REWIRE YOUR ANXIOUS BRAIN:

4 Books In 1:

How to Use Neuroscience and Cognitive Behavioral Therapy to Declutter Your Mind, Stop Overthinking and Quickly Overcome Anxiety, Worry and Panic Attacks

EDWARD SCOTT

PART I COGNITIVE BEHAVIORAL THERAPY

Table Of Contents

PART II OVERTHINKING

Table Of Contents

PART III HOW TO DECLUTTER YOUR MIND

Table Of Contents

PART IV OVERCOME ANXIETY AND DEPRESSION

Table Of Contents

PART I

COGNITIVE BEHAVIORAL THERAPY

Introduction

Cognitive Behavioral Therapy is one of the types of psychotherapy that helps you change your automatic thought patterns, cope with your emotions, and challenge your negative behaviors. Has anyone ever told you that your actions have consequences? Well, they do! The way we react can stop us from moving on our lives, and that is why it is so important.

The basic principles of Cognitive Behavioral Therapy, lies within its name: Cognitive = This is because the whole idea of CBT focuses on our ideas, beliefs, thoughts and generally how we feel.

Behavioral = You already know that behavior is something we do, or the way we act. Sometimes this can be driven by how we feel or our emotions.

Therapy = If we are in therapy, we are receiving treatment for something we have, are trying to work out, or fix it. The therapy eventually remedies the problem.

Think about an occasion when you've been reactionary. Maybe this is because someone takes something for you or treats you badly. We can be reactionary as our thoughts or behaviors are automatic, and as a result as we struggle to control our gut-reaction. This means we can't always help the way we feel or act, so we get emotional and behave without thinking. This is the way we are programmed to think or react in a negative way instantly.

These basic principles indicate how this therapy is focused on what you think and how you act in relation to an event or trauma. They are like a gut-reaction as we respond without thinking. Our thoughts and behaviors occur regularly, but what are automatic or intrusive thoughts and how can we change them?

Cognitive Behavioral Therapy is a well-known psychotherapy treatment that is notable for its positive results and feedbacks. It has helped thousands of individuals overcome different things and helps them regain control of their lives. It caters to different aspects of how a person responds to a certain situation or problem.

CBT is quite a complex and intricate method to understand. It is composed of several different aspects that should be understood before initiating the process. The three main parts of CBT are: thoughts, emotions and behaviors. These are all interconnected and can significantly impact one another.

Emotions are also crucial since this greatly indicates what and how a person might feel. Behavior deals with how a person interprets their thoughts and emotions and how they want to handle as well as express themselves in any particular situation. In this chain, behavior is mainly dependent on a person's thoughts and emotions. It deals with the cognitive aspect of a person as well as their behavior.

In terms of their train of thought, individuals will learn how to accept the current situation they're in and, instead of choosing to think about the negative thoughts and emotions, they learn how to focus more on the positive aspects in life. They try to see the silver lining in each situation and try to shed light on favorable emotions.

As for the behavioral aspect of CBT, it alters the way that different individuals react to various scenarios. When faced with a problem or situation, CBT allows them to take control of how they might behave and respond to it and lets them be a better person.

CHAPTER 1:

What is Cognitive Behavioral Therapy?

Cognitive Behavioral Therapy was developed by a psychiatrist called Aaron Beck in the early 1960s. He came up with Cognitive Behavioral Therapy after noticing that his patients tend to talk to themselves as a way of expressing their internal problems. He also saw that the feelings of his patients impact their thoughts.

This system developed by Beck has been widely used in the treatment of a wide range of mental issues such as depression. Cognitive therapists use this system to help clients identify negative thought patterns that may cause them to be depressed.

The patients are also encouraged to challenge and question these erroneous thoughts. They are encouraged to seek alternative ways of thinking which are more productive and rewarding.

Aaron Beck believes that your reactions to negative thoughts contribute to irrational behaviors. You regularly deal with different situations in your life, some positive while others are negative. The negative situations thoughts can come back to haunt you later in life.

How Aaron Beck Came Up with CBT Model

While working as a clinical psychologist, Beck noticed a respite in the patient's symptoms, and he concluded that the patients were presented with negative recurring thoughts, some of which were activating or triggering mental disorders. He later termed these events as automatic negative thoughts.

Beck believed that when your thoughts are incredibly harmful, you end up depressed. These thoughts can persist even in the face of contradicting evidence.

CBT Models

Beck identified some models that explain what contributes to depression and other mental disorders in an individual. These models are:

- The Cognitive Triad
- Errors in logic
- Negative self-schemas

1. The Cognitive Triad

Beck came up with what he called the cognitive triad while working with people suffering from depression. These are the three kinds of negative thinking that characterize depressed people. They are:

- Thoughts the self
- Thoughts about the future
- Beliefs about the world

These thoughts tend to come automatically and suddenly to people suffering from depression. When these three thoughts interact, they cause you to be obsessed with negative thoughts. They interfere with your standard cognition processing leading you to have distorted perception and memory.

2. Errors in Logic/Cognitive Distortions

Beck identified several faulty thinking processes. If you have these illogical thinking patterns, which are negative, you end up with anxiety or depression or both.

These illogical thought patterns include:

Arbitrary interference – You draw conclusions based on flawed or insufficient evidence, e.g., you think you are worthless because your favorite soccer team which you support has lost to a rival team

Selective abstraction – This happens when you focus on one part of a problem or situation while ignoring the rest. You feel responsible when your team loses a game even though you are just one player on that team.

Magnification – This happens when you tend to exaggerate a problem. e.g., if you are responsible for a small dent on your car, you conclude that you are a terrible driver

Minimization – Is the opposite of magnification. You tend to underrate or underplay the importance of an event in your life. For example, you get recognition for excellent classwork, but you undervalue this as trivial and non-important

Overgeneralization – This happens when you tend to use one negative incident to draw a broad contrary conclusion on the whole process. For example, if in one instance you score an E in a subject that you usually get an A then you conclude that you are foolish.

Personalization – This occurs when you take the blame for the failings or negative emotions of others. You attribute negative emotions of others to yourself.eg if your teacher comes to class looking unhappy you conclude that you must be the cause of that unhappiness.

3. Negative self-schemas

Beck believes that individuals susceptible to depression have a negative self-schema. They tend to organize and interpret information negatively. Their expectations and feels about themselves are generally negative. They also view life generally in a pessimistic way.

Beck attributed this negative schema to a traumatic childhood occurrence that you might have experienced. The experiences, which might play a role and contribute to this negative schema includes:

- If you experienced the death of your close family members such as a sibling or a parent

- If you experienced parental rejection, neglect, abuse, overprotection, criticism,
- If you were a victim of bullying at school or dismissal/exclusion by your friends from their groupings
- If you have these negative schemas, you become susceptible to making irrational errors in your thinking patterns. You focus your thoughts on specific aspects of a situation while disregarding even more relevant information

Other Key Principles in Beck's CBT Model

1. Importance of Cultivating Good Relationship with the Patient

Dr. Beck believed in the crucial role that a personal relationship with the patient plays in ensuring the success of the therapy. A trusting relationship is essential for the therapist to access and explore the automatic negative thoughts of the patient. He observed that some patients were reluctant to admit these thoughts. He focused on the reframing of these thoughts, which resulted in some significant improvement in his patients' conditions.

2. Cognitive Restructuring

Beck put particular emphasis on beliefs, meanings, ideas, and expectations, and the role they play in memory, the formation of concepts, and processing of information in a patient suffering from mental problems.

To treat depression and other mental disorders, Beck recommended cognitive restructuring. Your way of thinking and the thought patterns are modified. This will enable you to develop new schemas that will allow you to see the world and your immediate environment more positively and beneficially.

CBT Techniques – Beck's Model

1. Socratic questioning

Beck encourages the need to question your thoughts and beliefs to examine the truth in them. This technique encourages probing of evidence, rationale, and reasoning. For example, a patient who believes his life is in danger will be asked to provide a justification for his/her beliefs. The CBT therapist is encouraged to use questions to get to know the patient's reasoning.

2. Reality testing

You are encouraged to actively look for evidence and use it to test your beliefs or assumption. For example, if you believe there are giant ants that eat people, you are encouraged to find evidence of their existence and prove that they indeed exist and eat people.

3. Behavioral Experiments

This is whereby a scientific experiment is conducted to test a specific prediction. If you believe that your neighbor is threatening your life by coughing, your therapist will help you change that pattern of thought by presenting you with contrary information. You watch a film in which people cough due to various reasons such as smoking or illnesses like flu.

Essential Points in Beck's CBT Framework

1. The cognitive organization

Becks model holds that you give an interpreted and evaluated response to a situation. Your behavior is not automatic. There has to be a situation that triggered it. There is a specific way you are processing information based on cognitive schemes and which significantly influences your behavior.

2. Cognitive schemes

These are the structures that carry the techniques of coding, storage, and retrieval of information. They include how you memorize, interpret, and retrieve store information in your brain. It is about how you perceive your surrounding and how you interpret it

3. Beliefs

Beck concluded that cognitive schemes are made up of opinions. Your expectations influence how you view your world, make sense of it, and experience it. These beliefs tend to influence your thought patterns

4. Cognitive products

These are the thoughts that result from your interaction with specific information that a particular situation presents. This is how you act according to how you see and view your surroundings. How do you behave when presented in a specific case?

Important Evaluation of Becks CBT

The following were the findings:

- The study concluded that over 80% of patients have benefited from the therapy
- It also found that the treatment was more effective as compared to drug therapy
- The study found that the treatment registered a low relapse rate in patients as compared to the other treatments
- The study concluded that the application of CBT was vital in improving the health of people suffering from depressions and other mental illnesses

Key Changes in CBT

Like all other fields, psychotherapy keeps changing, as new models are being added in order to improve its effectiveness. CBT has not been left behind. With intensive research, new ways of working have been developed. These new ways are what is popularly known as The Third wave.

What is the Third Wave?

These are new approaches that attempt to develop and improve the tools of cognitive therapy. It is a movement away from the way cognitive focus on what you think and feels and towards focusing on how you relate to what you think and feel.

While CBT focuses on dysfunction, the third wave therapies focus on helping you to accept yourself and your world. Unlike CBT, third-wave therapies are not seeking only to solve your problem. They equip you with skills that help you navigate your issues successfully.

Summary of third-wave treatments techniques as compared to CBT:

- Focus on health over dysfunction
- Process-based over solution based
- Increasing skills to manage rather than trying to get rid of a problem
- Identification of context over seeking problem

CHAPTER 2:

Phases of Cognitive Behavioral Therapy

Cognitive Behavioral Therapy involves 6 different phases and now we will learn of these in detail in order to understand things better. Each of the phases has to be carried out in the right order to achieve the intended outcome. It cannot be done in a haphazard manner.

Phase 1- Psychological Assessment

Psychological assessment is the first phase of Cognitive Behavioral Therapy. In this stage the focus is on the issues faced by the client. There will be a detailed study of their flaws and their errors. The psychiatrist can evaluate these with the help of conversations with the client or a list can be made by the person himself. Using psychological methods to assess the client is important in order to understand the main problem. The client has to be a willing and consenting participant in this process. They have to agree to being evaluated first and if they do, the psychiatrist will give them a series of questions. The questions may be casual at times and some may be much more intrusive. However, the client has to trust the therapist and answer honestly if they want Cognitive Behavioral Therapy to work on them. The client is usually asked to lie down or just relax during this initial session. They are then asked the questions in a nonjudgmental manner by the therapist. A professional and good therapist will not react to any of the answers in a way that may cause the client to feel insecure or uncomfortable. There are no comments or judgment or laughter involved. There are some clients who even agree to be hypnotized during this process but this is rarely done. However, those who do allow hypnosis should ensure that the therapist is experienced and an expert in the field or things can go quite wrong.

Phase II- Reconceptualization

The second phase of Cognitive Behavioral Therapy is reconceptualization. This phase involves restoring beliefs, emotions and feelings. The therapist studies the client in the first phase and determines what their real issue is. In the second stage they will start helping the client from the ground up. The human mind obviously cannot be instantly rebooted like a machine. However, with time and effort, the mind can be trained to return to a state of normalcy.

The second phase of Cognitive Behavioral Therapy is very pivotal to the entire process. This is because it helps the client learn new concepts that will help them shed any old habits or thoughts that might affect them negatively. The old structure of the mind is shattered in a way to make room for a better and newer structure. This process is particularly time consuming and hence the second phase will take different lengths of time for different clients. It will depend on how deeply the client is affected by their condition. For some this phase might last a few hours but there are some who might need weeks or even months. There is no fixed time limit with this phase of Cognitive Behavioral Therapy.

The therapist will only be able to estimate this time after the therapy is started and they assess the client. Therapists will usually refrain from giving the client an estimated time for this phase because it can elevate the level of anxiety in them. The client just has to trust the professional and dedicate as much time as required to facilitate real healing.

Phase III- Skill Building

The first phase is the assessment of the condition and the second is the correction of errors. The third phase involves the acquirement of new skills and this is a very crucial aspect of Cognitive Behavioral Therapy. This phase is aimed at improving the patient's condition and life to a higher extent. Solving errors alone cannot be effective. The person has to learn more than that. Just correcting the errors they have made will not be effective in completely curing them.

Learning and developing new skills is what makes a whole lot of difference. The therapy is not singularly aimed at restoration and controlling damage already done. Therapy should also help the client move forward and add value to their life. It should help them improve their persona in as many ways as possible. Like the second phase, the time required for skill acquisition cannot be determined in a generalized way. It will take different people different amounts of time.

The person himself has to be willing to make the required changes in their thought process and behavior. There are certain people who have a harder time accepting change while others may embrace it quite fast. The first type of people are difficult for the therapist to treat as compared to the second type who make their job easier. However, with time, the therapist can definitely help any person learn some new and helpful skills that will help them improve their lives.

Phase IV- Consolidation and Execution of Skills

This fourth phase helps the client learn how to consolidate their newly acquired skills and execute them efficiently. Anything new can easily rust and lose purpose if left unused for too long. This applies to skills as well. The more frequently you exercise these skills, the better you will get at it. The therapist has to teach the client different ways in which they can actually apply their new skills. If the skills are not consolidated well, they may wither away. It would be the same as not learning anything at all. You can strengthen your skills and get real experience only when you use them in some practical situations. Mere theory is not much use. The skills are further cemented by practical application. This is when they can be useful, in the long run, for the client. The therapist has to guide the client in a way that they not only use the skill but also do it as well as possible.

They cannot just direct the client to carry out a certain exercise. The therapist is responsible for being present and guiding them through this exercise so that they do it in the best way possible. This is the right execution of any newly acquired skill.

Phase V- Maintenance

Acquisition of skills and practical application of them is not where it stops although you may think it should. The fifth phase after this is just as important and maybe more so if you want the Cognitive Behavioral Therapy to work for you in the long term. Think of it as maintenance of a fit body. You may go to the gym regularly for three months and get a toned and muscular body. However, you will easily lose this state of fitness if you completely stop exercising after those three months. Maintenance is crucial if you want to remain or improve that state. Similarly, the fifth phase will teach the client the art of maintenance. It is not possible to optimally use the skills at all times. However, continuous practice will make you as good at it as possible. Your skill will get stagnant if you stop practical application for too long. Generalization is also important in this phase. Having worked on the smaller details of your condition, you now need to generalize them together for the sake of adaption. It is easier to accept all the different aspects of your personality when you generalize it as a whole. A single consistent pattern of behavior is easier to deal with than different forms.

Phase VI- Post Therapy Assessment

With the sixth stage, you come full circle. You now know that the first phase of Cognitive Behavioral Therapy assessment and the sixth phase is a type of assessment as well. This phase is mandatory as it follows up on the impact of the entire therapy. Post therapy assessment will help the client or the therapist evaluate if the therapy worked at all or how effective it actually was. Skipping this final assessment is like leaving loose ends that leave room for error. The sixth stage helps in preventing any mistakes and checking that nothing important was overlooked. Finally, you have to know that any Cognitive Therapy process will vary depending on the client and the therapist. There can be certain medications and additions to the process to further facilitate the recovery of the client. It is also important for there to be feedback after each session so that you are more aware of anything more than that might be required or improved in the future.

CHAPTER 3:

How CBT is Different from Other Therapies and Why That's Important

The most significant difference about CBT is that it focuses on analyzing your actual thoughts and determining their rationale based on a number of factors. Do you get angry in particular places or situations? There are roots to these triggers, and once identified, CBT can make a sudden fear or anger into a quantifiable thing, a manageable thing. For instance, perhaps there is a certain floral pattern that drives you into a panic attack. Analysis of the thought and graded introduction to the pattern might teach you that it was a pattern worn by a lost loved one. Maybe you get angry if someone throws you a surprise party for your birthday, and through CBT analysis, you find out that you had a really bad one in your past. This sort of analysis may be done through a thought journal. Before getting into more detail, however, it must be stated that some of this information may be triggering for some readers. Don't push yourself to a point where you have a panic attack. Take steps at your own pace and don't be too hard on yourself if you do need to stop.

Medication

It is important to remember while you are reading through this book that it is not anti-medication. Some people do need medication to survive and have found that their lives are better for it. This book simply provides you with other effective ways that can be used for managing your anxiety and depression.

Medicines like Ativan, Xanax, and Valium are all very helpful in reducing anxiety and eliciting a feeling of calmness. They will often be prescribed for short-term anxiety and should be taken when having a panic attack as they rapidly reduce symptoms. Other drugs like Lexapro, Prozac, and Celexa are also helpful for those who suffer from depression and chronic anxiety. Many people find that combining some is helpful and upping the doses, under doctor supervision, improves symptoms as well.

But medication can also be challenging for some people; there are many patients who feel like they aren't themselves when they're on mind-altering drugs. Some people have moments when they aren't as creative, and others notice a reduction in their sex drive. Certain medications may also cause an increase in some symptoms while reducing others.

And when these medications are hard to access, it is very challenging to find an alternate method that works as quickly as taking a Valium or Xanax. For example, if your prescription ran out and you can't get to the pharmacy for three days, you might even experience withdrawal symptoms from not having these pills.

Cognitive Behavioral Therapy differs from medication because it lasts longer. There won't be times when you have to go through withdrawal, because the tools lie within your mind, not the tablets inside a pill bottle. You won't have to worry about becoming addicted to CBT, and it is not something on which you can overdose. Cognitive Behavioral Therapy is a personalized way of overcoming certain mental health ailments, and you won't have to go through dangerous and mind-altering prescriptions trying to find the right one.

Other Types of Therapy

Some treatments require you to spend six months or more talking to a stranger about every skeleton in your closet. Cognitive Behavioral Therapy, however, can empower you to learn fast, proven techniques that can help minimize the effects of triggers, and with a little work and patience, understand them for yourself. Complete self-awareness is hardly required. Even if you aren't certain what is triggering an event, if you can quantify the conditions so that you know an event is coming, then you can prepare for it.

These techniques are about structuring your thoughts and responses, not about dredging up the past. For one thing, simply knowing what caused the problems to begin with doesn't necessarily mean that you won't be triggered anyway. That is where CBT comes in.

Another difference between CBT and other treatments is that CBT is not a medication-based therapy. Some of us have had experience with psychiatrists, and while there may be powerful results, it is a slow process that involves trying medication after medication for months with the hopes that something will help.

Cognitive Behavioral Therapy is faster than these methods and is effective for a large number of people worldwide. Your therapist knows you, but only so much, so they won't be able to have all the answers for you. With CBT, you are the one who is doing the most providing, so it is important to remember that you won't have to depend on anyone else to actually start feeling better.

CHAPTER 4:

Is CBT Right for You?

With that said, now is the time to ask yourself whether or not trying Cognitive-Behavioral Therapy (CBT) is right for you.

With so many different options available when it comes to counseling and therapy, it's easy to feel confused and not know which one is best for you. Sometimes it's all a matter of trial-and-error, or simply asking your therapist for their professional opinion about which treatment option is suitable for you. However, if you can't afford to do either, then you can at least do some research and try to find out for yourself.

Thankfully, over here most of the research has been done for you. Now all that's left is for you to do some confirmatory checks and ask yourself the following questions:

What is My Diagnosis?

Most of the people seeking therapy have been diagnosed with serious mental illnesses, and while it's not entirely necessary, it's best to ask for a psychological assessment from a trained professional to determine whether or not you have one, too.

In a nutshell, most mental illnesses are gauged on the level of the distress they cause the individual; how much they deviate from normal behavior; whether or not they cause significant impairment in a person's social, mental, or emotional functioning; and whether it makes the individual a danger to themselves or others.

Knowing your diagnosis can do a lot to help you figure out which type of psychosocial treatment is best for you. Each mental illness responds differently to each kind of therapy, so you should go for the one that's been proven to be the most effective in treating your specific diagnosis and symptoms. However, if you don't have a diagnosis or feel that you don't have a serious mental illness, but rather, personal problems that you would just like to resolve, then CBT is the best answer for you. And as we've said many times before, it's also effective in treating anxiety disorders (i.e., GAD, social phobia), eating disorders (i.e., bulimia, anorexia), and mood disorders (i.e., bipolar disorder, major depression).

How Serious Are My Problems?

Similarly, to the last question, the seriousness of your mental health problems is an important consideration of whether or not CBT will work best for you. If you suffer from panic attacks, general anxiety, insomnia, specific phobias, substance abuse and addiction, relationship problems, anger management problems, or any of the disorders we've mentioned above, then CBT can help you.

However, if your problems are far more severe or complex, then you might need a more long-term treatment plan, especially if you have multiple diagnoses or have a chronic or recurrent mental illness (like a personality disorder or an intellectual disability).

Are my problems rooted in my thoughts?

The fundamental idea behind CBT, as you already know, is that our thoughts direct our behaviors, so it's specially designed to treat problems wherein the individual's automatic thoughts play a crucial role in their problems.

With the help of CBT, you will be better able to control and reframe your mindset ("I'm no good, I'm always messing up. Nobody likes me."), which will alleviate your emotional distress and help you eliminate your dysfunctional behaviors.

However, if your problems are not rooted in your thoughts, but rather your environment, physiology, or things that are out of your control, then CBT might not be the best choice for you.

Someone suffering at the hands of an abusive partner or family, for example, can only be able to change the way he/she feels and responds to the abuse, but still have to live with it. So as long as there is abuse, there will always be some degree of unhappiness in his/her life.

Do I have a clear problem to solve?

CBT is a solution-focused, goal-oriented therapy. Because it's only a short-term process, there needs to be a specific problem that you want to work on resolving. If you want to quit smoking or get over a bad breakup, for example, then CBT is a suitable option for you.

On the other hand, if you're just generally unhappy or dissatisfied with your life but can't think of a particular reason why (i.e., no past trauma, no abuse, no significant failures), CBT might not help.

If you're also interested in exploring your dreams and unconscious memories, or want to understand the meaning of life, then you're better off with psychoanalytic therapy than CBT, because CBT is more practical, direct, and focused on the here and now.

Am I ready to confront my personal issues?

Because you will be spending so much time analyzing and understanding your thoughts, you might learn or recall some uncomfortable things about yourself and your life. You will need to talk about your problems openly with your therapist in order for them to help you, as therapy cannot work if the client isn't ready to be emotionally vulnerable and confront their personal issues.

So before you go on, ask yourself: am I comfortable thinking about my feelings? Can I handle my emotions and my anxiety? It might be a bit upsetting at first, but if you really want to solve your problems, the best way is to face it head-on.

Can I dedicate time to therapy?

Even though CBT is quite a short-term process by most standards, 5-6 months is still a considerable period of time. You will have to go to hourly sessions once a week and sometimes even come home with homework and exercises for you to do outside of these sessions. This can be time-consuming, but you have to commit to the process in order to benefit from it.

Some people go to therapy and don't come back because they feel it wasn't effective or that it wasn't worth the time. People like this often expect therapy to happen overnight, but it doesn't. Not only will you need to dedicate your time to the healing process, but you must be emotionally invested, committed, and motivated as well.

So if you can't dedicate this time to allow yourself to get better and work through your problems patiently, then CBT will not be effective for you.

Do I believe in the power of therapy?

Before you try CBT, be honest with yourself first about whether or not you really believe that it can help you. If you're not at least open to trying it or willing to commit yourself to the process, then therapy is really not for you. Therapy is a collaborative effort. Your therapist will help you get better, but they're not going to do the work for you. They're only there to guide you and help you through it. You need to be the one who wants to change and actually make an effort to improve your life with more positive thinking and healthier behaviors. If you don't, then you're just wasting your time. There really isn't any magic pill to be the cure-all for the mental distresses that we may face over the course of our lives.

So is CBT really right for you? It's a difficult question to grapple with. In summary, those who will benefit from CBT are generally people who: know what their problem is and want to fix it; are willing to work hard and put in the effort to do so; and know that their issues can be solved with a more positive and constructive mindset.

Getting the Most Out of CBT

After you've pondered on all the questions above and have decided that CBT is the right choice for you, it's time to discuss how you can make the most out of it. Here are a few ways you can do that:

Embrace change

CBT is all about creating positive change in your life through your thoughts. Sometimes change can be hard — maybe even painful — but it's a crucial part of life. Trust in the process, even if you feel a bit of distress or discomfort. You have to understand that sometimes, things can get worse before they get better, just like how it always seems the darkest before dawn.

Set a time table for yourself

It's good to have goals and then set deadlines for yourself when you want to achieve them. This gives you a sense of urgency and makes you more committed to getting better through CBT.

Be honest with yourself

It can be scary to think about your deepest fears and insecurities, but it can also deliver the best possible results CBT has to offer you. So, no matter how painful may think it be to think about what you're going through, you need to allow yourself to be completely honest and emotionally vulnerable in order for you to succeed in therapy.

Reflect on what happened

Everything you do in CBT has a purpose, so reflect on it. CBT helps you understand the dysfunctional thoughts, feelings, and behaviors you experience and redirect you towards a better, healthier way of life. Taking the time to ponder on what these problems are will give you a better idea of how to resolve them and enhance your therapeutic experience.

Integrate CBT into your life

Perhaps the best and most effective way to get the most out of CBT is by integrating it into your life. This ensures that, even after this book's 21-day step by step program with CBT is over, it can still create real and lasting change in your life.

You may do this by setting aside some time from your schedule to practice CBT techniques and methods on a regular basis or coming back to them every time you encounter a personal problem. Remember everything you learn here and apply it as often as you can to yourself, your life, and even as advice to others. Only by doing this, then you will be able to reap all the benefits and rewards of Cognitive-Behavioral Therapy.

CHAPTER 5:

Myths about CBT

While most therapists and others in the psychology field have a good understanding of cognitive-behavioral therapy, it is easy for the layperson to develop misunderstandings of the method due to commonly purported myths. Along with myths about CBT running around, there are also a few common mistakes people might make during the therapy process. Thankfully, CBT is a simple and straight-forward therapy to complete when you are prepared with the proper tools. However, it is still possible to make simple mistakes.

Common Myths

These myths spread easily, as people who have only a small understanding of cognitive behavioral therapy might misunderstand what they have learned. However, as this person believes that they understand the matter, they begin to spread their misunderstanding that creates these common myths. Let's have a look at the most common myths and the true facts behind them.

Myth #1: CBT is a Rigid One-Size-Fits-All Approach

One of the beautiful aspects of CBT is that it is a fluid approach with many techniques that can be applied to a variety of disorders. While a person with depression will use one set of techniques, a person with post-traumatic stress disorder will use another set of techniques. While these two individuals will use two different types of techniques, they will also use a few of the same methods, such as journaling, to help restructure their cognition. The truth is that CBT is in no way rigid or one-size-fits-all. It is a highly customizable method that can be altered depending on a person's diagnosis, age, and individual circumstances.

This approach recognizes that every person is unique, and therefore require a unique and tailored approach.

Whenever a person sees a highly trained cognitive-behavioral therapist, this therapist should be able to tailor their treatment plan for their specific needs. A person can also customize their own plan if they are using this book without the help of a therapist. Of course, it is always advised to seek professional help.

Myth #2: CBT Only Focuses on Replacing Negativity with Positive Thinking

Cognitive-behavioral therapy focuses on restructuring the cognition to be more balanced and less negative. Although, this is different from positive thinking. With typical positive thinking, a person is simply saying something positive to cover up their negativity. For instance, the person may say "nothing is wrong, I'm happy," even though they just got bad news and are devastated. This type of insincere positive thinking is like trying to get rid of deadly mold in your house by simply painting over it. This does not correct the issue, it only coats it with a temporary distraction, which will likely lead to disastrous results later on.

Instead of covering up your life with insincere positivity, cognitive behavioral therapy teaches people to learn to see their lives as realistically as possible. This means that you see both the good and the bad, with neither overshadowing the other. Having balanced and realistic cognition allows you to enjoy the good and address and fix any problems. Along with looking at the world and yourself more realistically, CBT also teaches people to think more flexibly. This means that if a person is feeling nervous about giving a speech in front of a crowd, they can think about the situation flexibly. If the individual told themselves "I won't mess up, so I shouldn't be worried", this will not help, as messing up is certainly possible. Instead, the person is taught to think about other perspectives. For instance, they can think "Even if I make a mistake, I can still do well and succeed."

Myth #3: CBT Ignores Emotions

Nothing could be further from the truth than to say that CBT ignores emotions. The truth is that emotions are a very important part of the therapy process, they are just handled with a different approach from other types of therapy. With CBT, instead of dealing with emotions on their own, they are dealt with hand-in-hand with a person's cognition. This is because cognition is what affects a person's thoughts, behaviors, and emotions. Therefore, if you want to deal with difficult and troublesome emotions you must first understand the cognition behind them.

Myth #4: CBT Ignores the Past and Childhood

This myth has only some truth in it, but only partially. The truth is that CBT usually focuses on the here and now, the problems currently affecting a person. Yet, when needed, a therapist will look back at their patient's past and how that past may be affecting their cognition and causing problems in their current life. For some instance, if a person is suffering from post-traumatic stress disorder, the therapist will look back to see what caused the trauma. If a person has a social anxiety disorder, the therapist may discuss situations in the past that might have triggered the fear of social interaction.

A person may be able to look back at their past and see how it has negatively altered their cognition, but this can only be done reliably by a trained therapist. The truth is that we are unable to look at our own pasts completely accurately, especially when trauma is involved. However, a therapist specializes in understanding these difficult pasts and traumas with the ability to learn how to restructure the cognition into a balanced and healthy manner.

Myth #5: CBT Only Treats the Symptoms, Not the Problem

Cognitive-behavioral therapy, by definition, treats a person as a whole without reducing them to a setlist of symptoms. This is because before it can even treat symptoms, it must first change a person's cognition. As a person's cognition improves, the person will find themselves thinking in a more balanced way about the world around them, others, and themselves. This, in turn, creates a change to the person's thoughts, emotions, and behaviors. The result is a therapy that focuses on inside to outside fixes, rather than only covering up the outside symptoms while ignoring the inside problems. You can feel confident in knowing that CBT makes a real and lasting change to a person as a whole, allowing them to continue to experience improvements even after the course of therapy has ended.

Myth #6: There is Limited Scientific Evidence Supporting CBT

The truth is that cognitive-behavioral therapy has a high degree of science-based evidence supporting its success, especially when compared to other forms of psychological therapies. A meta-analysis published by the University of Boston analyzed over one-hundred studies on the use of CBT for various disorders, addictions, sources of stress, and other possible circumstances. This meta-analysis revealed that in almost all cases, CBT was more effective than other forms of therapy used. Not only that, but it was found to be especially effective for individuals with anxiety disorders, general stress, anger control problems, bulimia, and somatoform disorders.

Myth #7: CBT Requires the Person to be Motivated

It could be hard to get started on therapy, even if you know it will help. When you are depressed or anxious it can be difficult to do anything, even what is best. Does this mean that CBT won't work for unmotivated people? No! In fact, most people are struggling with these issues when they begin and are, therefore, unmotivated. Yet, despite this lack of motivation, time and again these people find success and find themselves experiencing amazing benefits.

Even for a person lacking motivation, CBT works if they have a schedule with goals, preparation techniques, and a person to stay accountable to. By having someone on your side - even when you lack motivation - you have the aid of someone pushing you to stick to your schedule and continue using the techniques you have learned. This person is a therapist for most people; however, you can also stay accountable to a friend or a family member. The person who you are accountable to should understand the premise behind your therapy, so try to get them to read this book or show them highlights that will help them understand the process. If the person understands what you need to do, they can ensure you are doing it well and can give advice when needed.

Myth #8: It's Long/Short-Term Psychotherapy

People tend to believe that CBT is either long or short-term. The truth is somewhere in the middle. Unlike some forms of therapy that require a person to go multiple times a week for an unforeseeable amount of time, CBT has a set number of sessions. By the end of these sessions, the patient should find that they have restructured their cognition and are ready to go back out into the world on their own. They may still use the techniques they have learned to maintain healthy cognition, but they will no longer need to practice extensive therapy daily.

On the other hand, it is not necessarily a short-term therapy, either. While there are usually a set number of five to twenty weekly sessions, the number of sessions will go on as long as is needed for the individual. You don't have to think about only getting five sessions if your case needs twenty.

Myth #9: CBT is Really Easy

It is true that cognitive-behavioral therapy is a simple and straight-forward therapy that anyone can accomplish, adult or child. However, it is important to keep in mind that no therapy is "really easy." All forms of therapy will have their own struggles, as it is not easy to overcome our inner pain, trauma, or habits. Restructuring your cognition into a more balanced and healthier worldview will take diligence and hard work. You will have to ensure you use your learned techniques on a daily basis and remain honest with yourself.

At the end of your time using CBT, you will find that the effort and work you put into restructuring your cognition was well worth it. Sure, it might not always be the easiest thing, but it is one of the most worthwhile ways you can spend your time. With diligent daily effort, you can find yourself much more balanced, satisfied, and happy.

CHAPTER 6:

The Relationship Between Cognitive Behavioral Therapy (CBT) and Body Language

Emotional Empath

The most common kind of empathy is emotional empathy, and it is just what it sounds like. In general, the book you are reading pertains mostly to an emotional empath. An emotional empath will pick up someone else's feelings easily, to the point that you experience them as your own and have difficulty distinguishing the difference between yours and theirs.

It is a much deeper emotional sensation, leading to having emotions that don't belong to you, simply because you are near someone who is feeling a strong emotion.

The most important thing for an emotional empath is to learn the difference between your feelings and someone else's so that you can better help others with succumbing to emotional exhaustion.

Physical Empath

A physical empath is able to "read" the emotional energy of someone's physical body, meaning that they can interpret what is ailing someone on the physical level.

This can be like seeing someone's posture and immediately being able to sense that they have chronic pain in their low back. Some physical empaths choose to work as doctors, healers, and medical professionals because of their skills. Sometimes, this ability will be felt, or "picked-up" in your own body and can lead to chronic fatigue symptoms if you are not aware of your ability to pick up on this kind of energy.

Geomantic Empath

For someone with this type of empathy, they can understand the energy of the environment. This could be like walking into a room or a building and automatically being able to discern the energy of the space. Many people associate this form of empathy with environmental studies and those with an urge to help with ecology and sustainability might be a geomantic empath. Usually, you sense this ability if you either feel incredibly uncomfortable or alternately, incredibly at ease and at peace, in certain environments. It's like "reading the room." With this type of empathy, you will likely feel a deeper bond to certain locations, landscapes, buildings, or places in nature. It is possible, too, that you are sensitive to the historical and cultural history of a certain environment and are able to feel that energy as well.

Horticultural Empath

For this type, it is all about plants and how they "feel." Someone who is a horticultural empath will likely feel drawn to flora and how plants exist in relation to you and the space that you are in. For those who are drawn into working with plants and gardening, you might have this type of empathic tendency.

Animal Empath

An animal empath can feel the emotions of their pets, as well as those of wild animals in nature all around the world. You will likely know just what an animal wants or needs if you have this skill and can offer the pet another kind of support and comfort based on your emotional understanding of their reality.

Intuitive Empath

An intuitive empath can understand something about a person simply by being in their presence. This can come with a lot of practice if you are any of the other kinds of an empath, and it is something that naturally occurs if you are open to your gifts and skills. With this type of empathy, you can usually tell immediately if someone is lying to you, or if they are hiding their feelings behind the words they are choosing to express themselves with. This can create an issue of being too open to others and requires that you understand how to guard and shield yourself well.

We will mostly focus on the emotional and intuitive empath, but all of the techniques and guidelines will be beneficial to any kind of empathy you may have or experience. You can look forward to knowing and understanding all of the various ways that being an empath can be a wonderful gift and before you get further into those aspects of empathy, it is important to look at how there can be challenges to working with this type of emotional availability all of the time.

Understanding the Empath

None of us comes into the world while knowing what empathy is—it is modeled, taught, and learned. It is also something that can naturally occur in the brain function simply because we are all human—and when we see another person in need, a lot of times the urge is to aid them or offer them some kind of consideration.

It begs the question: where does this ability really come from, and how are some people more of an empath than others? There are certainly a variety of ways that these skills can manifest or become a part of someone's regular personality and attitude in life.

It isn't just a process of deciding to become an empath and all of a sudden you are one; there is a strong physiological and biological link to your empathic skills, as much as your genetics and early life history and environment play a part.

There is always some kind of debate in the scientific community about what can cause or create certain functions in our brains and thought processes, and in the case of empathy and studying how it works in all of us, there are certainly some specific, neurological causes that form these connections in mind, allowing someone to comprehend someone else's experience through a form of emotional mimicry.

Other components can also play an important role in your mental and emotional ability to practice empathy—for example, your quality of life as a child. There is the argument of what genetic factors might play a role compared to how you are nurtured and cared for by your family and caregivers. Early life is when you begin to form your personality, and so much of it is impacted by what you are shown by the people in your life. All of the above can be a major force in how someone becomes an empath.

To gain a better understanding of what it can look like from a neurological standpoint, let's learn a little something about mirror neurons and how they can become a part of your brain's ability to understand and practice empathy as an adult.

The Empath and Mirror Neurons

Have you heard that your brain is more powerful at computing than any computer in the world today? The 3 lbs of tissue in your skull is a massive machine that can process information in a way that no technology can (at least not yet). The brain is still a mysterious organ that many neuroscientists today feel mystified by as they continue to delve deeply to understand our cognitive abilities and other brain functions.

As with any type of scientific research, the neurological studies that were conducted a decade ago, might not hold sway anymore as we have continued to discover new ideas about what the brain can actually do, through several different studies. Today, scientists are still discovering how empathy works in correlation to your mind matter, and there are a lot of valuable theories that seem to explain some of what can go on in mind to establish an empathic sense.

Recent research in the field of psychology, neurology and empathic studies has looked at the connection between mirror neurons and empathy. The human brain has trillions of neurons, and only some of them are considered to be mirror neurons. These are the only neurons that have been linked to empathic behavior, and as such, have been studied in specific cases to try and understand how empathy really plays a part in your mental cognition and function.

Mirror neurons are located throughout the brain and are not confined to one specific location, so you will find the same mirror neurons in the temporal lobe—where you process language, hearing, and memory—as you would in the frontal lobe, where you produce speech, control motor skills, and solve problems. There may still be a lot of research being done to understand how these mirror neurons function and how they are linked to certain aspects of our growth.

Current research describes the following functions of the mirror neurons of your brain:

Understanding Language

This will relate to your ability to learn language by hearing what your parents say to you as a small child, mimicking their words and language to learn it through watching and copying mouth movement, as well as auditory response and reaction.

Imitation

This is an automatic reaction that will often occur starting at a newborn's age when a child will copy, or mimic what the person is doing, such as smiling or sticking a tongue out playfully. Children continue to do this as they get older.

Reading and Understanding Intentions

This relates to knowing when someone wants you to follow their lead, by copying what they are showing you how to do, such as during instruction or class, or if you are a child, being modeled how to use a fork to pick up your carrots.

This can also develop into reading someone's emotional or physical intentions without having to use words to understand what is being demonstrated.

Observatory Learning

This is how many people learn new skills and is just what it sounds like: watching someone doing a task and then repeating it, such as watching someone thread a needle and begin to sew, and then mimicking the actions on your own.

Developing Personal Awareness

This is how someone will determine what kind of a person they are, through witnessing or observing someone else's actions and then making a choice to either perform the same action or choose an alternative. An example of this would be watching someone jaywalk, and then deciding to use the crosswalk yourself.

This might all seem so simple, and it is when you think about it, but on a bigger scale, noticing that these mirroring functions are occurring from the moment you are born, determines what capability you might have to perform certain functions as you grow into adulthood. If you are "mirrored" by your caregivers to be unforgiving, disinterested, and incorrigible, then you might not become a very empathic adult. There is some research being done with regard to mirror neurons and how they are linked to disorders like Autism and that it is possible that Autism could be attributed to a lack of functioning mirror neurons. Other sources have indicated that your early guidance by caregivers and a lack of "healthy mirroring" is what can lead someone to become a narcissist, sociopath, or any other kind of person who lacks empathy.

The research is still coming forward to ascertain what role mirror neurons play in any person's ability to develop empathic skills and our general ability to experience love, compassion and generosity are what set us apart from our animal companions here on Earth. The studies on empathy are certainly offering a gateway to understanding more about this human condition, and even with this knowledge, there is still the process of determining if you are a true Empath, or just acting with empathy.

CHAPTER 7:

CBT and Cognitive Dissonance

Cognitive dissonance, also known as cognitive disharmony, refers to the state of conflict within our mind, in which we are faced with two conflicting beliefs at the same time. The human brain is designed to eliminate disharmony as much as possible, and it fulfills this role by changing the way we perceive or feel about certain things.

Understanding cognitive dissonance is vital for the effective use of CBT because this can be used as a powerful motivational tool to change a person's behavior.

All of us have already experienced a form of cognitive dissonance at some point. In general, if you believe that you are a good person, and you do something bad, the discomfort that you feel is caused by the dissonance. The common signs of cognitive dissonance include:

- Discomfort
- Guilt
- Shame
- Avoidance
- Rationalization
- Ignoring the facts

The disharmony within our minds is usually most intense if we are highly committed to a certain belief, but our behavior follows the opposite direction. For example, if 30-year-old Lucia believes she is an independent woman, but still lives with her parents, she will experience a great deal of tension. In response, the brain will try to justify or deny, or change her attitude and beliefs – all to reduce the tension:

- My parents will not allow me to leave home. They need me.
- How can I survive if I live alone?
- I am comfortable living with my parents anyway.
- I love my parents and I would love to spend more time with them.
- I help in paying the bills.

Even though Lucia might have not initially agreed with these reasons, her mind justifies her situation to weaken the disharmony she feels. Cognitive dissonance is such a powerful tool because no one really likes to feel disharmony.

Forms of Cognitive Dissonance

Cognitive dissonance can happen to anyone in almost all facets of life. It happens in relationships, in the workplace, in the supermarket, and even during your break time.

Cognitive Dissonance in Relationships

One common example of how cognitive dissonance influences relationships is when people are dating. For example, Lawrence really likes a woman who values money. When he first met her girlfriend Stacy, she seemed practical and seemed to know how to save – but it turns out that she's spendthrift who often requests for gifts that are expensive.

There were red flags like the fact that she always seemed to be chasing after some designer bag. "Maybe that designer bag is just a one-time thing." It was the first thing that he said when he noticed. She's also always worried about maxing out her credit cards, but Lawrence ignored them. She also tends to choose expensive restaurants but does not offer to split the bill. He told himself that perhaps, Stacy just likes luxury --

nothing wrong with a woman who has an appreciation for the finer things in life.

Ignoring the obvious facts is an indicator of dissonance. Rather than admitting that Stacy may not be a good fit as a partner for him, Lawrence convinced himself that she would eventually change.

Cognitive dissonance is quite common in relationships that are abusive. Consider the couple, Robert and Diane, who have been dating for six years, but only married for a year. The marriage is happy and filled with passion. But one day, Robert slapped Diane during an argument.

Diane -- who thought she really knew her husband – started to experience dissonance. She abhors violence against women and has sworn that she will never let herself be victimized. But she found herself trying to rationalize his behavior. "He was too drunk, and I might have pushed my limits," she said. "He is kind and cares a lot about me when he is not drunk."

It is common for people to rationalize bad behavior and choose to stay in the relationship despite the abuses.

There are cases in which people are oppressed in a narcissistic relationship, but they end up forming a codependent relationship with the abuser. Cognitive dissonance causes a phenomenon known as trauma bonding.

For instance, if someone is experiencing Stockholm Syndrome, they usually justify the abuse despite the fact that they are just rationalizing to cope with the situation.

The victim of the narcissistic abuse will start to believe that staying with the oppressor is important for survival. If they think the relationship is threatened, the victim may panic and become anxious.

This is why some people resist undergoing any form of psychotherapy session as they are often afraid to hear the truth.

Cognitive Dissonance in the Workplace

Many of us have already experienced cognitive dissonance in the workplace. Consider the following examples:

At the office, Mary (who sees herself as an honest person) is not monitored regularly. Each time she takes her one-hour lunch break, she actually extends for another 30 minutes. Instead of believing she's stealing money because she is getting paid for the time she is not working, she reasons that she is overworking anyway and uses the extra time to unwind before heading back to work again.

Fred is the HR manager for a large company. One day, he was ordered to fire an employee for misconduct without enough evidence to support the claim. Because Fred strongly believes in fairness and justice, the unsupported claim from his superior will result in dissonance. The dilemma will cause tension because he may even lose his job if he fails to execute the order, even though it is clearly unethical to do so.

Cognitive Dissonance and Consumer Decisions

As adults, we want to believe that we are capable of making good decisions. But a simple shopping experience can create a dissonance if something that we purchased which we thought was good turns out to be bad.

For example, Gary would like to believe he's an environmentalist. He likes to help in eco-friendly causes and even volunteers for a local group supporting environmental initiatives. He bought a new car, and it turns out that it was a gas guzzler. Upon learning this, he struggled with the idea that he will drive a car that is harmful to the environment. He decided to sell the car and instead use public transport.

Meanwhile, Wilma declared that saving more money is one of her New Year's resolutions, but she ended up buying a designer coat using her credit card. Admittedly, she spent way more money than she can afford and so, she feels buyer's remorse. Instead of being bothered by the undesirable emotion caused by the dissonance in her mind, she decides that the bag will last longer than her cheap ones. Besides, the bag is really a nice bit of eye candy – maybe it will help her project a better image of herself at work (and during dates). Back in 2003, there was a study published in the Journal of Product & Brand Management, it was revealed that consumers usually use three coping mechanisms to reduce their cognitive dissonance when they think they are overspending:

Look for more information – they try to find more information that will justify their purchase

Change their attitude – for example, they may try to re-evaluate the worth of the purchase

Downplay the significance of what they spent – "I can earn the money again!"

Cognitive Dissonance and Addiction

With the sheer amount of information available today, people surely know that smoking and drinking too much alcohol can heavily damage their health. But still, people are into different forms of substance abuse. They are already informed and many of them understand the risks, but their actions create cognitive dissonance. Laura is a doctor. She understands how damaging smoking is. Now that she is already 35 years old, she smokes at least three cigarettes a day to cope with stress, and even takes cocaine occasionally especially when she is depressed. "My grandmother used to smoke a pack of cigarettes a day, and she lived to be 98!"

By justifying her behavior, Laura was able to reduce cognitive dissonance. Laura is trying to change her frame of mind so she will not feel bad about the discomfort caused by the dissonance.

Cognitive Dissonance and Spirituality

Religion and spirituality also play a vital role in the formation of the Cognitive Dissonance Theory. After all, religion influenced many of the values we hold dear today – love, honesty, charity, generosity, sacrifice, and service, among others.

For people who are strongly grounded in their life principles based on the tenets of their religion, intense dissonance could happen especially if they are faced with a situation and their behavior was against their belief.

For example, Joe is a pastor at a local Christian church. He believes in the sanctity of life, which is much in line with his pastoral career. When he watched the evening news, he learned about a group of drug dealers killed during a police operation.

He thought they deserved to be killed, but he felt a strong dissonance within himself. He justifies his thoughts by telling himself that the drug dealers have destroyed the lives of hundreds of people, and that they deserved to be punished.

Cognitive dissonance can also happen if we have seen certain things damaged our belief system. For example, Jesse really looked up to the humility and spirituality embodied by Buddhist monks. But on a trip to Thailand, he saw a group of monks drinking alcohol and was even rude when they were drunk. He experienced cognitive dissonance as he strongly believed that Buddhist monks should not behave that way.

The Role of Cognitive Dissonance in CBT

Psychotherapists specializing in CBT can use cognitive dissonance as a powerful motivational tool to influence the behavior of a person, especially those who are trying to cope with the effects of abusive relationships.

Cognitive dissonance is particularly helpful in cognitive restructuring. When we become more aware of the presence of disharmony in our mind, we can learn more about how the dissonance can affect our decisions. When we are aware that we have the tendency to reduce the dissonance, we may choose to challenge it and choose a more beneficial action. For example, if you are in the shoes of Pastor Joe and your thoughts (drug dealers deserve to die) are not aligned with your belief (life is sacred), being aware that your mind will try to justify the dissonance will allow you to strengthen your grounding on your principles.

Some CBT specialists also use cognitive dissonance as a vital tool in script playing. People that are suffering from anxiety disorder are often prescribed to work on their cognitive dissonance and visualize in their minds the consequences of their actions if they are not in line with their beliefs. Allowing specific scenarios to play out in the mind can help people to realize the possible outcome of their decisions if they continue with their behavior.

CHAPTER 8:

Identification of Problems

You already know that you need to make a change as there's some problem in your life, preventing you from moving forward or being happy. For the next part of the CBT process, we need to work out what your problems are exactly.

Working out what our problems are isn't always easy. When we've thought a certain way or believed a certain thing for a long time, it can be difficult to see that it's causing a problem because it feels natural and normal to us. Often, we know something is wrong, but we don't pinpoint the root cause and we only touch the surface. This means that the problem recurs because the problem isn't solved. There are a range of strategies we can use to identify problems and then we can use CBT to work on solving those problems. It's time to get to work!

Strategy 1 – Identifying your irrational and negative thinking patterns and actions with Journaling.

Identifying your problems takes time so it's important to be patient at this stage. It can take up to a number of weeks to identify your problems. One simple way of doing this is to use Journaling. Journaling is an important CBT technique if you need to work out what the problem is. It can help you to evaluate the extremity of your problems and how deep they run.

So, what is journaling?

Journaling is a commitment. It's a record of your thoughts, actions and feelings. It's a great way to work through your emotions, actions and thoughts. At first it is a simple record, but as you start to move on in the CBT process, it will prove to be a useful tool to reflect on how you feel, and it often helps you to pinpoint the root cause of your problems. You can describe how you feel, what caused it and how you coped with the situation (Ackerman, 2019) and then when you have enough information, you can analyze this and use a range of further tools to help you take action against those.

Journaling should be continuous, but you can start to work through and assess your diary after the first 7 days. Once you have already gathered enough information, you can start with the next part of your CBT process, and identify your cognitive distortions.

What are Cognitive Distortions?

Cognitive distortions are a collective name for your irrational thoughts and negative thinking patterns. As we explore them, try to relate some of the problems you have identified as part of your journaling process to the different categories. Cognitive distortions are key when we start to make sense of our problems as they help us to make sense of our thoughts, feelings and actions. This works hand in hand with journaling and we can identify the negatives problems and then we can start to categorize them. Once we categorize them, we can plan a course of action.

Defining Cognitive Distortions

We've already established that Cognitive Distortions are different types of automatic thoughts. When we have thought in a particular way for so long, we believe it and we continue to think in that way. That's why people who suffer from anxiety and depression have a relapse, because those thoughts and feelings don't just go away. Even if we've worked through our previous issues, we can still have doubt in the back of our mind that creeps through now and then to stir things up. That's why learning to cope is so important. It breaks the cycle, so we are less likely to fall back into old habits and patterns.

Have a read through the descriptions and examples of each below. Grohl (2016) suggests that there are 15, but in this book, we have formed ten based on his studies (Ackerman, 2019). We will use the ten we've formed below throughout this book:

- Filtering - The way we filter information is complex and often we filter out anything positive and concentrate on the negative.

- Polarized thinking - This is when we are defeated easily because we believe there only two options in life; we either succeed or we fail. If we don't succeed, we just believe we are failure and there is no room to grow or middle ground. Sometimes we simply don't get the desired effect because we need to work on one area, but polarized thinking again focuses on the negative aspects and accepts failure far too easily. This line of thinking acknowledges only perfection as success, but in reality, perfection is often rare.

- Overgeneralization - This is a cognitive distortion identified by Beck (1967) and this basically means that we allow one single experience to define us. This is like heading to the kitchen and a recipe not going to plan. From that point on, you generalize yourself as being a bad cook. As human beings, we can all learn from our mistakes and should never put ourselves in a box that prevents growth and development. Cooking is a skill and if we make a mistake, we can learn to cook better next time. That's the case with most things in life. Overgeneralization blocks us from experiencing that learning curve. This can escalate to a more extreme level and affect the way we think about or treat others. It can cause us to judge and mislabel based on generalized assumptions.

- Jumping to Conclusions - This is another way that we block, and it is similar to overgeneralization, except we make judgement calls and assumptions based on nothing. This is like assuming people do not like you or judging you do not like someone without really getting to know them, or that we don't like something without trying it first.

- Magnification or Minimization - This is another distortion that was identified by Beck (1967) and this is when we magnify or catastrophize an event and make it more dramatic than it actually this. This is the mother of blowing something out of proportion. When you magnify an event to this level it is difficult to see clearly because panic sets in. Does that tiny spelling mistake in an email really mean your reputation is ruined and that your boss will sack you for instance? Although this isn't professional, others do accept that it is human nature to make occasional mistakes. It is often how we react in response to the mistake that matters. Now, Beck (1967) looks at magnification and minimization as separate entities, but Grohl (2016) counts these two distortions together, because minimization can be just as damaging. Sometimes we minimize the positive things that happen in our life because we focus on the negative. These two distortions often work hand in hand, because we magnify the negative and minimize the positive.

- Personalization -This is another distortion identified by Beck (1967) and this is when we believe we are responsible for certain things that happen. We take things personally and think that something happened because of us or someone who doesn't like us because of something we have done. Often this is us thinking irrationally.

- Blaming others - We can sometimes blame others for how we feel. For example, we can say that we are in a bad mood because of the actions of someone else. In reality, only we are responsible for how we feel.

- Dwell on what we should have done - Often, we have certain expectations of ourselves and we dwell on certain situations in which we don't adhere to those. We reflect on what we should have done, rather than what we did do. We can be hard on ourselves and beat ourselves up over how we should have acted.

- False beliefs - Grohl (2016) talks about different fallacies that we have and how sometimes they are beyond our control, because we believe them to be true. We just accept that things are out of our control, life is unfair, or we have no luck, when something negative happens, without attempting to take back control or trying to rectify the situation. For example, if something bad happens and we think that we are fated to have bad luck and just accept that there is nothing we can do, we have already accepted defeat and we don't even try to win.

- Assuming we are always right - This is another attitude that can hold us back as we can miss opportunities due to being so closed-minded. We have to be willing to try new things but we also shouldn't be afraid to try again, even if something didn't work on the first attempt. Imagine you applied for a job last year, and you didn't get it. You now see the job advertised again and you really want it, but you convince yourself you won't get it because you didn't get it last time. This assumption holds us back because we've already accepted defeat. We've convinced ourselves we're not going to get it, so we might apply but not really put the effort in or we might not apply at all. This is because we assume we are right – we'll never get the job.

Do any of these distortions sound familiar? With CBT, we need to take a step back, to identify and assess our cognitive distortions in a logical way. Now you know what distortions are, you can start to categorize your problems with strategy #2.

Strategy 2 – Identifying and Making Sense of Cognitive Distortions

The first step to resolving cognitive distortions is to identify what is causing us these roadblocks. You may do it with or without a therapist, although if you are having trouble identifying and categorizing your problems at first, a therapist can help you to do this.

You should now have your journal and you will have filled this with at least 7 days of content, so you have an idea of the kinds of problems you face.

You then need to analyze these distortions in a logical way.

What evidence suggests this? What are my distortions based on? You might surprise yourself here because you may have more negative thought patterns than you think. You really need to put on your objectivity hat at this point and you can start to unravel those thoughts and begin to

set goals to change these beliefs. Once you identify the problematic areas and approach them in an objective and logical way, the easier it becomes to challenge them and change them. Don't hold back!

CHAPTER 9:

Description of The Main Problems Part 1

Anxiety is the sense of uneasiness an individual may feel about a certain person, object, place or situation. Sometimes taking shape as fear or worry, anxiety is such a common feeling that comes and goes in everyone's lifetime.

However, there are some that develop certain kinds of anxiety disorders that can lead them to have extreme and irrational reactions or behavioral responses. Anxiety disorders are actually psychiatric problems that can make an individual feel extreme negative emotions that may lead to unfavorable circumstances.

There are six major kinds of anxiety disorders, all of which involve certain types of anxiety and different ways of how it is triggered and addressed. Cognitive-Behavioral Therapy (CBT) is actually widely regarded by many mental health professionals as the preferred psychosocial intervention for most of them. Aside from its impressive effectiveness, it also helps the individual lead a better life even with the disorder, as it teaches them a lot of valuable skills that will help them cope with their conditions. Listed below are some of the anxiety disorders and how CBT can help individuals overcome each one.

Generalized Anxiety Disorder (GAD)

One of the most prevalent anxiety disorders around, Generalized Anxiety Disorder (GAD) is characterized by excessive worry about almost everything in a person's life with no particular cause or reason as to why.

Individuals who have GAD tend to make a big deal out of everything. They become anxious about everything in their life — be it their financial status, work, family, friends, or health — and are constantly preoccupied with worries that something bad might happen. They expect the worst-case scenario about everything and always try to look at things from a negative point of view.

With that said, it's easy to see how GAD can make it difficult for someone to live a happy and healthy life. It can come as a hindrance to their day-to-day life and become an issue with regards to their work, family, friends and any other social activities. Some of the most common symptoms of GAD include: excessive worry or tension, tiredness, inability to rest, difficulty sleeping, headaches, mood swings, difficulty in concentrating, and nausea.

Fortunately, however, CBT has worked wonders in treating all these symptoms and more. With the help of CBT, individuals suffering from GAD can change these negative thoughts into positive ones, which will ultimately change their behaviors for the better as well.

The term depression has been thrown around so lightly in today's culture that it has now come to mean any feeling of sadness or lethargy. However, depression is much more serious with that, as those who struggle with it already know.

Called the "common cold of mental illnesses" because of how prevalent it is, depression greatly affects an individual's thoughts, emotions and behaviors in a negative manner. Living with depression is like coasting through life, feeling unmotivated to do anything and drowning in self-loathing. Most people who are depressed struggle to even get out of bed in the morning, much less do anything productive with their day. There are over six major kinds of depression, each with different factors causing its arrival.

There are a lot of ways on how to tackle depression. An individual can opt to undergo different types of therapy in order to take their mind off matters and try to heal. Depression is an extremely difficult and complex topic to touch on. With a variety of options to choose from on how to handle this type of psychological problem, one of the most prominent and effective methods used by multiple individuals and therapists all over the world is psychotherapy.

There are different types of psychotherapy but the most used and found to be most helpful for patients is Cognitive Behavioral Therapy (CBT).

Handling the thought pattern, emotions and behavioral aspect of a person, it can allow them to get something greater in life and not dwell on the negative side of everything. CBT is a wide and complex treatment that is known to be helpful in treating a variety of mental illnesses, depression being one of them.

Listed below are some ways on how each individual can make use of Cognitive Behavioral Therapy when faced with different kinds of depression.

Major Depression

Major Depression is arguably one of the most common types of depression. There are approximately 16.2 million adults suffering from it in the US alone. Also termed as "Major Depressive Disorder", "Unipolar Depression", or "Classic Depression", this kind of depression is characterized by feeling too much grief or gloom, being overly fatigued most of the time, having a hard time sleeping well at night, losing interest in activities that once excite you, not wanting to eat as much as you can before, feeling hopeless or experiencing anxiety, and perhaps contemplating about self-harm or suicide.

However, this type of depression does not typically stem from a person's surroundings or situation. A person could have everything one may dream of and still have depression. Major Depression can last for as long as week or possibly throughout one's entire lifetime. Causing a hindrance between their social and personal life, major depression can keep you from enjoying everything you love about life and isolate yourself from others. Negative thinking patterns may lead you to an unhealthy lifestyle. So how do you overcome it?

Cognitive Behavioral Therapy could be one of the most common and effective methods used by therapists to help their clients overcome their depression. Enabling the individual to alter their thought patterns, Cognitive Behavioral Therapy (CBT) faces the problem head on and acknowledges the situation. This does not make any excuses or hide you from the truth, but rather allows you to accept the situation and think of a silver lining for it.

Most of its success comes from the fact that CBT touches on the most important aspects of a person's life, which are their thoughts, emotions, and behaviors. CBT helps you eliminate your negative thoughts and replace them with more positive ones. As it alters negative thoughts into positive ones, it also impacts the emotions and behavior positively like a domino effect. It also deals with dysfunctional behaviors and changing them for the better.

There are some specific CBT techniques that can help individuals deal with major depression, but most experts would agree that the most suitable technique to apply here would be cognitive restructuring.

Depressed individuals tend to have negative automatic thoughts. Through cognitive restructuring, they can deal with this and replace it with more positive ones that can help them function better, mentally and emotionally. Here's a guide on how to use cognitive restructuring on your own:

- Assess the situation. Find the negative aspect that's upsetting you
- Keep track of your negative emotions. Describe them in your journal and rate the intensity of each emotion.
- Pay attention to all the things you automatically think of whenever you encounter a difficult situation and keep track of how much you believe in each of them
- Examine these thoughts and see if they are realistic or not

- Generate better and positive thoughts that are realistic and seems more likely to happen when compared to your automatic thoughts.
- Evaluate the process and repeat as much as necessary.

These techniques can be extremely helpful in dealing with a depressive episode on your own. However, it is important to remember that professional help can sometimes be the better option, especially when dealing with depression which can often leave the person feeling unmotivated at all to complete their therapy and hopeless about ever recovering from their mental illness.

Major Depression can be treated with CBT in different healthcare clinics. Individuals have to assess their own state of mind and once things becomes too hard for them to handle, they should talk to other people, like therapists, about their problems. This way, they can live a better life and slowly regain control.

Persistent Depression

Also known as "dysthymia" or "chronic depression", persistent depression is the most recurrent of all types of depression, typically manifesting in episodes that last as long as 2 years and return throughout an individual's lifetime.

It may not come as powerful as Major Depression but it may still take its own toll on the one that may be experiencing it. The feeling of being sad and hopeless, having second thoughts about yourself, a lack of interest, and the problem of being happy during joyous occasions may be a sign of this type of depression. It can also change your perspective on how life works. Symptoms may fade out for a while before coming back as clear and powerful as ever, making it difficult for a person to feel like they have any semblance of control over their lives.

Therapy is one of the many ways to overcoming this particular type of depression. Cognitive Behavioral Therapy (CBT) can be essential in dealing with long-term depression such as persistent depression. This kind of depression may be an occasional experience for some individuals. There are moments when they seem to be normal and happy, and other times wherein they just can't seem to see the bright side to anything. When dark moments come, it is important to use CBT in handling this situation. CBT replaces negative thoughts with positive ones and changes the way one may behave around this kind of situation. It can possibly alter your trail of thought and behavior in order for you to see the better things in life. When people with persistent depression make use of CBT, it can be possible for them to get over this long-term illness and go on with their own life happily. A common CBT technique that can help you get through persistent depression on your own is problem analysis. Also known as "situational analysis", it helps people see the problem objectively and find a positive solution for it. Problem analysis starts with:

- Finding the problem
- Understanding the problem and how it works
- Dividing the situation into smaller parts in order to understand it better
- Finding out what your goal is and what you want to work towards
- Finding positive ways to reach your goal and move on from the problem

Problem analysis can be helpful in treating persistent depression since it can help them overcome the problem in a positive manner. This CBT technique will aid them in triumphing over depression and help them on their way to recovery. It can also teach them how to respond to specific situations that may be complex to handle.

CHAPTER 10:

Understanding Anxiety and Depression

Anxiety and Depression are two of the well-known common mental health issues people of today face. While it is normal to feel anxiety and periods of sadness at some level during life, pervasive anxiety or feeling down becomes a problem when it begins affecting day to day life and making it difficult to function. When you are feeling anxious over every little thing you face in life, or when you feel as if you are sad more often than feeling anything else, it may be time to consider whether you are one of the millions of people suffering from generalized anxiety or depression.

Anxiety on its own is a completely normal emotion with an evolutionarily motivating purpose. It increases awareness of one's surroundings and alerts the individual that vigilance is necessary. By being keenly aware that there may be a threat lurking nearby, the individual is ready to react proactively when presented with a threat. This helps with survival; one that is ready to react to a threat with the awareness that he may die if he does not is much more likely to survive than someone traipsing through a forest, completely unaware that he has been stalked by a saber-tooth tiger for the last mile. The one feeling anxiety feels more alert and aware of his surroundings because he feels his life depends on it. While there may be periods where one feels unwarranted anxiety, this still aids in avoiding dangerous, life-threatening situations. In fact, those with anxiety disorders are actually less likely to die in accidents than those with normal anxiety responses due to constant vigilance.

Anxiety is a normal emotion; it can very quickly become a problem when it becomes so prevalent and persistent in one's life that impacts one's general quality of life. The person suffering from a generalized anxiety disorder may feel a variety of symptoms, both physical and mental in response to their anxiety. Physically, they may feel their heart palpitating, or racing, along with chest pain that may feel like a heart attack. This could be paired with hyperventilation and dizziness or feeling faint. It may also manifest as chills, heat flashes, stomach issues such as diarrhea or nausea, pins and needle sensations throughout the body, or a dry mouth sensation. Mentally, the sufferer may feel a sense of imminent death or demise, along with irrational fears and dreading normal situations. They may experience insomnia and exhaustion, along with issues concentrating. Depersonalization is also common, which is a feeling of detachment from reality, as if one is not real, or is in a dream state. Someone suffering from an anxiety disorder may feel any combination of these symptoms. Some will feel only one or two, whereas others may feel the entire list, depending on the severity.

While the specific cause of anxiety disorders is not known for sure, it is believed that some people are born with a genetic predisposition to developing it, but it is triggered mainly by external factors. Most people who suffer from anxiety have suffered some sort of severe stress or trauma that triggers the disorder. This may be child abuse, an assault, surviving a war or natural disaster, or could be an accident. It could also be caused by general stress, whether at work, school, or personal relationships. Regardless of the cause, the effect is undeniably detrimental, and sometimes devastating, to those who suffer. Sufferers of anxiety are forced to live in a perpetual state of their brains and thoughts convincing them that there are frequently, or always, threats present, even when nothing around them will harm them.

During an anxiety attack, the brain triggers an unwarranted fight or flight response to some sort of stimulus. This could be anything from an impending work deadline to a crippling fear of speaking to strangers, to hating driving over bridges, to a phobia of spiders. Regardless of the trigger, the brain becomes flooded with cortisol, a hormone responsible for stress. This hormone increases one's heart rate, triggering an increase in circulation and reflexes to allow for quicker reaction time and reflexes. While this is useful in life or death survival situations, during anxiety, this reaction happens regularly and persistently, even in the absence of a real threat.

Anxiety, despite how debilitating and unending it may feel, is treatable. Cognitive behavioral therapy is effective at reducing symptoms of anxiety, showing a reduction of symptoms as you begin to learn the skills to cope. Through cognitive behavioral therapy, you will be learning how to reconstruct your thoughts. Just how Mia began to walk through her irrational fears, you will learn how to deconstruct your anxiety. While you won't be able to entirely reduce your feelings of anxiety, and you cannot control when you may feel anxiety, how you respond to it is entirely within your control. The thought that a spider may be hiding in your jacket is scary, but how you respond to that determines whether that one, scary-but-harmless thought spirals into a full-fledged anxiety attack. By recognizing that trigger for anxiety as simply a scary thought, but not necessarily true, you can relieve some of the distress you feel.

Depression, like anxiety, is a common mental health issue that many people misunderstand. Despite the fact that so many people see it as someone simply being sad and needing cheering up, it is, in reality, a much more complex mood disorder, and is completely separate from feeling sadness or grief. People who are depressed may find themselves unable to regulate their positive moods, feeling trapped in negative ones. Moods are the emotional states people find themselves feeling. How people feel influences how they respond to the world around them, and positive moods are important for normal human function. When in a positive mood, people are more likely to engage in problem-solving critical thinking and are more likely to help others. Conversely, a negative mood incentivizes engaging in mood-elevating acts, such as helping others to reduce one's own negative feelings. When unable to regulate mood, daily functioning becomes impaired.

A person with depression may find their mood permanently, or nearly permanently, negative, which then begins to impact their actions. They may largely feel uninterested in life or unable to feel pleasure. They may feel persistent anxiety, sadness, or emptiness, no matter what they do. Since they feel a persistent lack of pleasure, they begin to lose motivation to engage in acts that may elevate moods, such as helping others or personal care. As it does not help them regulate their mood, it is not incentivized, and therefore becomes a waste of time.

Someone suffering from depression may feel a variety of symptoms, ranging in severity. Commonly, a feeling of sadness, hopelessness, uselessness, or otherwise negative mood is reported, usually persistently or permanently. Those suffering may become uninterested in things they once loved, or find that they no longer feel pleasure when engaging in activities they used to enjoy. Sleep may be disrupted, though this varies from individual to individual; some may suffer from insomnia while others sleep too much. Individuals may feel a lack of energy or permanently tired, even when suffering from insomnia as well. Thoughts of suicide or death are serious symptoms that must always be addressed with members of the medical profession. It is a medical emergency if you or a loved one begins to think about suicide and it is imperative to seek appropriate treatment immediately.

Along with genetics, brain chemistry and structure play a large role in the development of depression as well. Imbalances of levels of serotonin and dopamine have been found in people with depression. Serotonin regulates physiological functions such as sleep cycles, metabolism, sexual behavior, and controls mood. Dopamine is a neurotransmitter that activates the brain's reward center; it makes people feel good to reinforce beneficial or pleasurable activities, such as eating food that tastes good, exercising, or bonding with family or significant others. When dopamine is reduced, pleasure as a whole is reduced, meaning motivation is. Likewise, when serotonin is reduced, functioning becomes disordered. Together, this culminates in the symptoms of depression.

Stress may also trigger an instance of depression, which may be called situational depression. This occurs when a person responds to extreme stress with a depressed mood. These stressors can be a single event or a long, drawn-out event, such as marital issues. Even events that would generally be regarded as positive, such as having a child or getting a new job may trigger depression if the stress associated with the positive event is more than the person can handle. This increase in stress can sometimes cause a temporary depression.

Depression, like anxiety, is treatable, often with a mixture of therapy and medication to try to correct for the physiological causes. In cognitive behavioral therapy for those with depression, the goal is to help people reconnect with activities that they may have given up during their depression. Through setting goals and learning how to follow through with them, people slowly ease into activities that they value, with the hope that they rediscover the joy in their hobbies. It also commonly involves analyzing thoughts for cognitive distortions and correcting these negative distortions in the hope of easing the feelings of sadness, hopelessness, or uselessness those with depression may face. Cognitive behavioral therapy may also adapt to the person with depression's situation and provide coping skills for their stressors.

For both depression and anxiety, some simple healthy lifestyle changes may aid in alleviating distressing symptoms. Exercise has been shown to lessen stress and increase endorphins, which regulate mood. Getting a full night's sleep also allows for neurotransmitters to be regulated and replenished, helping the brain function normally. A healthy diet helps foster healthy gut bacteria, which has been shown in mice to alleviate symptoms of anxiety. Avoid drugs that may alter the mind, including alcohol, nicotine, and caffeine, as these have been linked to increased feelings of anxiety and depression. By living a healthy lifestyle, you may see some improvement in mood and symptoms, but it will not cure your anxiety or depression altogether. It is important to look for treatment from a medical professional if, at any point, you feel as if your anxiety or depression is too much for you to handle on your own.

Depression and anxiety may seem impossible to handle for those in the throes of their symptoms. Despite this, you can get your symptoms under control through changes to your lifestyle, following the guides and lessons cognitive behavioral therapy aims to teach and, in some cases, medication. Cognitive behavioral therapy will prepare you to handle stressful, triggering situations, as well as how to rewire your thinking in ways that are more conducive to living a happy, mentally healthy life. Through effort, perseverance, and with time, you should feel your symptoms lessen and life should become easier. If you feel as if you cannot do it alone or with the guidance of a book, there is no shame in seeking help from trained professionals.

CHAPTER 11:

Description of The Main Problems Part 2

Anger

Anger is a state of emotion that varies from mild irritation to full-blown rage of fury. Anger, just like other emotions is accompanied by biological and physical changes. When a person gets angry, their heart rate and blood pressure shoot up. The energy and hormonal levels also increase with anger.

A person can experience anger from internal or external stimulants. You can be angry because of a certain person or event, or from being worried and brooding on your personal issues. Having traumatic memories of events that were enraging can also cause anger feelings.

How Do People Express Anger?

Naturally, a human being expresses anger by responding aggressively. Anger is a natural way to respond to threats. It results in aggressive feelings and behaviors that allow you to be defensive to protect yourself when attacked. For one's survival, a certain amount of anger is acceptable.

On the extreme, a person can lash out physically at any person or thing that annoys and irritates him. Social norms and law work to limit how far a person can express their angry feelings.

A person uses a variety of processes to deal with anger. These processes can be unconscious or conscious. There are three approaches to dealing with anger, through expressing your anger feelings, suppressing them, and finding calm. Depending on an individual, you can deal with anger by:

- Expressing your anger. A person should express their feelings of anger, not in an aggressive way but assertively. In order to do this, a person needs to learn how to clearly express their needs and formulate ways to get them met without hurting other people. When a person is assertive, it does not mean they are demanding or pushy; it is respectful of you as well as of others.

- Suppressing your anger. You can decide to suppress, convert, or redirect your anger. This is usually when a person holds their anger in, stops thinking about the situation, and focuses on something positive. This technique is aimed at inhibiting your anger feelings and converting those feelings to more constructive things or behavior. However, this approach can be dangerous. With no outward expression of anger, it can turn inward and can cause depression, high blood pressure or hypertension.

Anger that has not been expressed can cause many problems. It can result in pathological expressions such as passive-aggressive behavior. This is where a person decides to get back at people indirectly without their knowledge instead of directly confronting them. A person that suppresses anger can also develop a cynical and hostile personality. This is a person that will always put others down, criticize, and make cynical comments. A person with these strains most often fails in having healthy and successful relationships.

- Calming down inside is another way of dealing with anger. This involves taking control of your outward behavior as a result of the anger feelings as well as your internal responses. You can do this by finding a way to lower your heart rate, calming yourself, and allowing the feelings to subside.

Fears and Phobias

Phobias and fears are types of anxiety disorders. A fear develops into a phobia when a person is expected to change their lifestyle to manage it. A phobia can be explained as an extreme or irrational dread or fear that is aroused by a particular situation or object to the extent that it affects your life.

If a person is suffering from a phobia, they will go to great lengths to stay away from a situation or an object that other people may be considering it to be harmless. The superb news is that you do not need to live with phobias because they are treatable conditions. It is possible to overcome a phobia. Some phobias are easier for a person to live with because they do not affect their daily life. For instance, the fear of snakes, also called ophiophobia, will not affect your day-to-day life. On the other hand, agoraphobia, which is fear of open spaces, can make it difficult for a person to lead a normal life.

If a phobia starts to interfere with your daily life, it is time to seek help. In some severe cases, a person may have to stop working because they are unable to take public transport. This kind of phobia can interfere with a person's way of life, and statistics show that it affects about 8% of the UK population.

Fears and phobias can be specific. They may include fear of heights, spiders or dentists, or generalized kind of phobias or fears.

Some of the common phobias include:

- Social phobia – this is the fear of social interactions.
- Agoraphobia – This is when a person is afraid of open public spaces
- Emetophobia – this is when a person is afraid of vomiting
- Erythrophobia – this is the fear of blushing
- Driving phobia – as the name suggests, it is the fear to drive

- Hydrochodria – this is the fear of getting sick
- Aerophobia – this is the fear of flying
- Arachnophobia – this is the fear of spiders
- Zoophobia – the fear of animals
- Claustrophobia – the fear of spaces that are confined.

A person that suffers from social phobia may have started by being just shy. When this behavior is exaggerated to the point that it disrupts a person's life, it is time to seek treatment.

How Do Phobias Start?

No one knows exactly how phobias develop. However, certain phobias are believed to originate from childhood, mostly between the age of 4 and 8.

Agoraphobia and social phobia often start later in life. These phobias start mostly at puberty or in the later teenage years and earlier twenties. According to psychologists, a good way to eliminate some fears in children is by familiarizing them with the objects of their fear.

CHAPTER 12:

Description of The Main Problems Part 3

Defining Negative Thoughts

Negative thoughts are any problematic thought patterns that you may develop. Some are cognitive distortions, meaning they are false by nature, while others are simply thoughts that are so focused on the negative, they caused you issues. For example, if you think that you are a horrible person after dropping a plate on the floor and shattering it, you may have a problem with negative thoughts and cognitive distortions. These distortions should not be entertained and instead need to be banished and ignored, and yet people everywhere foster them.

By and large, these negative thoughts can corrupt your core beliefs, turning them into false; negative beliefs about yourself that then become problematic in general. When this happens, you frequently are stuck on those negative thoughts, so caught up in them that you cannot move past them. You instead fixate upon them and let them corrupt your image of yourself or those around you. This struggle can lead to all sorts of anxiety problems in the future.

Problems with Negative Thoughts

When you let negative thoughts rule you, you essentially allow yourself to root your mind in negativity. You are accepting the fact that your negative thought patterns run your life once you become aware of them and do nothing to change them. They can ruin relationships, destroy self-esteem, cause myriads of mental health issues, and yet, some people prefer to hide their heads in the sand and pretend there is nothing wrong despite the clear issues ahead of them. This is problematic—people are then so stuck in these negative thoughts that they cannot control the negative feelings or behaviors that follow as a result.

Remember, negative thoughts deserve to be challenged. You do not have to live a life rooted in negativity in which you are stuck living out the same negative cycles over and over again. You deserve to challenge those negative thoughts and develop happier, healthier mindsets that can benefit you greatly, ensuring more success and happiness in your relationships in the future.

Most Common Negative Thought Patterns

When you are engaging in negative thought patterns, you are likely unaware of it. However, there are 15 common negative thought patterns known as cognitive distortions that people regularly find themselves caught up in. Take the time to study these 15 distortions and identify whether you yourself suffer from them. If you do, you know that challenging them is within your future.

- **Filtering:** When you filter, you see things as entirely negative. You ignore the existence of the positive entirely, only allowing negative occurrences to define the way you see the world. For example, you decide that because you dropped a bottle of milk and shattered it once, you are a horrible, wasteful person, despite the fact that you rarely actually waste milk.

- **Polarized thinking:** When you engage in polarized thinking, everything is black and white with no in-between. Things are good or bad, right or wrong, perfect or failed. For example, you decide that you are largely undeserving of friendship simply because you accidentally betrayed your friend once and therefore, are not worthy of being a good friend.

- **Overgeneralizing:** When you overgeneralize, you assume that people and experiences are all the same based on small sample size. For example, you may have gotten sick once at an ethnic restaurant, and after that, you assume that all restaurants of that ethnicity are to be avoided.

- **Jumping to conclusions:** when you jump to a conclusion, you essentially state that you know exactly what will happen next without actually knowing or having any evidence to prove it. For example, you assume that your best friend will drop you without warning because you have a gut feeling it will happen.

- **Worst-case scenario thinking:** When you engage in worst-case scenario thinking, you essentially assume that the worst possible result has occurred before having any evidence that supports it. For example, when your friend does not answer the phone, she must have been kidnapped and murdered because she always answers the phone, but did not that time.

- **Personalizing:** When you personalize, you assume that you are the cause of all misfortune around you, at least in part. If someone is having a bad day, it must be your fault. If someone is angry, they must be angry with you. If there is a car accident, you must have done something to cause it.

- **Control fallacy:** When you have a control fallacy, either you blame yourself wholeheartedly for everything around you or you reject the fault of anything that happens around you. For example, you decide that it is your fault that your neighbor's house was robbed instead of yours because you left a light on.

- **Fairness fallacy:** When you act under a fairness fallacy, you assume that everything has to be fair and pursue utter fairness despite the fact that life is not, in fact, fair, and you may not be deserving of it. For example, you assume that if someone else got a job with the same degree and credentials you have, you should be able to get the same job and pay rate, with no regard to differences in personality.

- **Blame:** When you fall for the blame fallacy, you point at other people as the cause of problems rather than acknowledging that you could have some fault in the matter. You essentially refuse any accountability. For example, you may tell someone that you had no fault in the matter of a car accident because the sun shone in your eyes and you could not possibly see the other car because of it.
- **Should have:** You focus so much on whether something should or should not be, rather than seeing the world for the way it is. For example, you fixate on the fact that you should have gotten the job you interviewed for, so you refuse to try for other jobs, assuming that you will, in fact, be called back for that job.
- **Emotional reasoning:** When you engage in this, you assume that anything you feel must be true simply because you feel that it is true despite the fact that it is not the case. You may feel like you failed and therefore deem yourself a failure
- **Change fallacy:** Within this fallacy, you assume that other people will change for you to meet what you want or need, and when they do not, you get flustered or struggle to function.

CHAPTER 13:

Your Thoughts

Negative Thoughts

Not all negative thoughts are the same. Work on properly identifying the type of thoughts you're having by categorizing them into some of the following distinctions:

Blaming

Becoming a passive victim of circumstance can make it extremely difficult to act on changing your situation. Blaming others for the problems in our lives can ruin relationships and take away our personal power. Not taking responsibility also makes you powerless to change. Blaming thoughts include "How was I supposed to know that", and "That wouldn't have happened if."

Always/Never

Thinking in absolutes and categorizing your views into extremes is typically made up of self-defeating thoughts. These types of thoughts typically include words such as "everyone" and "always" or "never" and "none." Black or white thinking can stem from a constant need for approval, being a perfectionist, and believing that attaining something makes up your value.

Take time to examine those irrational beliefs you may hold. Aim to create a balance between your actual and your ideal selves that is more realistic. These thoughts can include "They never listen to me", "I'm the only one that takes care of things", "I'll never get a raise."

Focusing on the Negative

One example of focusing on the negative would be if you had an evaluation at work. Your manager could give you 9/10 excellent marks - detailing everything that's great about the work you're doing and give a few notes on the one area that may need work.

When you're focusing on the negative, you're more likely to dwell on the one thing that needs work than the nine excellent things. When we allow negative thoughts to flourish, this can turn into a situation where you feel bad about yourself, resent your boss, and ultimately let it affect the work you're doing. Seek to find the positive and allow your thoughts to help you find more balance and optimism in situations.

Generalizing is just as bad. "That was a bad day." Was your entire day bad or are you only remembering the bad parts of it? This is a sign of depression that you'll want to watch out for. Look for the things you enjoyed during the day - challenge yourself to find a few things.

Thinking with your feelings

Our feelings are very complex and are usually deeply rooted in powerful memories. Feelings aren't always telling the truth either. When you have a strong negative feeling, examine it - is there evidence behind the feeling? Are there real reasons to feel this way? Typically, these types of thoughts start with "I feel" like - "I feel stupid" or "I feel like they hate me."

Fortune telling

Negatively predicting the outcome of a situation can really influence how you feel.

Don't cut yourself off from changing because you don't think it's possible. For example, thinking "I never manage to make new friends at work events, I never have." Just because you never have doesn't mean that you never can.

This type of thinking can lead to inaction - stopping yourself from having experiences and participating in life because you think you already know how things will go. Remember, if you could tell the future, you'd more than likely be a billionaire right now.

Guilt

Guilt is not a productive emotion and it can cause you to do things you don't want to do. Be careful to not use guilt to motivate or punish yourself and when making commitments and promises, don't overextend yourself. Guilt can be insidious and destructive.

Although separate from shame, guilt can eventually lead to shame when it's irrational or not absolved. These emotions have the opposite effect of increasing empathy and self-improvement. It's easy to beat yourself up and prolong these types of thoughts, and we tend to simply brush them under the rug by rationalizing our actions.

These types of thoughts are repetitive and evoke progressively more intense feelings the longer we sit with them. Thoughts that include words like "have to, must, should" are typically associated with guilt. These negative guilty thoughts can manifest when we procrastinate as well.

We often don't want to do things simply because we think we must. A great way to combat those is by switching it from "I have to spend more time at home" to "I want to spend more time at home."

Mind reading/Projecting

These negative thoughts can manipulate your mind and are very common in those with social anxieties. Mind reading thoughts usually lead you to assume that other people are judging you negatively.

Projecting takes your own thoughts and transfers them onto others - convincing us that they are thinking the same thing we are. We would never really know what someone else is thinking unless they tell us.

Believing that you are a mind reader can damage the relationships we have with others, especially our significant others. It's easy to assume that you know what someone is thinking based on our past experiences, but we must be careful to clearly communicate with others.

When making assumptions, we can find ourselves misreading someone's negative look to mean something personal when it could be something completely unrelated such as them thinking that their lunch isn't agreeing with them.

Labeling

We can become unable to deal with people reasonably and recognize them as unique individuals when we attach negative labels to them. This is also true when attaching negative labels to ourselves.

For example, if you think of someone as arrogant, you're likely to mentally sort that person into the same category as all the arrogant people that you've ever known.

Be careful to challenge distinctions you are automatically assigning to people based on a situation, most times you'll find that someone's behavior in a certain situation isn't the only characteristic about them.

What will you feel if someone considered you the same way? Challenge any preconceived labels and work to discover a person's real character.

Personalizing

We can never fully realize why people do what they do. Personalizing the behavior of others and events with negative explanations can make us feel terrible and puts a distance between us and everything else.

These types of thoughts can lead us to become fixated on what's happening only to ourselves, and blind us from seeing the big picture. Assigning personal meaning to simple events such as a co-worker that didn't greet you can lead us to believe that they must be angry or don't like us anymore.

Personalizing thoughts go hand in hand with catastrophizing thoughts - getting negative feedback on something doesn't mean you aren't able to become better or that it questions your worth. Our connection to experiences can be completely impersonal and doesn't necessarily need a response from you. Break free from this type of self-imposed solitary confinement by letting "what is", go about its business.

Exposing a lot of these thoughts to a more rational light can lead you to believe that you're just lying to yourself. Just because you've believed something for an extended period doesn't make it true. Most negative thinking is automatic and will go unnoticed.

Write down these thoughts and respond to them. Take away the power they have on you and take control of your moods. Build that inner voice to keep these types of negative thought patterns in check. You can do this by catching a negative thought and identifying which type it is most like. Try to change the thought to something that is more realistic.

For example, "You never listen to me" would be identified as an "Always/Never" type of thought and we can change it around to "I get upset when you don't listen to me but I know you have listened to me in the past and you will again." Start to build positive thought patterns by approaching life more confidently and productively.

An easy exercise you can do when focusing on building positive thought patterns and associations is by using your 5 senses. Jot down a list of the happiest times you've experienced. Describe them in as much detail as you can. Where were you, what music was playing, what colors were around you, smells, people, setting, etc. Creating these paths of positive memory will strengthen your bonds. This will spur you to act lovingly and act on that feeling.

There are physiological and mental health issues that can seriously affect people above and beyond the bad moods and thought patterns most of us experience.

The Mind/Body Connection

Consider the mind-body connection. Our bodies, feelings, mind, and spirit are all interrelated and our mind-body wellness is in a reciprocal relationship with the way we manage stress.

If we don't feel good physically, the chances that we will feel good mentally are very low. Try listening to your body and treat it with respect by giving it what it needs to function properly.

There are easy steps to improve how you feel physically, and in turn mentally. Take note of your energy levels, what activities deplete you and what activities restore? Create a healthy balance by creating ways to nurture and care for all aspects of yourself.

CHAPTER 14:

Setting the Objectives to Be Achieved

As you have learned so far, having a negative attitude towards life keeps us from being happy and affects those that we interact daily with. Science has more than enough proof to show how being positive impacts your levels of happiness and terms of success. This is why making positivity a habit with the help of small changes can help you to drastically change your overall life and the mindset you have towards the world.

The life you are living is a mirror of your overall attitude. It can be quite easy, almost too easy, to be cynical at the world and see it as a mess of injustice and tragedy, especially thanks to the media that we all spend many hours a day on. Negativity is holding you back from really enjoying your life and has a great impact on your environment as well. The energy that people bring to the table, including you, is very contagious. One of the best things you can do in your life that is free of charge and simplistic is to offer positive attitude. This is especially beneficial in a world that loves and craves negativity. One of my favorite quotes of all time comes directly from the King of Pop, Michael Jackson: "If you want to make the world a better place, take a look at yourself, and make a change."

As humans, we are creatures of habit.

Smile

When asked who we think about most of the time, the most honest answer would probably have to be ourselves, right? This is natural, so don't feel guilty! It is good to hold ourselves accountable and take responsibility for ourselves. However, I want to challenge you to put yourself aside for at least one moment per day (I recommend striving for more) and make another person smile. Think about making someone else happy and that warm feeling you get when you receive happiness. We don't realize how intense the impact of making someone smile can have on those around us. In addition, smiling costs nothing and positively works your facial muscles!

Point Out Solutions, Not Problems

Embracing positivity doesn't mean you need to avoid issues, but rather, it is learning how to reconstruct the way you criticize. Those that are positive create criticisms with the idea to improve something. If you are just going to point out the issues with people and in situations, then you should learn to place that effort instead into suggesting possible solutions. You will find that pointing out solutions makes everyone feel more positive than pointing out flaws.

Notice the Rise, Not Just the Downfall

Many of us are negative just by the simple fact that we dwell too much on the hate and violence that is in our daily media. However, what we fail to notice is those who are rising up, showing compassion, and giving love to others. Those are the stories you should engulf yourself in. When you are able to find modern-day heroes in everyday life, you naturally feel more hopeful, even in tough times.

Just Breathe

Our emotions are connected to the way we breathe. Think about a time that you held your breath when you were in deep concentration or when you are upset or anxious. Our breath is dependent on how we feel, which means it also has the power to change our emotions too!

Don't Get Dragged Down by The Negativity Of Others

I'm sure you have gone to work cheerful and excited to take on the day ahead, but then your co-worker ruins that happy-go-lucky mood of yours with their complaints about every little thing, from the weather to other employees, to their weekend, etc.

It is natural to find yourself agreeing with what others are saying, especially if you like to avoid conflict. However, you are initially allowing yourself to drown in their pool of negative emotions. Don't fall into this trap.

Conflict may arise, but I challenge you to not validate the complaints of a friend, family member, or co-worker next time they are going about on a complaint-spree. They are less likely to be negative in the future, if they have fewer people to complain to.

Swap Have with Get

I am sure you often fail to notice how many times we tell ourselves that we have to go and do something.

"I have to go to work."

"I have to go to the store."

"I have to pay for rent."

"I have to mow the lawn."

You get the picture. However, watch what happens when you swap the word have with the word get.

"I get to go to work."

"I get to go to the store."

"I get to pay rent."

"I get to mow the lawn."

See the drastic change in attitude there? It goes from needing to fulfill those obligations to be grateful that you have those things to do in your life. This means:

You have a job to go to

You already have sufficient money to support yourself and your family to provide a healthy meal

You have a roof over your head

You have a nice yard

When you make this simple change, you will begin to feel the warmth of happiness snuggle you as the cold blanket of stress falls away.

Describe Your Life with Positive Words

The choice of vocabulary we use has much more power over our lives than we realize. How you discuss your life is essential to harnessing positivity since your mind hears what you spew aloud.

When you describe your life as boring, busy, chaotic, and/or mundane, this is exactly how you will continue to perceive it, and it will directly affect both your mental and physical health.

Instead, if you describe your life as involved, lively, familiar, simple, etc., you will begin to see changes in your overall perspective, and you will find more joy in the way you choose to mold your entire life.

Master Rejection

You will need to learn to become good at being rejected. The fact that rejection is a skill. Instead of viewing failed interviews and broken hearts as failures, see them as opportunities for practice to ensure you are ready for what is to come next. Even if you try to avoid it, rejection is inevitable. Don't allow it to harden you from the inside out.

Reframe Challenges

Stop picturing your life being scattered with dead-end signs and view all your failings as opportunities to re-direct. There are little to no things in life that we have 100-percent control over. When you let uncontrollable experiences take over your life, you will literally turn into mush.

What you can control is the amount of effort you put into things without an ounce of regret doing them! When you are able to have fun taking on challenges, you are embracing adventure and the unknown, which allows you much more room to grow, learn, and win in the future.

Write in a Gratitude Journal

There are bound to be days where just one situation can derail the entire day, whether it is an interaction that is not so pleasant or something that happens the night before, our mind clings to these negative aspects of the day.

I am sure you have read on multiple sites about how keeping a gratitude journal is beneficial. If you are anything like me, I thought this was total rubbish that is until I started doing it. I challenged myself to write down at least five things that I was truly grateful for every day. Scientifically, expressing gratitude is linked to happiness and reducing stress.

I challenge you to begin jotting down things you appreciate and are grateful for each day. Even on terrible days, there is something to be blessed about!

CHAPTER 15:

Identification of the First Obstacles

Anxiety can be dangerous for your health. Persistent negative thoughts and fear can cause a person to seek palliative elsewhere if they cannot find one. if the symptoms are not treated, then the symptoms can grow worse. Anxiety is linked in many cases to depression and the combination can lead to extremely poor decision-making when a person seeks to alleviate their discomfort. Some will turn to alcohol or drugs; others might cause themselves physical harm.

There are other ways that anxiety is manifest such as a decrease in social interaction which can be highly counterproductive to their Improvement. Indeed, reduced social contact can make their symptoms worse and they might have trouble noticing when they are worsening. It can also make relating to others more difficult if they fall out of practice and thus, they are enforcing their feelings of poor self-esteem.

Not every case of anxiety will go this way, however, in cases where there are high levels of constant anxiety, the person may seek assistance and relief in illegal substances or alcohol or other methods that might reduce their anxiety. Once someone starts to become addicted to something, they need to seek help immediately. That is why they need to seek the assistance that a CBT therapist may offer to go through exposure to anxiety-inducing things, as well as convincing them of the necessity of rebuilding their support network.

There are some primary and negative feelings associated with anxiety that one should be aware of. This includes a sense of impending doom that makes it difficult to concentrate because you are always having the feeling that something is going wrong. Anxiety attacks and the associated feelings of fear and stress may also make it difficult for the sufferer to re-establish a normal baseline.

Additionally, it is important to remember that depression is linked to anxiety but if one person has one it is likely they may have the other whether it is occasionally or over the longer-term. These in combination can cause high levels of stress which can lead to both high blood pressure and headaches and when this is the case avoidance coping may be seen as they try to manage what seems to be a very heavy emotional and mental load.

The feeling of constantly being uncomfortable may also make the person withdraw; they may seem quiet; or they may seem to be easily irritable. It may increase their reactiveness or their levels of anger if they are feeling discomfort or headache. On the other hand, there are others who will simply hide their pain and discomfort and not speak up.

In some extreme cases, anxiety may even lead to breathing trouble. Sufferers might experience rapid shallow breathing and their body may not enable them to take in enough oxygen to help them calm down. Constant stress can lead to stomach problems as well as ulcer gastrointestinal intestinal issues and other physical ailments. A balanced diet can alleviate some of these symptoms.

Anxiety will also affect the person's energy levels particularly, if they have panic attacks. If they are constantly feeling stress, they can wear down their immune system and tire out their body muscles and even their organs. They may require more rest to recover so they may also have trouble sleeping. This is associated with feelings of anxiety. it might also reduce their desire to have sex, possibly causing problems in their personal relationships.

Social anxiety is very common and it can affect the way a person relates to others as well as the frequency. It comes associated with a person having difficulty or feeling anxious at the thought of associating with other people speaking in front of groups or even having conversations.

These persons excel at avoidance strategies. For this reason, they may be afraid of making friends because they are suffering from low self-esteem and worried about what others think of them.

This is not simply a case of being shy and one should not accuse them of this. It is a high level of discomfort and fear that may even arouse the fight or flight Instinct in a person that makes them feel as if they were in danger. They might have elevated heartbeat sweating shallow rapid breathing and they may feel extremely uncomfortable. For those who have this severely and are not being treated, can even find themselves withdrawing entirely from society from support groups from those who love them and this does not bode well for their ability to recover. If they are not given the exposure and care that can encourage them to seek help, they will simply fall back into their emotional maelstrom.

The symptoms of social anxiety disorder will be very similar to those who have been reading about CBT and negative thoughts. Sufferers engage in negative interpretations of situations as well as negative thoughts and they worry about interacting with others. They engage in catastrophizing and predicting the future and they are strongly invested in believing that things will not turn out well. They may also engage in physical behaviors to express their anxiety; these may include such things as twiddling their fingers in repetitive motions or other physical exhibitions of discomfort.

Panic attacks can cause a debilitating amount of harm to the sufferer, including loss of income if they are too anxious to go to work. There is also a loss of physical activity as well as a loss of social activity. Additionally, the sufferer may even reach a point where they have suicidal thoughts or tendencies. In these cases, it is imperative that the sufferer see a physician or therapist immediately. Sadly, some sufferers of this disorder may not even know if they suffer from anxiety. They may wind up at a physician's office and hopefully the doctor will send them on to a therapist for treatment.

Sometimes people who are burdened with anxiety turn to other self-destructive ways of coping. They may choose to harm themselves as they feel a sense of relief from the thoughts that they have in their minds. Also, they may self-harm in an attempt to exercise some control over their pain when they feel that they have lost control over other aspects of their life.

There are two types of anxiety to be aware of when it comes to tendencies to self-harm: social anxiety and generalized anxiety. Those with social anxiety may choose to self-harm and feel in some way that they're feeling that they are punished. Those with generalized anxiety are generally more inclined to use self-harm as a way to release the stress and anxiety that they feel constantly. It may also be a way for them to silence the negative thoughts in their heads. In serious cases those who may have felt anxiety for quite a long time may in fact self-harm because they simply are looking to feel something because they are numb to more normal feelings. The most destructive type of this behavior can be seen in those who self-harm out of anger. These are those with social anxiety and they are also feeling angry towards themselves because they feel that they did not do enough to avoid being in this situation. They may also feel the need to be punished if they are not able to manage the symptoms they are experiencing. This is a dangerous form of self-harm and could hint towards underlying more serious emotional issues.

Risk factors for those who use self-harm as a coping mechanism include such areas as depression, anxiety and alcohol abuse. If you find yourself in such a situation please seek the help of a qualified professional to help you find assistance as soon as possible.

CHAPTER 16:

How Does One Go About Identifying Cognitive Problems?

To start with, cognitive problems, just as you may have noticed, are issues that an individual may encounter with their memory or thinking. There are two ways an individual can deal with such impairments.

Step 1: Recognizing the Thoughts Made, Physical Symptoms, Or Even A Change in One's Behavior.

Cognitive therapy usually calls for a step by step breakdown of actions to take with the motive of dealing with a cognitive problem.

This is because one can take some time while dealing with a problem. Therefore, the breakdown of a plan into small manageable goals is usually indispensable.

A good example to back up this statement is in the case of test failure by a student. It might prove hard for the student to move directly from the low grade to the higher one.

However, it is much easier and more realistic for the student to move from one rank to another until he or she reaches the desired level.

In this step, therefore, one has to evaluate their thoughts, physical symptoms, and even a change in their behavior.

Eventually, this will be able to help one ascertain whether or not they are under some emotional distress or not.

To understand this better, we are going to deal with one at a time.

Thoughts

These are the things that one thinks about during an event. When one is depressed or anxious, amongst other issues, various thoughts usually cross one's mind. These thoughts typically depend on the situation. The thoughts we are concerned with here are majorly negative thoughts. They may include:

- I am a failure.
- This is not meant for me.
- I will not succeed.
- I am weak.

When you realize you are having such thoughts, you are under some distress of some kind that is clouding your judgment. How does one even get to the bottom of it?

An example of this is by maybe considering:

- What happened?
- What did you think about the situation?
- Was it correct or not?

Try to answer what is causing what you are thinking. Finally, evaluate the answer and compare it to the judgment you would have made while in a sober state or rather the decision you would have come to if the current problem did not tamper the mind.

Physical Symptoms

The next thing one must check for is the physical symptoms. The physical symptoms we are talking about include:

- Restlessness
- Lack of sleep

- Increased heartbeat
- That feeling of butterflies in the stomach

A majority of people have encountered a number of these physical symptoms, and the leading cause is emotional distress. Therefore, when one notices such problems, be sure that you are under some difficulty, maybe stress or even depression.

A good example is like in the cases of anxiety and stress, where the adrenaline in one's body can cause one to feel hot, sweaty, an increased rate in heartbeat, difficulty in breathing, shaky, increased urge to go to the toilet, and even butterflies in the stomach which could lead to stomach discomfort. On the other hand, in the cases of depression, one feels tired, exhausted, lack of sleep, or even lacks interest in sex.

Behavioral Change

The last thing that will help one ascertain whether they are under some emotional distress is a change in behavior. Stress and depression are usually associated with some responses which include:

- Distancing oneself from friends and family
- Lack of self-motivation
- A tendency to eat less

A prime example of this is in the case of depression; one may decide to stay in bed and pull the covers over their head, choose not to go out, not to answer the phone, watch television, or even decline a friend's invite without a solid reason. On the other hand, in the case of anger, one can shout at someone, throw or aimlessly vandalize things around him, beat up someone or something.

Look at it this way, the thought that one is a failure can lead to increased stress which may cause the person to become restless and thus lead to a lack of sleep. The notion that one is a failure may even lead to one thinking that they are not meant to socialize with certain people, and therefore one distances himself or herself with the outside world.

Thus, there is a cycle of negative thoughts, changes in behavior, and physical symptoms that will keep you emotionally distressed. We will see as we go on.

Step 2: Define Your Problem and Set Goals

This step usually follows after one has determined whether he or she has a cognitive problem, as in step one. This step, therefore, helps one define the problem and, most importantly, accept the challenge. While describing the problem, one has to be sober and of clear mind to determine the problem precisely. Instead of just saying that the problem is the stress, one should go more in-depth and establish the cause of your stress. There are several parameters that one should consider while identifying the problem, like:

- What causes the problem?
- Why does the problem affect me?
- When does the problem affect me most?
- Where does the problem occur?

Once you ask yourself these questions, then you will have achieved the second step in dealing with cognitive problems.

Identification and evaluation, however, does not end here. One must accept the fact that he or she is facing a particular problem and thus face it head on to solve it.

The best way, therefore, is by setting goals. A goal, here, is what the patient wants to achieve by the end of the therapy. Remember that as you read this, you are the therapist helping yourself to get better. A goal helps in remaining loyal to the course and going ahead to check on the progress for some feedback.

The one trick with setting goals is that one must be as transparent as possible. Instead of saying that by the end of the therapy you want to feel less depressed, be more precise and note down what you would want to do once you are less depressed. Some examples are:

- I would want to go on a road trip somewhere
- I would like to have a cup of coffee quietly
- I would like to visit a specific place

These goals can be very enticing while writing them down, but it is good to have just a few goals so as not to make it cumbersome while achieving them.

CHAPTER 17:

The Best Techniques to Follow to Obtain Effective Results and Solve Problems Part 1

Develop Your Own Mantra

If you are an avid follower of Hinduism or Buddhism, you are probably familiar with the word “mantra." Essentially, a mantra is comprised of words or sounds that repeated during meditation to help you focus. The long and protracted “ohm” sound is one of the most popular meditation mantras. They are a really important part of the meditation routine. However, for this task today, mantra takes on a different meaning, as I am not really talking about meditation. The mantra I am referring to has more to do with shaping your life and mindset in this phase. And today, you and I are going to work on allowing you to discover your power mantra.

When you start up a company, one of the things you need to work on is branding. This helps your customers identify you from the sea of choices available to them. A lot of the work that goes into the branding of a company is more focused on the aesthetic aspect. These companies strive to influence the market's opinion of them with the help of logos, brand colors as well as the fonts used in all of their marketing materials. However, the element that really defines how the company operates is tan organization’s motto. In the same way, the clothes you wear, the way you wear your hairstyle and all the physical stuff are just aspects of personal branding.

You can use your style to influence people's perception of you. But the thing that really defines how you interact with people, reacts to situations and generally carry yourself is strongly influenced by your personal beliefs. When we hear beliefs, we start thinking about religion and culture. And to be honest, our religion and customs heavily influence our behaviors, but think about this for a second. If our beliefs were really rooted in our religion and culture, there is a very strong possibility that people who share the same customs would be more like replicas from a factory than the unique individuals that we are.

The differentiating factor for us is what we believe in personally. And even though you may not have given it a definition yet, the fact is you have a mantra. When you do not make the conscious effort to choose the mantra that defines you, life and everything else that happens to you will make that decision for you. And this can be one of the many reasons why many people are trapped in a destructive emotional cycle. There is a popular saying that if you don't stand for something, you will fall for everything. Today, you have to take the bold step of defining you. The mistakes you made in the past, the failures that you are living through in the present as well as your fears and concerns for the future are aspects of choices you make that have affected or will affect you. But they do not define you.

"You are what you eat" phrase. Like aesthetics, this premise influences one aspect of you. Mantras are like your mental food. And because we know how strongly the mind influences our behavior, this makes a great case for choosing to define yourself. And now that we have established that, we have to go to the next step which is selecting the mantra.

Mantras in this context could be anything from a favorite celebrity quote to something culled from your favorite ancient Greek philosopher. Whatever you choose, do not fall for that socially trendy type of quotes just because you feel people would think it is cool. Like with everything that has to do with your emotional wellbeing, the final decision is up to you. From the moment you were born, there are several voices competing for a space in your head. These voices are telling you how to do things, how to live your life and basically how to exist.

Now is your opportunity to create and horn in on your own voice. Let this voice take root deep within you and drown out every other belief system or voice that have held you down. The mantra should resonate with your innermost thoughts and desires. That is how you can tell that you are on the right track. It does not have to be "deep". But it should be something that every time you hear the words or speak the words, it changes your countenance for the better. Something as simple as "you are powerful" may be all you need to put your confidence together.

You can also have a mantra for different situations. Like if you are going to be making a presentation at work or maybe you have to speak in front of a crowd, you could choose a mantra that gives you the courage to step up and own the stage you are given. For times when you are experiencing some emotional lows which are very common if you are combating any of the negative emotions we talked about in the beginning, you can find mantras that will empower you to keep the darkness at bay.

When you find the right mantra, the next thing you might be thinking is how often you would need to say those words for them to have any effect. The only appropriate answer to that question is "as often as you need it." An old friend of mine used to reiterate, "motivation is like a bath: you need it every time you get dirty". And I agree with him 100%. Your mantra is not going to be a one-time-fix-it type of word that you just say, snap your fingers and everything falls into place. I wish there was some word or phrase like that but until that word has been discovered or invented, you are going to have to do your own behavioral conditioning every time you think you need it.

Say the words with conviction. Repeat them in sequence if you have to, but ensure that you are rooted in these words every day. As you evolve, you may have to take on new mantras. But you should always have that mantra that defines you no matter what. Find those words, speak those words, own those words and become those words. It would take a while for you to get there, but for today, let us settle for activating the inner victor within you with your own words!

CHAPTER 18:

The Best Techniques to Follow to Obtain Effective Results and Solve Problems Part 2

You may see methods that simply do not resonate with you, and that is okay. The important part is that you find several that you do enjoy. With the methods that you do find particularly enjoyable, you will still be able to see the same effects. Ultimately, the methods do not matter so much as the ability to utilize the methods that you do develop a preference for. With that said, however, keep an open mind as you learn some of these. While they may be off-putting at first, you may find that they really work for you after giving it a shot at least once.

Good luck as you begin the process of controlling your emotions, and remember that this is also a process that will take time. It may not be easy to learn to manage your emotions effortlessly, and you will slip up sometimes. That is okay. The important part is that you do not give up hope, and you continue to try to better yourself. If you continue to try, you will see results over time. They might start out small, but even a tree started out as nothing more than a seed at one time.

You can do this, even when you think you cannot. Remember that, and let's begin with the first emotional management technique.

Affirmations

Let's start off simply—Affirmations. Now, you have already learned about affirmations and utilized them in the cognitive restructuring process, so is it really that surprising to see them here in emotional regulation as well? When you are attempting to utilize an affirmation, remember that they have three distinct requirements:

They must be positive in structure to ensure you have a positive mindset. They must be personal in nature, so you are able to remain in control over the veracity of the process. They must be present-tense, so you know that they are true for you at the moment.

When you decide to use affirmations to manage your mood, the process is a little different than when using them to restructure your thoughts. This time, you will be focusing on how emotions, identifying them when you create your affirmation. The affirmation will be designed to be calming, reminding you to keep your mind in the present and keep your cool at the moment. If you are capable of doing this, you will find that your emotions will come and go without you overreacting to them in any way. Notice how this affirmation reminds you of exactly what you need at that moment. It cues you to pay attention to your own control of the situation first and foremost, while simultaneously reminding you to breathe. By giving you an extra action to do that you can control, this affirmation puts you in a position of control. That control makes it easier to control other actions as well, making this an effective affirmation for you to repeat. Other examples of emotional regulation affirmations could be:

As I breathe, I am able to feel my negative impulses leave my body

I know that managing my emotions is in my best interest right now

I am able to maintain respect for those around me, even when I am angry

These emotions I am feeling are valid, but I am still in control

Each of these affirmations seeks to ground you back in your own mind and remind you that you should not be responding emotionally. While they recognize that you will have emotions and that emotions should not be shamed or judged, they also recognize that you should maintain control and not allow yourself to act impulsively, using emotions to guide your behaviors.

Meditation

Meditation may seem intimidating, but it is something that you can manage with a little practice. It is also incredibly effective at encouraging you to develop a level of peace of mind that you may not have known was even possible. Through meditation, you will find that you are capable of relaxation and clarity you did not know existed. This is because meditation seeks to clear your mind, letting go of everything that stresses you out and focusing on that moment. You want to live in that moment, in that relaxed state. This can be incredibly helpful with your own emotional regulation—you can use it to ensure that you are calm and comfortable. You can use it to develop better emotional regulation skills simply because it is easier to regulate yourself when you are relaxed and calm. To meditate, you should stop and find a quiet, peaceful location and create a focal point of sorts.

You want to locate somewhere that you know you can spend at least thirty or forty minutes uninterrupted, in order to really get the full effects of the process. When you do find that place that you can meditate in peace, get comfortable, and settle into deep breathing. Take in a deep inhale for at least four seconds, hold it for two seconds, and then slowly release it for four more seconds, and hold it for two seconds before repeating.

This process will slowly settle you into a state of relaxation. As you reach that point of relaxation, begin to repeat your mantra to yourself. You will repeat it to yourself with every inhale, creating a slow rhythm to the process. Focus on the mantra completely as you repeat it to yourself, really envisioning what you are saying.

If your mind wanders, just gently redirect it back to your mantra and move on.

You let go of the thoughts that come to you and move back to your meditation every time you feel them. Over time, you will find that you can meditate longer with less distraction, and you may even notice your emotional regulation skills regulating.

Mindfulness

Similar to meditation, mindfulness will require you to settle down somewhere quiet and free from distraction. In doing so, you are capable to ensure that you will manage the process without someone interrupting you. In this process, you will be allowing your thoughts to occur, acknowledging them, but letting them pass with no judgment or attachment. You simply want to be aware of the thoughts that arise. This will teach you how to identify how your own emotions jump around, and you will even begin to identify chains of feelings as well. When you start mindfulness, you should again begin with the deep breathing, but this time, instead of repeating a mantra to yourself, you will be focusing entirely upon the feelings within your body. Notice the air enter your nose, travel down your trachea, and enter your lungs.

Really feel the oxygen as it flows through your body, following as it courses through your veins. Let the air escape your mouth and feel that as well. As you do this, allow your mind to drift and wander without judgment. Wherever it goes, you should acknowledge the thoughts there, but do not distract yourself. Do not try to change the subject or even try to change the thought patterns themselves. Simply acknowledge and continue to observe in a detached state.

Over time, you will begin to notice what your thoughts naturally want to do. If you notice any feelings that seem to occur in response to specific thoughts, that is okay too. You may simply want to be aware of how your mind works and what you need to do to understand them.

Positive Thinking

Remember how your thoughts are pervasive? Your thoughts will influence your feelings, which will influence your behavior. This process has been repeated throughout this book because it is so incredibly important. Using that process, however, you can understand why it is so incredibly important to use positive thoughts wherever you can. Because of that, you are challenged to play a little game. This game is very low risk—no one will get hurt, no one has to do very much, and the worst-case scenario is that you are stumped trying to think of something good in your life.

This challenge is simple: For every negative thought, you catch yourself thinking for the next week, you must drown that thought with three positive ones. This is the ultimate practice in appreciating your surroundings and finding things you are grateful for, even when you least expect it.

If you think negatively about a meal, for example, you are now tasked with creating three positive thoughts about that same meal. The three things that you come up with must be different, and they must all be about the meal, so there is no wording the same compliment in three different ways.

If you watch a movie with your best friend that you hate, if you voice a complaint, you must then say three things that you enjoy about the movie. If you complain about a chore or task you have to complete; you must identify three positive things about that. Be creative—you might realize you are grateful for more than you ever imagined.

For example, you complain that it is your night to do dishes after dinner. You are tired and frustrated and just want to sleep, and the complaint slips out. Now, you must come up with three things that are positive about the chore you are doing. Perhaps you come up with the following:

I am grateful that my body works well enough that I can do the dishes without pain or discomfort

I am thankful for the fact that I have fresh water to wash the dishes, to begin with

I am thankful for the fact that I have a home in which I have dishes to wash.

All of the three are related to each other, but they are distinct. You acknowledge that you are capable of dishes, which some people struggle with. You acknowledge that you have water, something not everyone has access to.

You acknowledge that you have a home and dishes to wash, something else not everyone is fortunate enough to have. In doing this, you slowly shift your mind to focus on the good in situations, seeking it out instinctively as you develop the habit of doing so. This will, of course, improve the behaviors you find yourself engaging in as well.

Journaling

Journaling is another way that is actually quite easy to utilize to regulate your own emotions. When you are going to be journaling, you must remember to focus on your emotional state in any given moment and simply write. There is no real guideline you must file during this technique—you simply let your mind wander and follow the process as it does. With your mind wandering, you will get a better feeling for your own normal trains of thoughts and emotions.

This is great—it enables you to understand how you can predict your own emotions later on. When you can predict how you will feel at any given moment due to predicting your triggers and signs that will usually come before an emotion, you will be able to regulate yourself. With knowledge comes power, and when you know when you are likely to blow up at someone, you will be able to stop it in its tracks simply because you will have the knowledge necessary to do so.

CHAPTER 19:

Maintaining Wellness

Putting in the effort to heal yourself from troubling emotions like depression and anxiety only matters if you will continue putting in the effort to maintain your wellness. If you immediately stop practicing your wellness approach the moment you start experiencing relief from your symptoms, or if you throw in the towel when you do not see the results you desire, you can guarantee that your results will not come.

You need to stay devoted and continue supporting your healing practices even long after you feel relief from the symptoms that were disturbing you for so long. By maintaining your wellness on an ongoing basis, you can ensure that you experience complete ongoing relief from your psychological disorders so that they do not slowly creep back in after you have spent so much time healing from them.

The following practices are things you can do to maintain your mental wellness on an ongoing basis.

Eating a Healthy Diet

When you are experiencing things such as anxiety and depression, staying on top of your dietary habits is not always easy. With both conditions, it is not uncommon to avoid often eating because you genuinely feel as though you are not hungry. The reality is that you are hungry; however, your increased stress has resulted in your appetite being suppressed so that the energy that would be used to digest food can instead be used to manage your symptoms of stress. Because of this suppressed appetite, it is not uncommon to completely avoid eating or to find yourself only eating the things you crave.

Eating this way for a prolonged period can result in your body becoming even more stressed because it lacks the nutrients that it needs to thrive. Once your body begins to lack vital nutrients, a whole slew of other issues can arise. This can lead to increased stress, increased anxiety and depression, decreased resiliency, decreased immune system functions, and many other troubling symptoms that can further impede your ability to experience a healthy life, including emotionally.

Focusing on eating a good healthy diet on a day-to-day basis is important if you want to nourish your body and reduce the number of stress hormones being produced inside of you. If you need to, start with eating small healthy meals or snacks throughout the day as a way to begin building up your appetite. As your body grows used to these smaller meals and snacks, you will be able to continually increase the amount of food that you are eating consistently until you find yourself consuming healthy portions daily.

In some cases, your diet can be used to help combat emotional disturbances, too. For example, if you are experiencing depression, increasing the number of B vitamins and omega fatty acids that you consume has been said to be very supportive in managing your moods. If you are experiencing anxiety, decreasing your caffeine intake and eliminating other stimulating substances from your diet can support you in gaining more control over your anxiety.

Drinking Plenty of Water

Like food, many people who are experiencing anxiety or depression will also forget to drink a healthy amount of water on a day-to-day basis. Dehydration is another factor that can increase the number of stress hormones being produced within your body. It can produce troubling symptoms that can worsen your mood, such as headaches. Raising your water intake and ensuring that you are consuming three liters of water every single day can support you in staying healthy and keeping your body happy. Water is essential in supporting your brain which is made up of 73% water so that it can function healthily.

If you struggle to drink enough water daily, try setting a reminder on your phone and carry a water bottle with you everywhere you go. Any time you experience even a small amount of thirst, take a sip of water. Begin training yourself to drink when you are thirsty so that you can ensure that you remain well-hydrated and that your body does not suffer increased stress due to dehydration.

Maintaining Your Personal Hygiene

Personal hygiene is another part of our daily routine that tends to be overlooked when we are experiencing intense anxiety or depression. Simple activities like brushing your teeth or hair or taking a shower may seem pointless when you are experiencing depression. You may find yourself feeling as though you lack the energy to stand up long enough to get the job done, so you simply avoid it altogether. Avoiding personal hygiene is not only unhealthy for you but it also worsens your mood and leads to you having an increased likelihood of remaining depressed or anxious.

For example, if you are too depressed to shower, you avoid going out because you feel as though your hair looks greasy and you smell somewhat like body odor. Because of this, you might decline going out with friends because you are embarrassed about the way you look. You declining offers to go out not only prevents you from engaging in experiences that could boost your mood but also keeps you feeling bad about yourself because the primary reason you declined was that you were feeling embarrassed about yourself. This can seriously lower self-esteem and self-confidence and makes it even harder for you to break your mood and feel free from your depression.

Even if you don't feel like it, make an honest effort to at least shower, brush your hair, and brush your teeth on a day-to-day basis. Keeping up your basic hygiene practices will improve your health and keep you feeling a small sense of pride and confidence in yourself even if you are feeling particularly low.

Caring About Your Looks

Your looks do not dictate your worth but they can have a massive impact on your mood overall. If you consistently look like you do not care about yourself, such as by not brushing your hair and wearing dirty clothes that don't fit or that don't go well together, you will feel like you do not care about yourself either. When people see you, whether you realize it or not, you will feel a sense of embarrassment because you have not taken pride in the way you look.

Paying attention to your looks does not mean that you have to be vain or that you have to apply makeup or put hours of effort into your appearance. Instead, it simply means that you take the time to clean yourself up and dress up nicely so that you look as though you care about yourself. When you behave as you care, your mind begins to recognize this and produces the feelings of caring too. You also begin to feel better because you can take pride in your looks when you go out in public. You do not have to avoid the public, look away in shame, or feel embarrassed about how you look. Instead, you can feel proud of yourself and your appearances.

Relaxing on Purpose

Many people fail to make time for relaxing on purpose and mistake it for sitting on the couch with their feet up as they scroll social media and read about stressful topics or listen to other people flock their negativity around. True and complete relaxation allows you to release stress from your mind and engage in peaceful activities. This can be watching a funny show on TV, reading, taking a bath or a relaxing shower, cooking or baking, or even just laying back on the couch with your eyes closed daydreaming about life.

Relaxing on purpose gives you time to unplug and tune out from the world around you. During this time, you want to release yourself from the pressure of the world and give yourself the freedom to just "be" with no strings attached. If you have a big family or find yourself living in a home that is not relaxing for you, you might consider engaging in some other form of relaxing activity regularly. Going to a spa, sitting at the park, attempting float therapy, or engaging in other solo activities can be extremely helpful. You don't have to be alone if you would prefer to relax in the presence of someone else. Even having a quiet evening on the couch with your family as you watch a funny movie can be relaxing. The goal is to engage in any activity that will allow you to release yourself from the stressful demands of life and simply enjoy yourself.

Getting Proper Sleep

Rest is powerful and unfortunately many people do not get proper sleep every single night. When you are depressed or anxious, you might find yourself feeling extremely tired regularly. You might even find yourself sleeping more often than not, potentially causing you to wonder why you are still so tired despite getting plenty of rest. The reality is that when you are depressed or anxious, your mind is often too stressed out to allow you to experience true and deep rest. Instead, your sleep is typically light and restless or filled with nightmares that result in you feeling stressed out and uncomfortable in your sleep.

Getting a healthy sleep every single night is important, and it can be accomplished even if you do not feel as though you are getting enough rest right now. It does take some intention and practice; however, especially if you have been having particularly restless sleep each night. Using nightmare rescripting can be a great way to relax the state of your sleep even if you are not necessarily experiencing nightmares. This CBT practice can support you in restructuring your sleep state in general so that you can experience greater relief from any stress that may be following you into your sleep each night.

You can also improve the state of your sleep by meditating before bed using aromatherapy, keeping your bedroom clean and tranquil, ensuring that your bedding is comfortable, and keeping your house at the right temperature. Avoiding stimulants in the evenings and night times and switching to relaxing foods and teas like those infused with chamomile and lavender is a great way to relax your body from the inside out.

You also need to make sure that you are sleeping on a consistent schedule. Even though you may feel exhausted every single day, maintaining a consistent bedtime and wake time and avoiding sleeping in between can ensure that your body becomes used to sleeping on a consistent schedule. This can help you overcome feelings of chronic exhaustion and support you in feeling better rested when you wake up. Be kind to yourself if it takes a while to level out your sleep. These patterns do take time and are not something that will change right away. Several weeks on a consistent sleeping schedule with a strong and consistent bedtime routine can be extremely helpful though.

CHAPTER 20:

How the Exercises Help in Anxiety

There are several benefits that exercising brings to an individual's life. The potential fighting diseases and improved levels of a person's physical condition are some of the benefits that have been established for a very long time. Physicians have been among the spreading awareness of encouraging people to stay physically active.

Exercising has been found in helping an individual to stay mentally fit. It has the potential of helping an individual reducing his or her stress levels. There are several studies and research that have been done on other benefits of exercising. They have been able to release that exercising helps a person to improve a person's alertness and level of concentration. The process helps a person reduce his or her fatigue that with the day to day activities an individual does in his or her daily life. The cognitive functioning of the brain is helped by exercise. There are several times that stress has the potential to deplete a person's energy.

It is alright for a person to experience stress and anxiety in his day to day life. Therefore, it is important for one to choose a perfect form of exercise he or she likes. It can be dancing, weight lifting or jogging. During these sessions, an individual's heart palpitations tend to rise as the heart palpitations. The phenomenon helps a person's heart to function even better than earlier on. This brings about the result of resting heart rate becoming slow during the exercise sessions. When a person has an improved functioning of the heart and lungs, he or she is associated with having an overall great sense of general well-being. This is very important when a person is offsetting the feelings of anxiety.

30 Minutes of Exercise a Day Would Be Adequate

One of the most recommended ways that help in dealing with anxiety is exercising. This is common advice that people who have faced anxiety and stress have been given by several people. However, it becomes a difficult task for a person to identify which type of exercise is fit for the purpose. It is obligatory for an individual to know which exercises can be effective to achieve the goals set for the exercise. An individual is not supposed to be worried about there being only one form of exercise. The good news is that the exercises that can help break anxiety and stress are several. This means that a person has a variety of options to choose from.

There are three major exercises that stand out over the years. They are effective because they are a combination of several activities in them. A third of the total forty million are the only ones that tend to seek treatment and help for anxiety. Therefore, the rest of the people are advised to at least do exercises whether at the gym or an open field. However, an individual is warned about having a bad perception that exercise has the potential of healing completely anxiety attacks. If an individual feels more overwhelmed during any of this exercise is supposed to seek professional help.

The exercises need to be done for an amounting time of thirty minutes for an individual to gain optimum benefit from them. These exercises are favorable to anybody who is either a beginner in exercising or someone who is used to exercising. These exercises include;

- **Aerobic, Cardio-Heavy Exercises**

These forms of exercise are majorly known for pushing an individual's heart rate to go up. Some people describe them as busters of anxiety in the psychology field. This form of exercise includes several exercises that are widely practiced by people across the globe. The process includes exercises that have high intensity such as running, team sports, swimming, and weight lifting. These exercises have the ability to make someone breaking a sweat, increate an individual's heart palpitations and hard forms of breathing.

A good explanation of this process can be drawn from the study conducted by Harvard Health. They have the potential to reduce the stress hormone in an individual's body. The most common stress hormones are cortisol and adrenaline. They help the body to produce a special kind of hormone that kills pain in an individual's body which is known as endorphin. This hormone also helps a person to have elevated moods during the day while active. Endorphin hormone aids human beings to find forms of relaxation and be optimistic which are resulted from intense moments of working out. After these exercises are over, an individual is supposed to have warm showers.

Aerobic exercises have heightened success when it comes to dealing with anxiety. There are several studies that have been conducted on aerobics by several scientists. A study conducted on people ailing from anxiety found out that their symptoms dipped for ninety minutes after an intense exercise of thirty minutes. The workout had a person have a maximum oxygen intake of eighty percent which made breathing become very hard. The massive amounts of energy an individual put on aerobics help to lower the levels of stress hormones and improve the levels of endorphin hormone.

- **Intense Work Outs**

An interesting study done in the year 2010 found that intense work out to up to thirty minutes has myriad benefits to an individual. A session of work out can be accompanied by several exercises that are beneficial to an individual in reducing the levels of anxiety.

The results are optimum if an individual is able to push him or herself to doing vigorous exercise for thirty minutes and not less than. An interesting discovery of the study was that it was easy and less anxious if the activity is done for a period of three months and not a lifetime.

There was an intriguing reason that was behind the stated point above. The basic thought by the researchers was that continuous exercise was responsible for increased palpitation and the breathing of an individual. These were things that helped a person have a comfortable feeling and helped a person when he or she was ailing from anxiety. It is because the symptoms of anxiety are characterized by an increased form of breathing and racing of an individual's heart.

The thirty-minute dedicated by a person to this exercise is supposed to be able to help him or her to maintain low levels of anxiety. It is because the signs and symptoms of anxiety reduce overtime when the process is continuous. The intense workouts are supposed to break into three parts for it to be sustainable and efficient to beginners. The intermittent breaks are supposed to be after ten minutes accompanied by a five-minute rest.

- **Yoga**

This is one of the ancient practices developed by human beings. Yoga practices have the potential of lowering the levels of anxiety in an individual's life. It is because the technique has the appearance of modulating the response systems that are responsible for stress management. The practice of yoga has the potential of reducing the levels of psychological arousals. There are several forms of psychological arousals which include respiration, heart rates, and blood pressure. The act of yoga centers its advantages on two key pillars which are its ability to calm an individual and the physical strength it gives him or her.

Intense sessions of yoga which are done up to thirty minutes have the potential of calming an individual. This is because it reduces the stress and it improves the palpitation of an individual's heart. The process helps the brain of a person to relieve from racing thoughts since it is shifted to focus on a person's breathing and the gentle and fluid movements. One of the common symptoms of anxiety is a person having low levels of continuous worrying and being too vigilant. These are some of the anxiety symptoms that yoga constantly alienates in an individual's life. These stated benefits of yoga were proven on three people in the year 2016. The research was conducted by Georgia State University which found yoga decreases general worrying in a person's day to day life.

Benefits of Exercising on Anxiety

There are several people who think that exercising is all about muscle size and aerobic capacity of a person. The thinking angle is straight since exercising can help a person to trim his or her waist, add years to a person's life and improve one's physical life. However, these factors cannot be described as elements that help a person to stay active in his or her day to day life.

There is another reason that makes people get motivated to exercise. This reason is that exercising has the potential of making a person have high levels of well-being and self-awareness. The process of exercising helps a person who does it constantly to have enough energy that can get him or her throughout the day. The practice goes ahead to make an individual have a better sleep when he or she rests after finishing his or her daily activities. The other advantage exercising brings to an individual is it improves his or her memory and make people have positive feelings about themselves. This makes them have a relaxed form of life that is very important for a person when he or she wants to achieve his or her goals. One is supposed to realize that exercise has various importance in an individual's mental health

The benefits exercises bring to an individual's brain come after a long span of days. The process can take an individual thirty to sixty days for the results to be depicted. A person is then advised to be very patient while he or she is exercising out. The call of being patient goes in hand with an individual being consistent with the type of exercise he or she is doing until the results start popping out. The results from these exercises are very advantageous because they relieve stress; promote quality sleep and betterment of an individual's mood. Researchers have shown an individual does not have to a workaholic on exercising since a modest amount of exercise can produce required results.

Exercising is a natural way that is effective in aiding to cure anxiety. The first step the process does is that it relieves tension and reduces stress levels in an individuals' life. It goes not higher in boosting a person's mental and physical energy. The process of exercising helps to control anxiety by producing an endorphin hormone that is responsible for the well-being of human beings. It is great for an individual to put his focus on things that keep him or her moving. It will be a waste if a person zones out while this process is ongoing. There are several ways that an individual can keep and maintain his or her focus while exercising. A person can choose to count his or her steps while walking; he or she can decide to concentrate on the sensations that are produced when he or she is running such as the rhythm being produced or he or she can choose to focus on the flow of wind. This helps to improve a person's awareness in turn.

CHAPTER 21:

How to Get More Productivity

Productivity is probably one of the most important things to know about when it comes to building your self-discipline. When you have a high level of productivity, you will find that you are getting more done each day and that greater self-discipline simply becomes natural to you. The key is fully understanding what productivity is and what it takes to improve it and get it to the most beneficial level possible.

What is Productivity?

Productivity is essentially a measure of efficiency. It shows how much you are able to get done and how quickly you can finish tasks without making mistakes. When you are not able to complete tasks, it forces you to get further behind. As you fall behind, you start to become overwhelmed, and this can make you want to give up on your goals.

When you are being as productive as you can be, the following benefits are often enjoyed:

You will have a greater sense of clarity since when you are productive and staying on top of things, you are making things simpler. Simplicity leads to productivity.

You will find it much easier to focus when you are productive since you will be able to work on a single thing at a time.

You can more easily eliminate things in your life that just lead to congestion, allowing you to worry more about the good and positive things that are present.

You will find that your life is more peaceful and less stressful since you will not constantly be fighting the clock to get things done.

You will find that you are more effective in all areas of your life because you will know what has to be tackled first and what is able to wait until you have some free time.

It is easier to achieve a work-life balance when you are productive since scheduling becomes far easier.

You can find more time for yourself to reflect and plan so that you can lessen the chances of unexpected, uncontrollable emergencies happening.

Techniques to Improve Productivity

Improving your productivity will feed you with an increase in your level of self-discipline. Because of this, it is important to work on your ability to be productive while enhancing your personal discipline. The following are commonly used techniques that will work to take your productivity to the next level:

You want to start by tracking how much time it takes you to do individual tasks so that you can see where all of your time is going.

Take breaks regularly to avoid burnout.

Create deadlines for yourself, but make sure that they are reasonable or else you will fail.

Know when you can say “no” to something that would simply distract you from being productive.

If something will take under two minutes, do it right away and check it off of your to-do list.

Keep a schedule where the most important tasks or meetings are written in a bright noticeable color so that you never accidentally forget about them.

Stop trying to be perfect, because it is only a distraction that keeps you from getting things done as efficiently as possible.

Stop multitasking and focus on a single task at a time, because if you try doing several things at the same time, you are only giving each task limited focus and thus reducing your efficiency.

Use your commute to your advantage and use it to reflect, get things done or something else that is productive, but safe, depending on how you commute.

Turn off notifications on your phone and computer, since they are designed to be a distraction and give nothing positive to your overall productivity.

Do everything that you can to minimize interruptions throughout the day.

Exploring How Spies Achieve Ultimate Productivity

Spies are fascinating, and they are the topic of some of the most popular action movies and books of all time. They are agile, and they always seem to win even when the odds are clearly not in their favor. This is solely due to their ability to be productive. Spies view their mind like a gadget. It works like a fishing pole when you are fishing, per se. It is a tool that can be used for just about anything, and there are ways to manipulate it to ensure that it can be beneficial for every situation.

When spies are thinking through a situation, they are able to do so without missing a beat. Coming to the right conclusion seems almost automatic to them. In a matter of seconds, they determine a course of action and immediately use it. First, they compile data, then they analyze everything that they have collected, then they make a decision and lastly, they take action. This seems like something that would take forever, but due to the intense training spies get, they can do all four steps in seconds.

Now, you do not have to be a spy in order to learn how to think like them. You need to start by learning how to rapidly collect data about a situation and how to analyze it immediately. You also need to ensure that your decision-making skills are sharp so that you choose to take the right action.

For example, you are in an interview for a job that you desperately want. You are asked why you want the job. You need to quickly collect data concerning the personality of the interviewer, the tone of the question and the personality of the company. Then, you will analyze this and make a decision about how you will respond. The action is your decision. As you practice in this process, you will find that you can get from data collection to action in a matter of two to three seconds every time.

Building on Your Good Habits to Eliminate the Bad

Once you develop good habits, you have to make sure that they are strong enough to outweigh your bad habits so that you are not tempted to go back to the bad ones. You want to work on both improving them and expanding them. These are two different actions, and it is important to differentiate between the two since there are different steps and types of work involved in each.

Improving Your Good Habits

Once you adopt a good habit, you have to make it stick. There are methods that you can start using now to ensure that your newly-gained good habits are ones that are going to be a part of your life for the long-term. In order to improve your good habits:

Commit yourself to using the good habit for just 30 days to start with.

Make sure that you give your new good habit some attention on a daily basis.

Approximately 14 days into your new habit, remind yourself about your goals and check to ensure that you are on track.

Keep it simple, and only focus on improving a single good habit at a time.

Make a trigger that will remind you to give attention to your new habit.

Ensure that you are consistent in how you choose to go about improving your good habit.

Find a person that will help you to stay accountable to yourself so that if you slip up, you can immediately get back on track.

If the bad habit you replaced the good with makes you feel like you have lost something by this point, find something positive to replace it with.

Get rid of temptations that are threatening to throw you off track.

Do not be afraid to be imperfect. In fact, embrace imperfection.

Think of this process as an experiment because when it seems less permanent, you will find that it is easier to work with.

Expanding Your Good Habits

You know how to develop good habits and even how to make them stick, but do you know how to expand on them? Essentially, expanding your habits means that you are using the good habits you have solidified to serve as a foundation for building additional great habits.

For example, you have created a good habit concerning how you eat. You have reduced your intake of junk foods to just once a week as a treat and all of your primary meals are healthy. To expand on this, start including exercise to your daily activities to further your health pursuits and habits. Consider taking a 15-minute walk after dinner or doing some basic calisthenics before breakfast. The key is to start simple and build just like you did when you improved your eating habits over time.

Another example is building on the good habit of improved time management at work. Once you master managing your time better at work, you want to expand this to improving your time management when you are at home. For example, create a schedule for cleaning and work on it until it makes the most sense and saves you time keeping a clean home.

Just make sure that you incorporate similar good habits at the same time. This will make it much easier to expand how many good habits you have.

Natural Good Habit Development and Incorporation

In working to develop and incorporate good habits naturally, there are six primary stages you want to follow. When you follow these exactly, you will find it much easier to engage in developing better habits naturally.

Start contemplating the good habits that you want to develop so that you can start getting used to the idea.

Start thinking of the exact good habit that you want to develop.

Start getting prepared to make the changes needed to naturally adopt the good habit.

Take action to make the needed changes

Work on maintaining your success.

Terminate your efforts and start the steps over on a new good habit you are looking to develop.

Eliminating Stress

This is an important element of working on your good habits since stress is the one thing that can quickly throw you off track. Stress is among the most common challenges that people face since making life changes can automatically cause stress. In fact, it is estimated that approximately 70 percent of adults in the United States experience anxiety or stress on a daily basis.

Try the following stress-reducing technique as you journey towards building good habits:

Take some time to exercise when stress gets too great, even if you do it for just a few minutes, as this helps decrease the stress hormones in your body.

Utilize certain scents that work to make you feel calmer, such as rose, bergamot, neroli, sandalwood, orange, lavender, vetiver, chamomile, frankincense, ylang ylang and geranium.

Reduce how much caffeine you are consuming, since caffeine can induce anxiety, especially when taken in large amounts.

Take a few minutes to journal what you are feeling and experiencing during a particularly stressful situation.

Know when to say "no" so that you do not get too overwhelmed with your responsibilities.

When stress is high, sit comfortably and breathe deeply for a few minutes to reduce tension and center yourself in focus with a task or responsibility.

Consider meditating when you start to feel overwhelmed with what is happening in your life.

Find a trusted person who will listen to you vent and help you find ways to solve problems that are causing you stress.

Every day take 10 minutes to decompress and reflect on your day. Put your stresses behind so that they cannot affect you tomorrow.

CBT & Mindfulness

Over the past decade, mindfulness has become increasingly popular in the world of psychology. Mindfulness-Based Cognitive Therapy (MBCT) combines some of the principles of classic CBT along with breathing exercises, meditation, and mindful practices that help you accept reality exactly as it is.

What is Mindfulness?

Mindfulness is the act of paying attention. If you're in a mindful state, you are fully present. You aren't obsessing about the past or fretting about the future. You are just in the moment, processing events as they happen. When we are mindful, we're more likely to feel calm and safe.

Why is Mindfulness a Good Thing?

Compare mindfulness with its opposite, mindlessness. If you are having a busy life, as most of us do, you probably go through your days on autopilot.

Have you ever driven to work and realized you have practically no memories of your trip? Or perhaps you've sat down to eat a meal and read the paper, and suddenly discovered that all the food on your plate has mysteriously vanished? Living mindlessly means missing out on your life. Mindfulness helps you re-engage with the world, moment by moment.

Understanding the Differences Between Classic CBT & MBCT

You can use CBT and MBCT exercises, mixing and matching as you like. However, you need to be aware that they are based on different approaches to mental health. There are some people that may find it easier to stick to one paradigm, and that's fine!

Here are some key points you need to know:

1. CBT invites you to challenge your thoughts, whereas MBCT encourages you to accept them.

Throughout this book, you've been weighing evidence for and against your thoughts. You've been identifying your cognitive distortions, asking other people to share their experiences, and generally gathering a lot of information. Research shows that these techniques work well in people with mild to moderate mental health problems.

MBCT takes a different approach. Like CBT, it involves noticing destructive thoughts. However, MBCT practitioners teach their clients how to accept them. MBCT techniques help you stay grounded in the present moment and wait for the thoughts to pass, instead of fighting them. You don't actively reframe your approach to the world as you would in CBT.

CBT works because it teaches you how to process your thoughts in a new way and arrive at different conclusions. MBCT works because it teaches you how to move through the world mindfully, accepting and letting go of whatever thoughts come to mind. In simple terms, CBT is analytical and MBCT is more immediate, with a bodily focus—meditation and breathing exercises both involve your senses.

2. MBCT focuses on preventing depression and anxiety, whereas CBT is used as treatment for people who are already experiencing an episode.

MBCT is recommended for people who have experienced multiple episodes of depression. MBCT is also used to help people living with chronic pain or another physical condition.

3. MBCT does not focus on problem-solving or behavioral activation.

Problem-solving is a key skill in CBT because it gives you a framework for taking action. Behavioral activation serves the same purpose; it helps you start taking steps toward changing your environment. MBCT doesn't emphasize these techniques. MBCT is about "being," whereas CBT is about "doing."

4. MBCT is usually delivered in a group therapy format.

MBCT is normally given in a structured group format that lasts several weeks. Clients are asked to practice new techniques they have learned between sessions. During their classes, they have the opportunity to practice together and receive guidance from an MBCT-trained therapist.

However, this doesn't mean you can't benefit from some of the techniques used in MBCT. Some research suggests that group and individual therapy are equally effective.

Here are three exercises you can try. They are similar to those practiced in MBCT classes.

MBCT Exercise #1: Body Scanning

Sit or lie down somewhere that is comfortable. Some people find this exercise so relaxing that they fall asleep, so if it isn't bedtime, do your body scan in a chair rather than on your bed.

Start by bringing your awareness to your entire body. Notice the weight of your body against the chair or the bed. Inhale, hold the air in your lungs for a few seconds, then exhale. Repeat this several times.

Next, focus on your feet. Notice their weight and temperature. Move your awareness up to your lower legs, then your thighs, buttocks, and trunk. Pay attention to how your back feels against the bed or chair.

Pay attention to your shoulders. If they are hunched or tight, take a deep breath and let them drop. Is your jaw clenched? Let it soften. If your hands are balled into fists, exhale and deliberately relax your fingers.

Repeat this exercise at least three times per day, for 20 minutes each time.

MBCT Exercise #2: Mindful Eating

For this exercise, you will need a small piece of food that you can chew or suck for a few minutes. A raisin or a piece of hard candy is perfect.

Find a place that's quiet where you won't be interrupted for five minutes. Hold the food in one hand. Notice how it feels against your skin. What temperature is it? How would you describe its texture? Put it up to the light. What color is it?

Now, smell the food. How would you describe it? Does it have a strong or subtle smell? Is it appetizing? When you really pay attention to your usual snacks, you might be surprised to find that you don't enjoy them quite as much as you think!

Next, put the food in your mouth. Don't bite into it yet. Roll the food on your tongue. What does it taste like? Is it sharp, sweet, salty, bitter, or a combination? What is the texture like? Is it dry, moist, or somewhere in between?

Finally, swallow the food. Do you feel it go down your throat and into your stomach? Was it easy to swallow?

This is a great exercise to do at the start of a meal. You will automatically eat more slowly, which will help you digest your food. It's also useful if you have a problem with overeating or binging; checking in at the start of a meal or snack helps you pay attention to your body's cues. It is not an immediate magic cure, but it may encourage you to stop eating when you start to feel full.

MBCT Exercise #3: Walking Meditation

Most people assume that you need to sit still for hours in the lotus position to meditate. Fortunately for those of us who find it hard to sit in silence for longer than a couple of minutes, this isn't the case! You can try a walking meditation instead. If possible, choose surroundings that are green and calm. Try relaxing your shoulders and maintain a good posture. As you step forward, notice how it feels when your foot meets the ground. What does the earth or tarmac feel like beneath your feet?

As you walk, use all your senses. What can you hear? What can you smell? What can you see? Look up—are there any interesting clouds in the sky? Take a few deep breaths. What does the air feel like in your lungs?

If it's raining, you can do this exercise indoors if you have a large room or clear hallway.

What to Do If Your Mind Just Won't Stay Still

Mindfulness and meditation practices are simple, but that doesn't mean they are easy. The biggest obstacle will be your mental chatter. When you slow down and let yourself notice your thoughts, you'll notice just how noisy your brain is. It's normal for your mind to leap from thought to thought, worry to worry, or idea to idea. The Buddhists have a term for this phenomenon. They call it the "monkey mind," because the mind is rather like a monkey, swinging quickly from branch to branch.

No one, however long they have been practicing mindfulness exercises, has a completely still mind. The difference between novice and expert practitioners is that the latter have long ago accepted that their brains will always be hyperactive. However, they aren't bothered. They know that being mindful isn't about removing every thought from your brain. Neither is it about becoming a cold, stoic robot.

Radical Acceptance

Radical acceptance is a technique developed by psychologist Marsha Linehan. In the 1980s, Linehan devised a form of therapy called Dialectical Behavior Therapy, or DBT.

Lots of people struggle to confront reality, but this is the first step to change. To help her clients make peace with difficult situations, Linehan suggested they try radical acceptance.

Exercise: Radical Acceptance

You can either wait until you are in a difficult situation to try this technique, or you can practice it by deliberately thinking about something that makes you angry or upset.

Start it by telling yourself that you don't have to approve of a situation to accept it. Take a few deep breaths. Say to yourself, "OK, this is happening. I don't have to do anything about it. I don't have to fight it. I just have to let it be, and it will pass."

This exercise is harder than it sounds! Your mind will still spin, and you'll still feel the urge to blame someone or something for your misfortune. When this is happening, try to bring your attention back to your breathing.

When you feel a little calmer, ask yourself these questions:

1. What led up to this event?

2. Did you play a part in causing this problem?

3. What part did other people play?

4. Did luck come into it?

5. You have a choice—you can choose to accept the situation or rail against it. What approach do you think is more constructive?

We like to think that, with enough effort, we can control everything in our lives. It's a comforting illusion. However, it just isn't true. Life is unfair sometimes, and it's often unpredictable. Learning to accept life as it is may be scary, but it is enormously liberating.

Summary

CBT therapists are increasingly using mindfulness in their work with clients.

Mindfulness exercises ground you in the present moment, freeing you from worries about the past or the future.

Mindfulness Based Cognitive Therapy (MBCT) is a structured form of therapy that draws on principles of both CBT and mindfulness.

You can use some MBCT-style techniques for yourself, such as mindful eating and meditation.

Mindful acceptance or "radical acceptance" can help you make peace with whatever life throws your way.

CHAPTER 22:

Coming to Your Senses

If you hope to come to your senses with cognitive behavioral therapy, it helps to have a scientific understanding of how this therapy can benefit you. By seeing the facts, you will better be able to believe in the results, which will then help you to consistently stick to using the CBT practices on a daily basis.

There are many amazing benefits of CBT which can help you come to your senses and better live your life. It can do this as it not only affects the way you think and your mood, but this therapy changes your brain and the way in which it operates at a cellular level. The truth is that many studies have been done on this, proving its effectiveness.

One such study, conducted in Sweden at Linköping University, was conducted on a group of participants who were all receiving CBT online with a therapist. The researchers focused on social anxiety disorder for this study. This form of anxiety is one of the most prominent mental illnesses, with approximately fifteen million Americans living with the condition.

The researchers conducted magnetic resonance imaging (MRI) tests on the participants both prior to beginning their therapeutic treatments and after the completion of the full CBT treatment schedule. The first set of MRI scans found that people who have social anxiety disorder also have a change in the volume of their brains, as well as increased activity in the amygdala. This is important, as the amygdala is the portion of the brain which humans primarily use in emotional responses, processing memories, and making decisions. It is obvious as to why increased activity in this area of the brain would cause problems for people.

Thankfully, despite this problem affecting our biological and cellular state, there are ways in which CBT can help. Amazingly, the second set of MRI imaging found that after nine weeks of therapy the scans had improved. The participants experienced both a decrease in their brain volume and a reduction in their amygdala activity, returning it closer to normal levels. These scans also revealed that the patients who experienced the most improvement in anxiety also improved the most on the scans.

This study proves that cognitive behavioral therapy is more than positive thinking; it, in fact, has the power to change your brain matter and activity. The change can affect your mood, decision making, reactions, thoughts, feelings, how you perceive the world, how your brain reacts to stimuli, and even the volume of your brain.

Cognitive behavioral therapy does not only show beneficial effects on anxiety but other mental illnesses, as well. In fact, people diagnosed with clinical depression found twice the help with CBT than they did when using antidepressants. This greatly helped prevent depressive relapse in participants. The light was shed on this shocking improvement when researchers conducted brain scans on the participants. It turns out that CBT and antidepressants both target different areas of the brain when treating depression. While antidepressants reduce activity in the emotional center of the brain (the limbic system), CBT was found to calm the area of the brain required for reasoning (the cortex).

The results of this study found that while antidepressants may help to reduce emotions, CBT helps in a more proactive manner. This is because this form of therapy enables the brain to better process information and make decisions in a more healthy manner.

While CBT certainly involves replacing negative thoughts with positive, it is not a simple matter of "thinking positively." If you have shared your struggle with doubts and anxieties openly, then you have most likely heard the exasperating phrase "just think positively." This may turn some people away from CBT, but let me assure you, cognitive behavioral therapy is about much more than positive thinking. This form of therapy requires many thoughts, feelings, and lifestyle changes that work together to change the brain at the cellular level. This change, in turn, improves your thoughts and feelings at a base level.

Don't worry, if you are depressed, anxious, or suffering from the effects of long passed trauma you are not alone. CBT has traditionally been used to help people just like you improve their lives and find contentment. This therapy does not recommend phrases as "just think positively," as if you try to use such a simple approach you won't find any long-lasting benefits.

Instead of simply thinking positively, it is important to focus on using your mind in the same way as you would a tool. As you learn to use your mind in this way you can better analyze information flexibly from various angles while controlling your emotions. When you use your mind in this way to calmly and flexibly consider situations from all sides, then you can gain a sense of understanding, find solutions, and lower stress.

Imagine using your mind in this way. If you are someone with anxiety about driving, then simply telling yourself "I'm fine, I don't have anxiety," wouldn't benefit you in any way. This statement is unrealistic, untrue, and only going to fail you when it counts. As you drive and begin to feel more anxiety, you will only believe that you have somehow failed. Once you begin to feel more negatively, it is a downward spiral, and you will only continue to become more stressed until you believe that positive thinking is useless.

Therefore, rather than giving yourself a quick clique you can step back and calmly see the situation from all sides as you come to your senses. Doing this allows you to be present, rather than in a state of worry about what might happen. While you are analyzing the situation, find a solution as to what you will do if you begin to feel anxious while driving. If possible, think of multiple options to counteract various sources of anxiety. After you have a solid plan, you can then affirm yourself with positive affirmations and deep breathing exercises to calm your mind. You will be able to trust in yourself and your plan to get you through the situation because you know that even if you become anxious you have a plan that will help.

You can come to your senses out of a negative cycle by calming your mind and body through meditation or mindfulness, analyzing the situation, testing out your thoughts, considering alternatives, and then prioritizing making truthful choices rather than emotional reactions. This will naturally lead to improved mental health, especially when combined with other CBT methods and components. Keep in mind that CBT is not one single component, but rather many components working together as though they were pieces in a clock.

One powerful way to come to your senses is to conduct behavioral experiments. This practice is regularly recommended by CBT specialized therapists. Behavioral experiments are used in order to test out your negative automatic thoughts. After testing the thoughts, you can then re-evaluate the beliefs behind them and your assumptions. Over time, you may find that as you test these thoughts you can become more aware of your surroundings and have a better more solid perception of the world around you.

Behavioral experiments can help in other ways, as well. For instance, they can decrease wallowing in negative thoughts, encourage consciously processing your thoughts and assumptions, increase the memory of positive experiments, and move the mind away from negative ideas. This entire process helps a person better see the objective truth rather than their preconceived assumptions, fears, and ideas.

You must be wondering how you can participate in using these powerful behavioral experiments. One of the great things about these experiments is that while they can have a profound effect, they are a simple process that you can do anywhere and at any time. Let's look at a couple of behavioral experiments to have an idea of what they can look like for different people in various situations.

Centering Yourself and Silencing the Inner Critic

Lastly, a vital piece in the puzzle when coming to your senses is silencing your inner critic. This critic is more severe than any critic on the outside may be, and we often allow it to control our lives. By listening to this critic, we are allowing it to smother our potential and decrease our belief in ourselves. But, if you learn how to silence this critic you will find that you can see yourself and the world around you in a new light. You will be able to see more clearly as if waking up from the night for the first time and seeing the light of the dawn.

In order to silence your inner critic, try following these four steps:

Notice Your Critic

If you hope to gain control of your inner critic, you can only do so if you first notice it. We are always having a conversation within ourselves. This happens whether we are asleep or awake, but we can most control and be aware of this critic while we are awake. Many people are unaware of this critic, as the conversations we have with ourselves are rapid and automatic. These happen so quickly that you hardly have the time to notice one thought before your brain moves onto the next. Therefore, in order to notice your inner critic, you must make a conscious and steady effort to be aware of your thoughts. Slow your pace down and focus on paying more attention to your various thoughts and emotions throughout your day.

Separate the Critic from Yourself

Your inner critic can thrive when you don't notice it. For this reason, it wants to remain hidden, slowly causing you harm. Yet, you were not born with this critic. Babies are free from this critic, and only develop it after they have internalized outside voices. For instance, they might internalize expectations placed on them, standards, and the criticisms they hear from others. Therefore, work on separating your inner critic from yourself.

Confront the Critic

You can't simply ignore your inner critic; you must confront it in order to take away the power it holds over you. How can you do this? Simply by telling your inner critic, whatever you have named it, that you don't wish to hear from it can have a surprising amount of power. Whenever you hear this inner critic begin to nag at you, try telling it to be silent or go away.

Replace the Inner Critic

Rather than simply ignoring your inner critic, you can best fight it off if you completely replace it. This way, rather than struggling with self-critical thoughts, you can create a strong and powerful ally that focuses on self-compassion. Imagine this new voice as being a close friend who supports and cares for you. But, in order to create this new compassionate inner voice, you must learn how to recognize your positive traits.

Don't be discouraged if it takes time to notice your positive traits. Our brains work with selective filtering, which always looks for evidence to support what we believe about ourselves to be true.

CHAPTER 23:

Developing Your Problem-Solving Skills

As long as you are walking the planet, you cannot run away from the fact that problems will always be there. No one ever has an "easy life." But then again, the human spirit possesses tremendous power, and there's no problem we couldn't overcome.

For most people struggling with mental illness, usually, they are overwhelmed by problems, yet they have no idea how to overcome these problems.

When you develop your problem-solving skills, you equip yourself with power, and not only do you afford to keep off mental illness, but also it helps you attract more success to your life.

The more you increase your problem-solving skills, the more utility you have both at the personal level and toward society. It's no secret that people want to be associated with those that have power.

The following are some incredible tips on boosting your problem-solving skills.

Active Listening

This is one of the most important skills for solving problems. Whenever there's a problem, you will notice that there are disputing sides, and this means there's a disagreement.

By improving your active listening skills, you stand a better chance to understand what the other person is saying, and this promotes cordial relations. On the other hand, when you are hardly paying attention to what the other person is saying, there are bound to be issues, and then it makes the problem even worse.

For instance, at your workplace, there might be a problem with your department head, and whether or not this problem gets resolved is up to whether the involved parties are willing to listen to one another and arrive at a common ground.

No matter what field you are involved in, you will always need active listening skills so that you may improve the quality of your life.

In some instances, mental illness could very well be fueled by the absence of active listening skills. For instance, if someone is battling depression, they are likely to have less concentration, and this would cause them to pay less attention to what other people are saying, and as a result, it would affect their capacity to reach a solution.

Some of these exercises include:

· Facing the speaker

It will sound simplistic, but it is a very critical exercise. This is whereby you face the other person and hold their gaze.it might feel invasive at first, but with more practice, you will get into a position that you are comfortable with that. Facing the speaker, you are in a position to understand precisely what they are saying, and it eliminates misinformation. Facing the speaker also boosts your confidence, and you are less likely to be afraid of stating your opinion.

· Watch nonverbal cues

If you are unsure about where someone stands in regard to a suggestion, just watch their nonverbal reaction. Most of our communication is nonverbal. When we can read a lot of information from the gestures that someone makes, we only need to be attentive and insightful.

· Stop judging

One of the problems with many people is the tendency to judge others. This makes it hard to have an objective conversation. For instance, when someone is battling extreme anxiety, they tend to not concentrate on what another person is saying, and they will rush things, thus making it hard to have an objective conversation. If you are looking to improve your problem-solving skills, ensure that you stop judging what other people say.

· Stay focused

It is also important that you stay focused on the task at hand. Allow yourself to be in the moment. You have to channel all your energies to what is going on. If you are wishy-washy, you will only limit your capacity to solve the problem. Also, when you stay focused, you win everyone's respect.

Research

Research is a very critical component of solving problems. When you are faced with problems, it is incredibly important that you understand everything about the situation, as this is a way of arming yourself with important knowledge so that you may make the best decision ever. In the modern world, we have the internet, and it makes research far easier. In the course of the research, ensure that you are sticking to the topic and that you are well aware of the impact of the information were it to be applied. Being good at research requires more practice. You must never tire of doing research in order to solve your problems and achieve your important life goals. One important thing about research is that you must stick to your circle of interest. Extensive research doesn't mean you wander away from your interests. Uniformity in research interests not only helps you acquire extensive knowledge but also you gain an understanding of your potential.

Creativity

If you look at most people who battle serious mental illnesses, you will realize that they have been pushed to the edge of the cliff by their day-to-day troubles. Creativity is one of the things that would help us overcome our troubles pretty easily. When we are creative, we have power, and we don't let our troubles overwhelm us.

But then again, not everyone is sufficiently blessed with a creative bone. Most of us cannot think on the spot and come up with a solution in the face of a problem.

To some extent, creativity is an inborn trait, and some people have generous loads of creativity, and that's good for them. You might not be tremendously creative, but you still have the capacity to grow your creativity.

For instance, if you are experiencing financial challenges, there's the risk that you have developed a mental illness like insomnia, but if you tap to your creativity reservoirs, you should find a way to get out of your financial challenges, and by extension, overcome insomnia.

One of the ways to boost creativity is through engaging in physical activity. When you get blood to surge through your body, you are in a position to get new ideas and solve your problems.

Ever realized that you receive the best ideas when you engage in physical activity like walking or during gym sessions? Another way to boost creativity is by working with a team. You weren't designed to go it alone. Make it a habit to collaborate with other people. You get inspired.

Also, ensure that you are in the right mood. When you are in a great mood, you are not only in a position to think up great ideas, but you also have the drive to act.

Also, ensure that you seek assistance from your superiors. Your superiors, in this case, are people with experience. You will see that most successful people have been mentored. Don't have too big an ego to imagine that you don't need any help.

Relaxing also ensures that you become a more creative person. When you relax, you let your mind relax, and your worries go, creating an enabling environment for creativity.

Increasing your creativity is, without a doubt, one of the best ways to boost your problem-solving skills.

Communication

Effective communication is one of the most important skills for success. Without great communication skills, you hardly ever inspire confidence in yourself. Thus, ensure that you develop sharp communication skills. For someone who's struggling with mental illness, communication skills are an especially important skill, for they help you express yourself. When you go out to make friends, you need to present yourself well, so that you may create a great first impression, and people will want to be associated with you. If you have poor communication skills, you won't be in a position to get what you desire, and that can be quite sad. When you decide to improve your communication skills, the one thing you need is practice. When you make a habit of practicing how to present yourself, eventually you will master that skill, and moving on, it will flow seamlessly. The following are some incredible tips on improving your communication skills:

· Make it a priority

Make a point of always practicing your communication skills. Never tire of wanting to be the best. You can decide to practice every morning. The more you practice, the easier it becomes to master your communication skills, and you have a pretty easy life. But when you practice infrequently, you might fail to make significant progress, and this could stop you from hitting the peak.

· Simplify your message

The thing that you need to keep in mind is about communication is always to simplify your message so that the other person understands what you are on about without a struggle. If you don't simplify your message, there's a risk of being misunderstood, and then it breeds even worse outcomes. Still, on the subject of simplifying your messages, ensure that you express yourself using a simple language. When we talk about simple language, it means you must use easy-to-understand words.

· Great body language

We perform most of our communication through our body language. Thus, ensure that your words are in agreement with your body language. When you have great body language, it allows you to make a good impression, and also your message is received as intended. Great body language is also advantageous for the purposes of boosting self-confidence.

Team Building

If you look carefully at most people who are accomplished, you will realize that they didn't do it alone, but rather they had a great team behind them. The capacity to work alongside a team is an extremely understated gift. No matter how you look at it, working as a team is beneficial than working solo. Most people with mental illness need help, even though they may not admit to themselves. Thus, such people need to collaborate with other human beings in order they may have a team that works effectively, and this will enable them to reach their important life goals.

When it comes to team building, you need to be frank with yourself and with other people. But most importantly, you need to have good intentions. When a person learns to combine their efforts with the efforts of other people, they acquire the power to accomplish their important life goals.

If a person doesn't have the capacity to work together with other people, this can become a great hindrance; it can stop them from utilizing their full potential.

Decision Making

At the end of the day, bear in mind that our decisions amount to our reality. If we make poor decisions, we end up with a terrible reality. Making quality decisions is a tremendous gift.

For someone who's battling mental illness, it is important that they become great at making decisions.

This will help them recover from their mental illness, but more importantly, it will help them stay away from the condition. When you make quality decisions, it means you have taken the power to control your life. It only means that you are not at the mercy of fate.

Not every last of your decisions will sail as expected, but in such times, you must have the fortitude to stay your ground. Always ensure that you are making quality decisions, and you will greatly improve your life.

CHAPTER 24:

Creating Positive Associations

Studies show that people can break bad habits quicker if they replace the bad habit with a good one. For example, if you want to stop smoking, then every time you would have taken a smoke break, take a music break. Put on some easy-listening music and practice some deep breathing exercises to calm you. Or, play your favorite Motown and put on your dancing shoes. You'll learn to let the music and deep breathing relax you instead of the cigarettes. Or, you'll dance until you're downright happy. Who can be sad when they're dancing to Motown, right? Anyway, you do it, you're replacing a bad habit with a good one.

You've allowed negative self-talk to become an automatic reaction to some event or situation that didn't go your way. Okay, so to break this bad habit, you'll need to create some positive associations that are strong enough to help you change your beliefs and behaviors. Once you have experienced some success with this step-by-step guide to assist you with your depression and anxiety, you'll be able to make positive associations between past experiences and present beliefs. These positive associations will help you to dispel the false talk that keeps you from moving forward and enjoying a future free of depression and anxiety.

What are Positive Associations?

Positive associations are when you take two or more objects or feelings and connect them together. For example, in Pavlov's experiment with the dog he would offer his dog a tasty treat, and as soon as the dog got the treat, he would ring a bell. After a period, the dog didn't need the food; he would merely hear the sound of the bell and begin to slobber in anticipation of the treat. He had made a positive association between the sound and his goodie.

Likewise, you can learn to make positive associations with your feelings and something positive. Without intending to, you have come to make negative associations with your sense of depression and sleep. You know that you can escape depression by sleeping. So, when you begin to feel depressed, you automatically want to curl up and sleep to get relief from the depression. Wouldn't it be wonderful if you could make some positive associations that would allow you to change your thinking? For instance, you also know that exercise helps to alleviate stress and anxiety. So, when you begin to feel depressed, it would be so much better if you automatically wanted to get active. It's an association that offers physical and emotional benefits.

At first, the positive association feels much like you are rewarding yourself for feeling depressed, but this isn't the case. You're creating an opportunity to link something positive to do that will occur automatically, taking the place of the automatic negative self-talk and need to isolate.

How Do You Create Positive Associations?

Creating strong positive associations doesn't happen right away. In fact, to make connections that can change your behaviors and perceptions requires lots of practice, repetition, and an intentional routine. The positive motivator also should be something you love, something that makes you feel incredible, free, and happy.

Many people think positive associations always deal with money, but it is the things that money can buy that create the connection. Having a lot of money sitting in the bank doesn't give most people a thrill, but enjoying an exotic vacation or buying a boat might be unbelievably powerful. Don't think that positive associations need to be expensive. Like Pavlov's dog, a routinely given treat might do the trick.

The longer you have suffered from depression and anxiety, the more frequently you will need to create positive associations to help you change your behavior and perceptions. Don't wait until you are in the middle of a deep depression to try to apply positive associations. Work on replacing the negative with positive when you're feeling good. If you're having an especially good day, use your journal to write down a description of the day. File in your mind how good it feels to be mentally strong and healthy. Enjoy the weather, and think to yourself how you're going to remember how warm and cozy it felt to enjoy the sunshine and be out in nature. Then, the next time you start feeling down, or you begin to hear negative self-talk, give your mental files a going through and search for those positive memories.

If your subconscious doesn't know the difference between imagined feelings and real emotions, then keep believing you're enjoying that incredible day. Try to think about how feels to walk in the sunshine or lay in a hammock on the beach. Remember what it was like to take that dream vacation with your loved one. Smell the salt in the air, and feel the sand beneath your toes. Soon, your mind will focus on positive thoughts and associations rather than the negative ones. Your mind will try to focus on what you tell it to, so make your automatic "go to" thoughts be the positive ones.

Once you've got your ideas headed in a positive direction, you'll need to follow that up with immediate action. Don't just think about something positive, do something positive. Again, if Pavlov had not continued to follow the sound of the bell with treats, soon the conditioning would have disappeared, and the dog would have stopped slobbering when he heard the bell. The ringing of the bell created the physical response of slobbering, but the food was the reward. The thoughts encourage you to begin to change your thinking from negative to positive, but it's the repeated affirmative action that helps you to change your behavior.

Rewarding your positive thoughts with positive actions will keep the associations strong. Merely distinguishing the behavior cannot be the association or the reward. Sometimes the reward is not to change. When you've experienced depression and anxiety most of your life, you could just be too darn comfortable with your feelings. You have accepted that you are a rather sad and anxious person, and then you think about ways to justify your feelings.

"I'm supposed to be the thinker in the family, the serious one, the one who has their feet firmly planted." See, how you begin to talk yourself into your sad way of thinking so you won't have to go through the pain of change? Change brings with it newness, unfamiliar territory, and many people believe they are more uncomfortable with change than with their anxiety or depression.

Did you see what happened there? You made a connection between change and something negative. If you feel that change is hard and requires sacrifice, then you have created a negative connection. If you are depressed or highly anxious, you need to change your thinking. However, if change represents a negative experience for you, then you won't change that negative self-talk because change is negative. Getting caught in a negative cycle can be devastating for the depressed.

Passionate Memories Create Strong Associations

To create connections strong enough to change behavior you need to learn to live life passionately. Whatever you do, do it with heart—with emotion—with desire. When you're living life to the fullest, there's no time for depression and anxiety. If you are doing things that you love, you will enjoy it every day. If you're not doing what you love, you may be paying the price with depression and anxiety.

I know what you're asking yourself about right now: "But, what if I cannot support myself doing what I love?" Okay, I'll buy that. If you love horses, but you live in the city and only get to ride twice a year, how are you going to do what you love and support yourself? Residing in these circumstances doesn't mean you can't be passionate about what you love and find enjoyment from horses every day. It just requires you to be a bit more creative.

You could surround yourself with equine art so that you could enjoy the beauty of horses every day. You could visit the equestrian center outside of town once a month and smell the hay and groom the horses, and then take a pleasant ride in the country. You could put pictures of horses on your screen saver or your phone. You could volunteer with an equine rescue on the weekends. You could read about them and watch movies and documentaries about the magic of horses. All these things feed your passion.

Whatever you do, make every positive association remarkable. Build the positives in your life to be bigger and brighter than a full moon. It will take the wind out of your negative sails. If you can say "so what" to the negatives, then you take away their power. "Okay, so I did poorly on that test—so what? I have five more this semester to do my best. One bad grade isn't going to make me fail the course." Or, "I wrecked my car—so what? I walked away unscathed, and nobody else was hurt. It could have been so much worse." By minimizing the negative, you take away its influence over your thinking.

Thinking positive is like performing magic on your thought process. You may not even know how it works, but you sure are enjoying the show!

CHAPTER 25:

Dealing with Persistent Negative Thoughts

Regardless of the issues that you are dealing with, having to force yourself to keep it up despite ongoing and extremely persistent negative thoughts can make an otherwise inoffensive therapy session and subsequent related exercises seem almost impossible to work through. Overcoming negative thoughts can often be extremely difficult, simply because you are using your mind to change something that your mind is doing, but the following exercises should serve to make the whole process more manageable.

Anchors

The anchor model can be used to replace any negative belief with a more positive alternative instead. To understand the basics, picture a hot air balloon being held down by an anchor at each corner. This balloon can be thought of as the negative thought that is holding you back, and the anchors are the social consensus, emotion, logic, and evidence that are holding it in place. In order to release the balloon once and for all, you must replace the offending belief once and for all as well. To get started, the first thing you may need to do is find the offending thought; this is as easy as running through the thoughts that are always with you and stopping when you get to one that hurts when you think about it. For example, if you are an artist, then a common pervasive negative thought is that you don't have what it takes to be successful at your craft. With the thought identified, the next thing you should do is determine what anchors are currently keeping it in place. To do so, consider the following:

· What events or evidence anchor the thought?

· What emotions are tied directly to the thought?

· Who are the people around you that reinforce the thought?

· What logic is locking in the thought?

Once you have tracked down all the specific facets of the thought in question, the next step is choosing a new belief to replace the old one. For example, if you're feeling depressed because you believe you will never achieve your dream, changing your thought directly to "I'm going to achieve my dreams" is too big of a shift. Instead, something along the lines of "If I continue to work hard and persevere, I will reach my dreams" is both motivating and not so much of a major shift the mind will reject outright.

Avoid Negative Thinking Traps

One of the hardest parts of removing negative thoughts from your mind is the fact that the more you try and remove them, the more you think about them, and the more entrenched they become. Many people think of negative thoughts sort of like unwanted hairs that can simply be plucked out of the system and discarded at will. Unfortunately, the truth is that when you actively pursue negative thoughts with this goal in mind, all you are really doing is giving them the full benefit of your focus, making it easy for them to take control of your perception and distort it as needed to justify their existence.

To understand why this occurs, it can be helpful to think of a negative thought as you would a coiled spring. The more the spring is compressed, the more energy it is going to displace and the more resistance it is going to place against the downward force against it. Likewise, when you make the mistake of trying to suppress a negative thought in this way, all that occurs is that an even greater counterforce is generated as a result. Thus, the more you try and not think about it, the more prevalent it becomes.

Instead, a better choice is going to be confronting the negative thought head on and determining whether or not it has any legitimacy behind it. By confronting the thought head on, you take away its power and can finally see it in an accurate light. With this done, you can then either disregard it completely as it isn't worth thinking about, or come up with a more effective solution than simply worrying about it all the time.

Practice Acceptance

For this exercise, rather than trying to get rid of your negative thoughts, you are going to accept them as a part of you and respond accordingly. For starters, it is important to try to understand that your mind is generating these thoughts outside of your control and without your consent. While at this point you can't always control what thoughts you want to have, just realizing this fact is a step in the right direction.

For now, it may be enough for you to take a stance as an observer; simply being aware of all of your thoughts is a big step forward. At this time, you don't need to react to your thoughts or pass judgement on them; simply use the mindfulness meditation techniques to separate yourself from then as much as possible. This goes for negative thoughts as well; let them enter your mind and then leave again on their own accord. Do not resist them.

You can find it useful to think of each thought as an arrow; once the arrow is released, it accelerates until it reaches maximum speed before then losing momentum moving forward. When it comes to negative thoughts, not resisting them will let them continue to fly until they have burned themselves out, while holding onto them, even to resist, will only cause their power to grow.

Change your Belief System

Another key step to banishing negative thoughts is to get at the source and remove negative thoughts that you have identified from your belief system. Specifically, this means divorcing your negative thoughts from your personal version of yourself by understanding that they don't reflect reality and are simply a product of an untrained mind. To start changing these thoughts, you need to express them in such a way that they are separate from yourself.

For example, if you are feeling anxious, you would verbalize this fact by saying, "I am having thoughts that I am anxious; this is just my mind telling me to feel this way and is not an accurate reflection of the current situation." You could then list all the reasons that come to mind as to why you do not currently need to feel anxious. By making an effort to complete this mental shift, you are distancing yourself from the negative thoughts, which then shifts them from automatic thoughts to active thoughts, thoughts you can more easily do something about.

Once they are in the realm of active thoughts, you can then use what is known as rational coping statements to plan how you are going to deal with these thoughts and feelings when they arise.

Writing and Destroying

If you find that your negative thoughts are typically linked to strong emotions, then you may find success, and catharsis, by getting them out in writing. For this exercise, you should get out a pen and some paper and physically write down everything that is bothering you.

Then, once you have gotten out everything that you care to, you can then destroy the paper in any way you deem appropriate. If you aren't a writer, any type of physical representation of your negative emotions will do. This exercise is about getting the negative emotions and their related thoughts out of your mind and into the physical world where they can be dealt with more easily. Destroying the representation of the feelings shows your commitment to moving on and living your best life.

CHAPTER 26:

Facts Versus Opinions

This simple exercise revolves around a common misconception plenty of patients make. It can't be helped, though. Sometimes, panic and depression attacks make it almost impossible for us to get our thoughts straight. When that happens, we are prone to making very big distortions about our beliefs. One thing becomes another and the next thing falls out of place entirely. This same thing can be said about the difference between facts and opinions. When we are quick to register one as the other, we end up with unrealistic expectations about ourselves. If left uncorrected, they spiral downward and drag us into a slump.

The Remedy

Fortunately, the difference between fact and opinion is a very solid wall. This wall has very visible characteristics to the trained eye. The job of the therapist is to help the patient see these characteristics in order to make the proper distinction.

For this technique to work, some journaling and writing are needed.

First, sit down with your therapist to write all your sentiments about certain events. Don't worry about categorizing them immediately, what's important at this point is to make a complete list of your cognitions.

These could be about a recent argument you had with a relative. It could also be another panic attack after your boss gave you additional work to finish. It could also be an interesting discussion you had with a friend. The good thing about this exercise is that it's applicable on a daily basis, making it an ideal therapy-cap after each session.

Procedure

Once you and your therapist have a list of your cognitions, go over each item to determine under which category they belong. Are they mere opinions? Or are they undeniable facts?

One good way to test each item is to give yourself reality checks. Ask yourself if each item can be refuted or not. See if you can present a counter-argument for each item. Here, you put on your thinking cap and question your own cognitions as if you were self-treating.

- He hates you. Who told you that? Did he say that explicitly?
- You made a mistake. What criteria did you fail to meet?
- You're weak. What standards? Who gave you that assessment?
- You won't finish that report this week. According to who?
- There is something amiss about you. What could it be? What makes it wrong? Does everybody think that way?

By questioning the items from your journal, you can determine which ones are opinions and which ones are facts.

Of course, facts are unchangeable, but you can change the way you view them. Opinions on the other hand, are more flexible and change depending on your current perspective.

The main purpose of this exercise is to show you the truth that thoughts are not facts. The way you think may be an honest representation of your internal conflict, but that is in no way empirical evidence of any of your anxiety disorder or depression-based claims.

The exercise is designed to teach you to reevaluate yourself from time to time, double-checking your cognitions to make sure that you don't run into any unhealthy distortions along the way.

For severe cases, this exercise becomes a life-saving technique that tells you to slow down and think things through with a rational mind. Coupled with the other techniques mentioned in this workbook, this exercise can also serve as a good test to see if you can distinguish fact from opinion in real time.

CHAPTER 27:

You Are Your Own Cure

Living with anxiety means that you are always on edge. You're living a life that is propelled by fear. This makes it difficult to find any type of enjoyment to speak of. Your mind is always stuck on the "what ifs" that may happen. Rather than being able to live in the moment, all your thoughts are directed either on the guilt of the past or the unpredictable prospects of the future.

This is no way to live, and the fact that you're reading this book shows that you don't want to continue to live that way. An anxiety disorder is a common occurrence in today's modern world. Things move fast, and the ability to take your time and relish at the moment is fleeting at best. However, it may be somehow comforting to know that you don't have to live that way. You can do something about it.

While there has been a great success in many of these methods, we encourage you to attempt to resolve your anxiety issues in a more natural way. Your brain and your body have already been designed to heal itself. It's just that most of us have lost touch with how to communicate with our bodies effectively. We go through life like an automaton, jumping from one function to another, void of feelings and confidence. However, if we learn to tap into our own natural resources, health and recovery are often just a short distance away.

The reality is that we have the power to heal ourselves within us. In most cases, you won't need expensive medications, pay hundreds of dollars to therapists, or buy into a lot of expensive gadgetry. There are steps you may take right now; techniques you can apply the moment you feel your anxiety start to rise to help your mind to reach a calmer state so you can add more positive feelings to your life.

Getting in the Right Frame of Mind

Regardless of the method, you choose to aid you in recovery, keep these basic points in mind before you begin any treatment. We are living in a world full of negativity, so it's no wonder that we drift off into the negativity of land without effort. The challenge, though, is to redirect our thoughts to the more positive. This will be the first step in getting our thinking back on track.

Show Gratitude: Cultivate a spirit of gratitude. Believe it or not, this is a mental exercise that all of us can practice. This is not just to be thankful for the basic things in life but to think deeper in that same line of thought. Making a concerted effort to at least feel gratitude inside can start us thinking more positively.

We are all accustomed to using cues to remind us of other activities, so it just makes sense that we would use similar cues to remind us to think on more positive lines. You do this enough, and eventually, your mind will start to do it automatically, without thought or planning. Sometimes, we do this without noticing. For example, if you know you're going in to work with a very demanding boss, you mentally prepare yourself. The same should also be true when you are preparing for a potential anxiety attack. The more you prepare your mind ahead of time, the easier it will be when you are actually working to push that anxiety back down to a level where it belongs.

Reject Your Inner Sensations: Next, you want to recognize those negative thoughts and refuse to accept them. A gift is not a gift until you accept it. So, if your mind is flooded with negativity, realize that you have the option to say no to it. You are not compelled to accept any thought or feeling just because it came from your subconscious mind.

Our subconscious mind has been trained by our myriad of experiences that we go through from the moment of birth. Sadly, we have often forgotten the experiences that led to our negative thinking, and only the thoughts remain.

Find Your Safety Zone: No matter how bad we think something may be or has the potential to be, it is always easier to handle when we know that we have a safe place to run to. Our safety zone doesn't always have to be a physical place to hide. It may be a form of meditation or just a mental state you train your mind to run to when things get to be too much.

Because of the crazy world we live in, most of us have forgotten how to live. We are constantly on the watch, focusing our attention on potential dangers, threats, malice, and pain. When we fall in that line of thinking, we feel everything is not safe. Start making it a habit to look for the good things in your life. They are there, hidden away behind a wealth of badness, but you can find them. You just have to look for them.

Change Your Inner Critic: We all have that inner voice that is constantly reminding us of our past failures and our faults. This is not a characteristic exclusive to those with an anxiety disorder. It is common to all of us. But we have to start making a concerted effort to put that voice on mute. Those harsh words were drilled into us years ago, possibly when we were children and couldn't do everything right. But now, we have to change that dynamic. Up until this point, it has been a one-way conversation with that tiny voice buzzing all these negatives in your ear. It's time for you to start talking back to it.

Our inner critic acts like a judge that has nothing good to say. It judges the quality of our work, predicts how other people will view you, and makes you feel guilty and ashamed at every turn. It mocks you, teases you, and bullies you. He doesn't care about your feelings and has only one goal, and that is to prevent you from having any joy in life by questioning and doubting everything you do.

Identify the Judge: One of the first things you need to do is to identify your judge. If his voice is constantly ringing in your ear, the chances are that you are not even noticing it anymore, you are automatically responding to it without even being aware of its negative prodding. You'll notice when your judge is there when you finally get up enough nerve to tackle something, and suddenly your energy starts to wane, you begin to question yourself, or without reason, you lose interest. Recognize, at this point, that it is your judge, your inner critic who is taunting you.

Whatever you hear, your inner critic tells you, don't remain silent. It's a tiny little voice in all of us. But the important thing is to acknowledge what he says, respond to it, and then keep moving on. That voice inside our heads is there to serve as a warning to keep us from getting into danger. But there are times when he gets a little too much power and warns us even when there is no threat.

The inner critic rarely goes away. As long as we're living, he will continue to find ways to sabotage our efforts. However, if we can identify who he is, acknowledge that we've heard him, and forge on in spite of the fears he creates, his voice will not have as much power over us, and we won't have to worry about succumbing to our fears.

A Word at the Right Time

Another mystical tradition that has done wonders to help relieve anxiety is the use of affirmations. Because you're caught in a loop of negative thinking that has undoubtedly taken its toll. The use of positive affirmations is a great way to interrupt those negative thoughts with something positive verbally. It aids your mind to shift its thinking and gradually guides the mental process in another direction.

You can use these affirmations in three different ways.

•You can say them silently to yourself.

•You can write them down.

•You can say them out loud to yourself.

Anytime you feel overwhelmed with negative thoughts, and you can use these positive affirmations to redirect your thoughts in a more positive direction. Some therapists recommend that you make it a habit by attaching your affirmations to something you do every day. For example, every time you get in your car, you repeat your affirmations. Or you could say them every time you go to prepare a meal, put your children to bed, or hang up the phone. The trick here is to make it a regular habit to use these affirmations until they become a natural habit to you.

Some examples of positive affirmations you could make to fight off anxiety:

•I am calm.

•I am a good person.

•I value who I have become.

•I am healthy.

•I feel relaxed.

•I can let go of this stress.

•I am a good student.

•I can make a new friend today.

•I have a great relationship with my family and friends.

There are some basic rules about creating your own affirmations. First, all affirmations must be used in the present tense. Do not make mantras about the future or refer to something in the past. Second, they must always be about you. You are battling your own inner thinking process; therefore, your affirmation should be there to redirect your thinking. This is the only thing you have any control over. And third, they have to be 100% positive. Do not use any words like can't, don't, won't, etc. Instead of saying I can't get angry anymore, rephrase it to say, "I am going to stay calm today."

While they are commonly used when negative feelings and thoughts pop up, some have found they do better if they start their day with affirmations, letting them start on a more positive note. However, you choose to use them; they can help ease your anxiety in many ways:

•They are a positive distraction and can keep your thoughts from running wild.

•They give you a positive belief. In time, the brain will adapt to the positive phrases and start to adapt to the belief that you are trying to form.

•And they serve as a constant reminder to insert more positive things in your life.

Make sure that the affirmations you choose have personal meaning to you. While you can find a long list of suggestions online, one that you create yourself will have more influence over your thinking than something written by a total stranger.

You need to be committed to using them regularly. It may take time before you begin to see the benefits, but if you continue, eventually, you'll start to feel the anxious discomfort fade away replaced by something more positive and actually makes you feel good.

CHAPTER 28:

How Long Should Your Therapy Take and Results

Different people need different levels of therapy. No one will be better or worse than someone else because of the amount of time their therapy takes compared to anyone else's. It just depends on the person's circumstances. If a person is required to have psychological treatment in an in-patient setting, they need to attend therapy sessions upon being released. You need to give yourself the time you need. Now, that being said, you need to be working towards making this time come to an end. There are those who need lifelong therapy, but these are extreme cases, and this tends to apply to a person who has an incurable personality disorder that can only be managed on a day-to-day basis, such as narcissistic personality disorder. Either that or they have such extreme trauma that it has become too much for them to handle on their own.

You get the results you work for. Therapy is helpful, but it is not magic. You must be the strongest driving force in your recovery. This is more than just wanting to get better. Everyone wants a better life with better things in it. You can want to be more confident but never take any chances. You can want to have a fuller social life but never say hello to a new person. When you have the drive to accomplish something, it means you are willing to achieve it, even if that means doing things that are difficult or outside of your comfort zone. Believe it or not, overcoming a cognitive distortion is going to cause feelings of discomfort. You may feel like you have lost an aspect of your identity. You may wonder what comes next. Humans are creatures who prefer the things we are familiar with, even if these things are unpleasant. This is why some people continue to choose toxic relationships and friendships. While they do not enjoy the chaos and drama that comes with these relationships, they are familiar with the way they work after having come from a tumultuous household. This is why going through cognitive behavioral therapy and taking steps to change the way you think takes so much courage. Not everyone can go through with it.

Good things will not just come your way. You will not wake up one day, and your life has changed. You have to earn these things. In the movies, a person gets a letter in the mail saying they are actually royalty and then are whisked away to a mansion where they meet a royal suitor. In real life, waiting for something external to change your life is not going to work. If you want to attract outside forces that are positive, you need to exude positivity.

Healing your mind starts with healing your body. In fact, if you have physical health problems, it will make your mental health suffer. Have you ever had that time where you were feeling physical pain that you were not also feeling irritable? This also goes for being tired and sick. It probably has been some time since you have taken a real look at how you are doing physically. Perhaps you haven't been feeling well but have just kept pushing through because you do not feel like you have time to get yourself better. If you are refusing to take care of yourself, you will actually lose time. If you ignore poor physical health, eventually, your body will let you know it is at its limit, and you will have to take time off from work, which will hinder your ability to make money.

An average person needs to have around 8 hours of sleep. This means if you need to wake up at 6 in the morning, you need to be in bed by 10 p.m. at the latest. You will be surprised at how much more productive you feel when you have gotten a good night's rest. You should also have breakfast. Even if you are not the type of person who likes to have a big breakfast, you need to eat something in the morning. Otherwise, somewhere in the middle of the morning, you will start to feel tired because your body is craving nutrition. Actually, you will feel the highest energy levels when you eat a lot of small meals throughout the day. Also, if you can afford to have, a 20-minute nap during the middle of the day will restore some energy you lost during the morning rush. There are times you will need to push through something, for instance, if you need to finish a paper within an hour. However, if there are no deadlines pressing on you and you are beginning to feel like you need to turn your attention away from whatever you are working on for a moment, feel free to do so.

Your mind and body will sometimes be vocal about what they need. Let's say you work at a computer and spend most of your day typing emails and other papers. You hit a point where your eyes are swimming, and you can't concentrate. When this happens, put the computer down for just a minute. You might think it is better just to keep pushing through, but this will actually only make you feel worse, and if you cannot focus, the work you put out will not be the quality it would if you just gave yourself a break. Try having a snack or another cup of coffee. If you give yourself this small break, you might find that you can return to your work later, feeling renewed and better able to complete the tasks you need to. It is not always about pushing yourself even if your body is saying no. Sometimes, you need to be kind to yourself, and you will end up getting a better performance out of yourself for that.

You also might need to broaden your definition of progress. You may feel like you are behind because you are still feeling poorly. However, there might be some progress you are making already but are no noticing it. For example, a person who has been suffering from panic attacks might feel like they are not getting any better because they are still experiencing them. However, if you are able to recognize that what you are having is a panic attack, you are a step farther from where you were. In the past, you might have felt overcome by this sensation and did not know what it is. You might have thought something horrible was happening to you that was out of your control and could even kill you, such as a heart attack or stroke. It was hard to breathe, and you thought you were going to lose control and even get yourself or someone else hurt. When you get to the point where you can realize "I have a panic attack," you have taken a step-in regaining control.

There is a shift in power when you begin to give the disorder less importance. In the case of anxiety and panic, when you stop thinking of it as something horrible that ruins your life, the less power it will have over you, and this will actually make it lessen in intensity. It might be difficult at first to not think of anxiety or depression as something horrible, considering the amount of distress and emotional pain it has caused you, not to mention the ways it has limited your life. However, think of it this way. Nothing in your actual life has changed. No true crisis has happened. No one has been injured. Your life is not in danger. Your house is not crumbling apart. Nothing bad is happening outside of your own mind. Nothing terrible is happening. You are experiencing temporary sensations in your mind and in your body that are uncomfortable.

Even when thinking of events in the real world, limit how much you use the word "terrible" or any other word to that effect. If you are afraid someone you work with doesn't like you, your fears might not even be founded. That could simply be how their personality is. They might be shy or the type of person who only says what needs to be said. However, even if the worst-case scenario happens and they do not like you, how bad of a thing is it really? Nothing in your life will change just because of this. You will still have your job. You will still have your family and friends. You will still have your house and your health. The only consequence you will suffer because of this is that someone you do not often see and does not play a big role in your life will dislike you, most likely in secret because, more often than not, people want to avoid unpleasant conversations. When you think of it like that, even if this fear your mind has conjured up is true, it will really not be that bad.

You will feel better when you accept where you are right now and do not think in terms of good, bad, worse, or better. This is how you are doing today, and it is acceptable. You might compare yourself to other people who seem like they have it all together and feel inferior to them. However, you need to keep in mind that you are only looking at their life from an outside view. They might be going through things you are not aware of. Everyone you know is dealing with stress on a daily basis. They also make mistakes on a daily basis. For all you know, they might be going to therapy to work on their own set of challenges.

While the span of time a person spends in therapy is up to them, there should also be a progression. The conversations should not be circular. A circular conversation is one where the same things are said over and over again, and there is no moving forward. Essentially, it is the same conversation happening repeatedly. If you feel that you have hit a place where you are stuck in your therapy, you are not doomed to this. You just need to figure out why you have run into this roadblock. There might be a fear you are no ready to overcome or an aspect of your life; you are not excited to face.

Cooperating with a Therapist

There's nothing wrong with asking for help. Sometimes when you are on the quest to improve your life, it is wonderful to have a steady ally, someone who will be your champion and coach as you dismantle old thoughts, emotions, and beliefs that no longer serve your purposes. Even if you were taught as a child that asking for help makes you a major annoyance, it is no longer true in your adult life. You deserve to receive help if you want it.

CHAPTER 29:

How to Find a Good Cognitive Behavioral Therapist

One of the first criteria for selecting a therapist is whether or not you like the person. Do you feel comfortable? Do you imagine that you could talk about difficult things and cooperate with this individual? There has to be that feeling of esprit de corps or you will be wasting your money.

Where to Begin

It can be quite helpful to ask all the professional people in your life if there is a particular therapist that they know or would recommend. Ask your minister, accountant, attorney, massage therapist, beautician, barber, and physician. Carefully take down the name and pertinent details, as well as a note about who referred you to that person, as you will want to mention that when you first meet. If there is a particular characteristic that you absolutely need to have in your therapist, a Google search might bring forth some possibilities. If, for example, you require a particular location, insurance acceptance, or expertise in something very specific, such as adult giftedness, sexual abuse, sexuality, or alcoholism, it could be easier to find someone with an Internet search. Follow up on each possibility with an e-mail or phone call and make an appointment. You will decide rather quickly upon meeting the person whether or not it might be a fit, but either way, take along your list of questions about the suitability of the arrangement. Does this therapist practice Cognitive Behavioral Therapy? What are the therapist's educational background and credentials? Any specialties? What professional organizations does that therapist belong to? What insurance is accepted?

Types of Mental Health Professionals

Any mental health professional may be well-equipped to help you with CBT. In other words, her primary title may be something else, like a clinical social worker. It is unlikely that you would find a person who is strictly a cognitive-behavioral therapist without those skills being in the context of a broader profession.

Some of the professions that might also offer CBT are as follows:

Psychiatrist

Psychologist

Master's level counselor (marriage and family therapist)

Social worker

Advanced practice nurse

Depending upon your personal preferences, you may be interested in a type of help that is not licensed, such as the following:

- Spiritual counselor
- Hypnotherapist
- Life coach
- The Initial Sessions

The first session will be a consultation, in which you decide whether or not you wish to continue with that person. You will want to ask about the values and orientation of the therapist in order to determine the presence of like-minded compatibility. It will be important to discuss costs, insurance, and the estimated length of the course of treatment.

The next few sessions will probably be centered on assessment. The therapist wants to get to know you, what brought you to therapy, and what changes you want to make in your life. The two of you will work out some plans and agreements in order for you to get the most out of your commitment to Cognitive Behavioral Therapy.

Limitations of Family and Friends

Although those people closest to you are highly interested in your life and experiences, they may not be the best source for advice or support. Undoubtedly their perceptions are skewed, and they may even be threatened by your hopes of making major changes in your life. It might be a better use of your resources to turn to people who can efficiently zero in on your difficulty. Those are the people who are in your inner circle who know you very well and could perhaps push your buttons when you are making significant efforts to change your life. Just be careful about what you disclose.

Conclusion

If you do engage in therapy with a skilled professional, you can expect your sessions to be short and structured. Despite the brevity, the structure maximizes the benefit to you. Typically, the beginning of the session is addressing any concerns you may have had during the time since your last session has ended, as well as reflecting on whatever assignment you were given to practice during the week. You then begin to learn a new skill that will be useful for you, followed by practicing your skills in a controlled environment with the therapist. You are then given the assignment to complete before meeting once again before you are sent on your way.

By having an assignment that requires you to use the skills taught to you during the week in real time, in real life situations, you can really reinforce their use in your life. As the skills taught prove themselves effective, you then use those skills more often, which gives you more benefits and positive reinforcement, which leads you to use them even more. This positive reinforcement cycle really affirms that the therapy process is working while also making waves in your life.

Just think; one tiny negative thought can derail your entire day, just like a single stone tossed into the water can ripple out to affect water quite a distance away from it. Now imagine, instead of just a pebble of a negative thought, you pick up a fistful of positive thoughts and behaviors turned into rocks and threw them into the water. It will make a much larger effect and will be much more noticeable. Those big waves are what you create as you utilize the skills taught in therapy in real time to be compliant with therapy.

This, in part, is one of the reasons CBT is so effective. You are literally tasked with changing thoughts behaviors in real time and taught the skills to do so rather than spending your time dwelling on the past to eventually try to come to terms with whatever has happened in your past. As you begin to utilize the skills and find them effective, your therapist can walk you through why the process worked in the first place, leaving you with a thorough understanding of how your mind works. This leaves you understanding the changes in your life at a fundamental level, which leaves you much more likely to continue including them in your life in various aspects. If you discover that repeating affirmations helped you overcome your crippling anxiety by helping you remain grounded during an exam, you may decide to try implementing them when you are stressed about a big deadline at work, or to stop yourself from lashing out at people when you are upset. Likewise, if grounding techniques worked when you felt overwhelmingly angry, you may try using them when you feel other overwhelming feelings as well. As these skills and techniques build upon themselves, they inspire you to continue making changes, and you end up with a domino effect. One change leads to another, then another and another, until you barely recognize the life you are living now but in a good way. The changes that will happen will seem almost too good to be true, but they are entirely possible if you put in the dedication.

Without a therapist to guide you, this process is a little bit different. There will not be anyone trained to help you or bounce ideas off, nor is there anyone who can walk you through the steps and explain why things work. You have a book telling you what to do, but unable to provide real-time feedback to help you better if it is necessary. You have a cookie cutter assignment designed to help the general public, and you must tweak it to make it specific to you and your situation. This does not mean it is impossible. However, if it were, this book would be pointless! You absolutely can attempt CBT on your own since the process itself is so relatively simple and action-based, if you have the dedication and fortitude to do so. However, if at any point, it is too much for you to handle, you should absolutely seek out professional help.

PART II

OVERTHINKING

Introduction

What precisely overthinking is? Specialists review overthinking disorder as investing a lot of your energy considering anything that causes pressure, dread, anxiety, concern, fear, disquiet, and numerous others. This state of mind isn't just about an excessive amount of considering something yet in addition to spending more considerations, which influences the individual's ability to work and finish the day by day errands.

If you stress over your companions, your activity, your family, your sweetheart, and sweetheart, your life and whatever else; You accept that you don't have the overthinking condition; No issue your thought process, maybe concerns you for a second, at that point a limited capacity to focus time, you proceed with different pieces of your life. Indeed, you return pondering now and then. Notwithstanding, you don't ceaselessly consider it, and you don't think it's disrupting the general flow with a special remainder. With this sort of state of mind, be that as it may, the disturbed is all that you can consider. Regardless of whether you may not be grieved on something very similar consistently and over and over, you generally stressed over something.

To know whether you are overthinking everything, you may encounter at least one of these cases:

□ Complexity in following and taking part in a conversation.

□ You are continually assessing yourself to people around you just as how you coordinate with them.

□ Giving such a great amount of time on most pessimistic scenario situations, including individuals from the family or yourself

□ Reliving past errors and disappointments on numerous occasions, so you are not fit for moving past them.

□ Worrying a lot about your items and dreams or future assignments until such time that they are not practical to accomplish

□ You are reliving your past horrible experience, which incorporates the loss of guardians, or any individuals from the family, misuse, which leaves you not equipped for managing it?

□ Lack of ability to hinder the progression of ambiguous considerations, feelings, and stresses

No two people will experience the ill effects of this sort of mental issue similarly. In any case, individuals who do endure overthinking will find that their daily living and way of life are in question. This is because of their absence of capacity of productively and effectively controlling negative considerations just like a feeling. Along these lines, they think that it's challenging to hang out and mingle. They likewise believe that it's challenging to make the most of their sidelines and interests. They are not useful and not innovative at work. All these occur since their brain spends a conflicting measure of vitality and time on specific lines of considerations. There is an inclination that they don't have total control over their feelings or brains that can be incredibly dangerous to mental prosperity.

You are making companions, getting together with somebody just as keeping companions is difficult with this state of mind as you oppose cooperating with them. You will encounter trouble in conversing with the individuals that encompass you. This happens because you are a lot of issues with the things you do or say. You are additionally a lot of worried about what will occur or how you will do it on the off chance that you, your companions, or any individual from the family experience, this psychological issue may battle to make general discussion with others. It will likewise be challenging to interrelate in a standard-setting. You may discover it so challenging to go out to the retail establishment or to your arrangement, which influences as long as you can remember when all is said in done.

This state of mind can influence practically all parts of your life. This can likewise change how you speak with others and how you perform. It can also affect your own life, public activity, work-life, and, obviously, your affection life. It can destroy your relationship to your accomplice, and individuals encompass you and your whole life by and large. Overthinking is an extreme condition. It can make issues throughout your life.

Overthinking is a disorder that can happen to anybody and whenever. If you have tension or any uneasiness issue, it can become overthinking tumult as well. The concern and the uneasiness which you have on various life conditions can become overthinking rapidly. You should consider what you should continue or how you prevent awful things from occurring. You can't avert awful and awful things from happening. Additionally, you can't stop yourself from every terrible choice. In this way, the best activity is to ask help from companions, relatives, and experts.

Do you stress excessively, even on fundamental things? Is it accurate to say that you are an outrageous worrier? Possibly you consequently feel that when you wonder,/fear enough, you can maintain a strategic distance from awful things from occurring. Notwithstanding, stressing and pondering can influence your framework from multiple perspectives, which may amaze you. On the off chance that you think a lot on everything or on the off chance that you have unnecessary stressing, it can bring about sentiments of high pressure and even lead to a genuine wellbeing condition.

CHAPTER 30:

Definition of Overthinking

There are numerous methods of portraying overthinking. It tends to be comprehended as a circumstance where one can't imagine quite agonizing and thoroughly considering things. Overthinking isn't a turmoil. It includes a dread that develops in you and overpowers you, yet you can't support yourself; however, let it do as such. Now and again, rather than dealing with it, you essentially select to keep down your tears. It's the dread of disappointment: falling flat at your particular employment, bombing a specific class, fizzling in your connections, overthinking drives you to buckle down for ridiculous desires. This may sound gainful; however, in all actuality, you will be depleted by keeping up this pace. You are thinking of an excessive amount of prompts depletion.

Sincerely and genuinely, you will feel depleted since your brain never stops. It is constantly overflowed with contemplations, and the most exceedingly terrible thing about it is that you accept there is no other viable option for you.

Overthinking is that inward voice that attempts to cut you down. It condemns you and devastates your certainty and confidence. You question yourself, yet you likewise question the individuals who are near you. It pushes you to re-think everything. Figuring an excessive amount of can is contrasted with a spreading fire. It torches everything that it finds on its way. Hence, you will endure because of overthinking.

Overthinking is the point at which your psyche sticks to the shortcomings that you have made and takes you through them for the day. At the point when you overthink, your life will be on steady interruption. You will consistently feel like you are trusting that the correct second will accomplish something. The issue is that this second never shows up. You're continually foreseeing that something could turn out badly. You will be excessively cautious while busy. This is impacted by the way that you are simply stressed things probably won't work out true to form.

Kinds of Overthinking

There are various types of overthinking that could influence the nature of the choices we make. Regular types of overthinking are compactly examined in the accompanying passages.

Theoretical Thinking

This alludes to a type of reasoning which goes past cement realities. For example, when you are attempting to plan speculations to clarify your perceptions, at that point, you're participating in dynamic reasoning.

Multifaceted nature

The multifaceted nature type of overthinking comes about when there are numerous elements to consider in your dynamic procedure. For this situation, these various elements could keep you from gauging the genuine significance of every last one of them. The impact is that it could keep you from settling on choices expeditiously.

Evasion

Evasion happens when one attempts to abstain from accomplishing something by utilizing the dynamic procedure as their reason.

Cold Logic

When utilizing cold rationale to think, you will, in general, abstain from depending on human elements, including language, culture, character, feeling, and social dynamics.

The result is that you wind up settling on one-sided choices that don't think about legitimate or social real factors.

Instinct Neglect

This happens when one neglects to consider what they know. As it were, one selects to overthink something that they know some things about. Rather than following your gut nature, you overthink and wind up settling on an inappropriate choice.

Making Problems

You may likewise wind up intuition in a manner where you are making issues that are not there in any case. There are sure circumstances or things that are not as perplexing as you might suspect. In standard circumstances, it would have taken you a moment or two to tackle them. It is fundamental to concentrate more on the master plan and not criticize the subtleties.

Amplifying the Issue

Typically, little issues require straightforward arrangements. There are cases where we intensify these issues, and we wind up concocting excessively complex answers to settle them. This is another type of overthinking. You wind up squandering your assets to think of colossal arrangements that don't coordinate the issues you are encountering.

Dread of Failure

The dread of disappointment is anything but another idea to the vast majority. This is the thing that rouses a large portion of us to buckle down. Rather than buckling down for a splendid future, you wind up drawing inspiration from the dread you have created inside you.

Superfluous Decisions

There are times when we settle on superfluous choices since we constrain ourselves to decide on these choices, yet we are not required to make them. For example, when pondering our future, there are cases where we wind up paying on unessential options dependent on suppositions.

CHAPTER 31:

How to Identify If You Are an Excessive Thinker

Signs You Are an OverThinker

Coming up next are clear signs that you think excessively. You may attempt to deny it, yet consider these signs and question yourself whether these are a portion of the things that you may have encountered.

You Overanalyze Everything

If you notice that you overanalyze everything around you, at that point, you are positively an over-thinker. This implies you may attempt to locate more profound importance in all the encounters that you experience. When meeting new individuals, rather than taking part in profitable correspondence, you may concentrate rather on how others see you. Somebody could be giving you a specific look, and you may make a few suspicions simply dependent on that look. Overthinking devours you. You wind up squandering a great deal of vitality attempting to make sense of and understand your general surroundings. What you don't understand is that not all things have inherent significance.

You Think Too Much But Don't Act.

An over-thinker will be influenced by something many refer to as investigation loss of motion. This is where you ponder something; however, don't take care of business at long last. For this situation, you invest a great deal of energy gauging the alternatives you have available to you. From the start, you decide on what the best option may be. Afterward, you contrast your choice with other potential choices that you could take. This implies you can't quit pondering the potential outcomes and whether you settled on the correct choice.

At last, you end up not settling on a choice. You just end up in an endless loop where you think a great deal; however, there is little that you do. Maybe the best procedure to keep yourself from falling into a reasoning snare is to evaluate the choices you have. A straightforward choice to act will have an immense effect.

You Can't Let Go

Regularly, we settle on incorrect choices that could lead us to come up short. At the point when this occurs, it very well may be overwhelming to give up more so when you think about the penances you have made to arrive at the point you are at. You may feel that it is excruciating to give up after you have put away a great deal of cash on a specific business. The issue here is that you would prefer not to fall flat. In any case, understand that neglecting to give up just keeps you away from evaluating something different that could work. It likewise influences your life since you will ponder your disappointments. You have to proceed onward.

You Always Want to Know Why

Point of fact, the idea of inquiring as to for what reason can be useful to tackle issues. This is because this examining mentality finds you the solutions that you may be searching for. In any case, it can likewise be harming when you can't resist the urge to ask why consistently. Ordinarily, we are acquainted with addressing inquiries from kids. They simply love to inquire as to why about everything without exception. They won't spare a moment to ask you for what good reason you don't converse with your neighbor. Why youngsters are conceived or just why you love to walk. There's something novel about how kids are interested.

You Analyze People

How you see others can likewise say a great deal regarding you. As a rule, you get lost reasoning a lot about how others carry on. You may in appointed general authority, everyone that you run over. This one strolls entertainingly. That individual isn't dressed well. You wonder what somebody sitting at the recreation center is grinning about. At the point when these considerations fill your head, you will just deplete yourself. Investing an excessive amount of energy concentrating on others will just stop you from utilizing your brain beneficially. Rather than envisioning your objectives and your future, you squander your vitality considering seemingly insignificant details that increase the value of you.

Normal Insomnia

Do you think that it's difficult to rest here and there? You may get worked up over the possibility that your cerebrum can't close down and quit thinking. Unfortunately, this can deaden you since your cerebrum doesn't get the rest that it merits. Bit by bit, you will see a reduction in your efficiency. You are probably not going to like yourself since there is little that you accomplish. Agonizing a lot over not having the option to rest can make you uncomfortable, and you may wind up in a condition of bondage. On the off chance that this is something that you have been encountering, at that point, it seems like you may be an over-thinker. What might you be able to do about this? To start with, on the off chance that you are not dynamic, at that point, it is indispensable that you discover a method of keeping yourself occupied.

You Always Live in Fear

It is safe to say that you are apprehensive about what the future has available for you? If you answer yes to this inquiry, at that point, the chances are that you're confined in your brain. Living in dread could drive you to depend on medications and liquor as your best cure. You will pick up the discernment that by consuming medications, it will assist you with suffocating you're distressed and help you overlook. Sadly, this isn't the situation since medications and liquor are minor depressants. They hinder your cerebrum working. Subsequently, you will, in general, accept that they are helping you overlook.

You're Always Fatigued

Do you generally get up in the first part of the day feeling tired? This could be a consequence of stress or wretchedness. Rather than carrying on with a gainful life, you end up getting up late, drained, and unmotivated. The motivation behind why this happens is because you don't offer your brain a chance to rest. It has been working day and night. At night, rather than resting, you end up alert throughout the night since you are overthinking. Your brain can't labor for 24 hours in a row at a similar degree of working. You will just experience the ill effects of burnouts.

You Don't Live in the Present.

Do you think that it's hard to appreciate life? For what reason do you think you think that its overwhelming to sit back, unwind, and be content with your companions? The minor truth that you can't remain in the present suggests that you won't center around what's going on in the Present. Overthinking will dazzle you from seeing anything great that is at Present occurring around you. You will frequently consider the most terrible that can occur. The issue is that you are caught in your brain, and there is nothing outside your considerations that you can helpfully consider.

The inability to live in the Present denies you the chance to improve associations with others. You will live in dread that they will scrutinize you. In this way, you will just need to exist in your cover. Once more, this will prompt pressure.

CHAPTER 32:

The Relationship Between Excessive Thinking, Anxiety, Stress and Negative Thinking

It is intriguing to discover that our musings characterize what befalls us. From a mental viewpoint, it implies that we can control what occurs us by essentially figuring out how to control our musings. This is a ground-breaking procedure, surely. Realizing that you have control over what occurs, you are something that the vast majority are ignorant of. You become what you think. On the off chance that you look carefully, whatever transpires, positive or negative, comes from your considerations.

Your opinion affects your emotional wellness and prosperity. Your considerations lead to the passionate express that you may be encountering. Frequently, this will influence your wellbeing. On the off chance that your contemplations are engrossed with pitiful occasions, at that point, the chances are that you will continually feel dismal. If you are continually pondering the pleasant exercises that you participate in with your companions, at that point, you pull in a similar vitality to your life. This part investigates how negative reasoning is connected to uneasiness and stress. From this, you will collect a more profound understanding of why your contemplations could be distinguished as the reason for your decreasing efficiency at work, absence of rest, and your bombing social connections.

Nervousness; Stopping Negative Thoughts

Various elements bring about nervousness. Now and again, it is brought about by a mix of hereditary elements and natural components. Dread inside; you can, without much of a stretch, cause you to feel stressed over things that haven't occurred. In outrageous cases, this prompts alarm. Your brain can, without much of a stretch, intensify the feelings of dread inside you and cause you to accept that something awful will occur. In social settings, uneasiness will leave you in a steady condition of the stress of saying an inappropriate thing before others. Likewise, you may pick up the supposition that others won't care for you. Such negative musings just keep you from acting naturally. It keeps you away from carrying on with your life.

Normal Thoughts in Anxious People

There are sure unpleasant contemplations apparent in restless individuals. The following are a couple of instances of a portion of these considerations. Distinguishing these contemplations is useful as it guarantees that you figure out how to manage your nervousness. Instances of normal musings in on edge individuals are as per the following.

"I am bad at what I do."

Restless people will concentrate more on the antagonistic parts of themselves. In any setting, their brains will continually consider their shortcomings. It will be hard for them to consider their qualities and why they were picked for a specific job in their work environment. Uneasiness will cause you to feel like your manager will fire you whenever, for instance.

"I will overlook it."

Have you, at any point, felt that you would overlook something even before the genuine thing happened? This is an indication that you are on edge. Accepting that you will overlook something implies that you can't confide in yourself. You're bringing questions up in your psyche that you can't make sure to accomplish something either during the day, tomorrow, or sooner rather than later.

"No one loves me."

In the web-based life world, it is simple for an on-edge individual to reason that individuals don't esteem them since they are not getting any reactions to their posts. This attribute depicts somebody who thinks excessively. This is an individual who is constantly stressed over what others may state. Thus, they will be too worried about their web-based life posts and the reactions they will get.

"Imagine a scenario in which I am straightaway.

No ifs or buts, we live in a universe of vulnerability. You can never make certain about tomorrow. This can affect your demeanor towards the obscure. There are times when you may be frightened that the most exceedingly awful could transpire whenever. According to this, you ought to comprehend that it isn't unexpected to experience such musings. In any case, this doesn't imply that you ought to permit such contemplations to overpower you. Since you have some degree of control of your contemplations, you ought to figure out how to oversee them. Living inconsistent concern that you may falter at any moment is an unacceptable quality of life.

"My accomplice hasn't called; they should be distraught at me."

Uneasiness can likewise influence your connections from various perspectives. Consider a standard model where your accomplice neglects to call you during the day. There are numerous reasons why this could have occurred. Perhaps they were occupied, or their telephone was out of battery. Notwithstanding, your stressing nature will give you the supposition that your accomplice is furious about you for reasons unknown. Having this recognition will just demolish the delightful relationship you share with your accomplice.

"Did I leave the entryway open?"

The vast majority will stress a lot over the straightforward things that they may have neglected to do. For example, you may interrogate yourself regarding your entryway, apparatuses, or your light switches. You will discover your psyche meandering contemplating whether the machines were turned off. Doing this more than once will just prompt tension.

In light of these normal instances of on edge contemplations, unmistakably overthinking can prompt tension. The basic truth is that you can prevent yourself from speculation to an extreme. Your accomplice was neglecting to mind you; for instance, there are numerous reasons concerning why this could occur. Possibly they are occupied grinding away and that their cell phones are in quiet mode. It could likewise be that they are in a gathering. Consequently, there is no requirement for you to contemplate it. Grasp accepting things as they are without entangling them.

Uneasiness Triggers

There are various reasons why you will be on edge. There are sure occasions, encounters, or feelings which could intensify tension's indications. These components are named as uneasiness triggers. Coming up next is a short investigation of a portion of the normal triggers of tension.

Wellbeing Concerns

Wellbeing concerns can be a significant trigger of tension. Normally, this occurs after an upsetting clinical finding, for example, constant disease or malignancy. It is regular for individuals to be worried about the bearing that their life would take when experiencing a ceaseless infection. Fortunately, you can manage this uneasiness by changing how you think and see your life. Carrying on with a functioning life, for instance, will keep you from focusing on the sickness. Rather, you will acknowledge what life brings to the table and appreciate it.

Prescriptions

There are sure prescriptions that can likewise cause you to feel on edge. This is because of the way that these drugs have dynamic fixings which influence how one feels. Basic prescriptions could cause tension to incorporate weight reduction meds, anti-conception medication pills, and clog medications.

Caffeine

Caffeine can decline or trigger the side effects of tension. With social uneasiness issue specifically, it is prudent to bring down your caffeine consumption.

Skipping Meals

There are cases where you may feel jumpy on account of skipped suppers. This happens because of the drop in your glucose. Eating a fair eating regimen is suggested for various reasons. Advantageously, you endeavor to eat a solid eating routine consistently to guarantee that your body gets all the supplements it requires.

Topping yourself off with sound tidbits assists with keeping up your glucose levels. Like this, it diminishes the probability of you feeling apprehensive or upset.

Negative Thinking

There is a great deal that we have talked about concerning how your contemplations figure out what your identity is. Thinking contrarily will probably degenerate your psyche with sentiments of dissatisfaction. This implies you are probably going to feel on edge since you stress a lot over the most terrible that can occur.

Money related Concerns

With the unforgiving financial occasions, it is overwhelming to keep yourself from pondering your accounts.

This turns into a significant issue when you have obligations to pay, and everything shows up wild. To manage triggers identifying with accounts, you ought to think about looking for proficient help.

Stress

Notwithstanding the triggers talked about above, ordinary stressors can likewise cause tension. For example, while missing significant cutoff times, you will stress over the possible loss of something significant. You could wind up building up a dread of losing your employment. Without a doubt, this will worry you a great deal. Now and again, stress can negatively affect the nature of your rest. This declines your circumstance as uneasiness will, in general, exacerbate when you need more rest.

Damaging Thinking; A Common Cause of Stress

Negative reasoning will regularly prompt pressure. At the point when you continually harp on negative self-talk, this is the thing that your inner mind-brain will concentrate on. Rather than ruminating on how awful things appear to tail you, it is crucial to understand that such considerations can negatively affect your enthusiastic prosperity. To grasp how our considerations lead to pressure, we should consider how pressure works.

How Stress Works

Analyst Albert Ellis proposed the ABC model of seeing how stress works. According to this model, outside occasions (A) don't trigger feelings (C). In any case, convictions (B) can cause feelings. This is to imply that their outside surroundings do not straightforwardly impact individuals' feelings, yet they are influenced by how they process what occurs around them.

Seemingly, stressors will consistently be there. For example, stalling out in rush hour gridlock is a typical thing. It possibly prompts pressure when you handle it negatively. In such a manner, having a cynical view about traffic will cause tension and stress. Perceiving that you can control how you think should assist you with perceiving that you can, without much of a stretch, dodge pressure. Things being what they are, the reason would it be a good idea for you to whine about a gridlock when you are sure that there is no way around it? To successfully manage such a circumstance, you should keep your brain drew in with something different. Tune in to your preferred music as you trust that traffic will open up.

CHAPTER 33:

The Law of Attraction

On the off chance that you are worried about the course that your life is taking, at that point, the law of fascination might be a valuable apparatus to get you in the groove again. By all accounts, you may infer this is a law that causes you to draw in things around you. All things considered, similarly, as the name proposes, this is an amazing law that recommends that you pull in what you center on. Believe it or not, this law is continually attempting to shape your life. What individuals don't comprehend is that they are continually forming their lives deliberately or subliminally. The existence that you have today is credited to your opinion of years prior. On the off chance that you recently pondered maintaining a fruitful business, there is a decent possibility that you have made strides to attempt to accomplish the objective. Without a doubt, you probably won't get precisely what you needed. However, you will be in an ideal situation than speculation adversely.

Your future is molded by how you think and how you react to circumstances today. Accordingly, on the off chance that you feel that the coming months will be hard for you, have confidence that they are bound to be troublesome. Then again, on the off chance that you have the discernment that you will have some good times, at that point, you are bound to appreciate life as it unfurls itself to your desires.

The law of fascination depends on a straightforward idea. You draw in what you decide to concentrate on. Regardless of whether you decide to think adversely or decidedly, it's everything up to you. On the off chance that you decide to concentrate on the positive side of life, at that point, you will pull in beneficial things your way. You will consistently be brimming with satisfaction and abundance; you will carry on with your life feeling vigorous and prepared to deal with whatever comes your direction. Despite what might be expected, on the off chance that you decide to concentrate on the adverse, your life will be loaded with wretchedness; you will never be content with the individuals around you. Frequently, you will feel like you are worn out on living. Your profitability at work and home will be contrarily influenced. You will consistently be that individual that finds the negative in all things. The entirety of this is an aftereffect of what you decide to concentrate on.

You know how the law of fascination functions can open the entryways of accomplishment in your life. This law opens your psyche to the acknowledgment that we live in a universe of unending prospects, boundless euphoria, and limitless abundance. Consider it; you can place your confidence in your convictions and help change future results. Isn't it unreasonably astonishing? Sadly, barely any individuals comprehend the law of fascination and how to utilize it to change their lives successfully.

Your considerations and emotions will cooperate to fabricate a perfect future for yourself. Since you can choose what you need, you should demand a day to day existence that you've generally longed for living. Your concentration and vitality ought to be following what you need to pull in.

Step by step instructions to Use the Law of Attraction

After understanding the way that you are the maker of your reality, you should start thinking deliberately towards making a superior life for yourself. For this situation, this ought to urge you to think decidedly since your musings characterize what you need throughout everyday life. This necessitates you channel your time and vitality on contemplating the beneficial things you need throughout everyday life. It additionally implies that you ought to purposely deal with your contemplations and feelings as they affect what shows.

Ask, Believe, Receive

The law of fascination gives off an impression of being a direct procedure where you simply request what you need, and you will get it. In any case, the application procedure requires something other than asking and getting. If it were this straightforward, at that point, everyone would be carrying on with upbeat carries on with liberated from pressure and uneasiness. All in all, would could it be that makes the law of fascination basic yet overwhelming to apply?

Inquire

Individuals make solicitations to the universe consistently, either deliberately or unknowingly, through their musings. Whatever you consider is the thing that you center around. This is the place you have diverted your vitality. Utilizing the law of fascination, you ought to understand that you must take purposeful activities to deal with your considerations and feelings. In such a manner, you need to conclude that you need something deliberately. This likewise requests you should live and go about like you as of now have what you are requesting.

Accept

For you to show what you need in your life, it is basic that you genuinely accept that you will get what you need. Your musings ought to mirror the sureness that you have in realizing that you will get what you need. In this way, your brain ought to be liberated from questions. This is the trickiest piece of the law of fascination.

A great many people just inquire. In any case, they think that it's hard to accept that they can get what they need.

The part of conviction decreases when people understand that what they requested is taking more time to show than they anticipate. In this way, they direct their concentration toward negative reasoning.

They start to persuade themselves that it is incomprehensible. Life isn't simple. Such recognitions just influence what you are requesting from the universe. The most exceedingly terrible thing is that the pessimism predisposition starts to come to fruition. Without acknowledging, they draw in pessimism in their lives since they just neglected to accept.

Get

The exact opposite thing that you have to do is to get what you were approaching or seeking after. Maybe this is the least demanding part since it just expects you to situate yourself in an ideal manner through your feelings to get your blessing. Consider a customary circumstance where you are accepting a blessing from your friends and family. You express from your non-verbal communication that you are upbeat. Feelings of affection and gratefulness ought to be clear while getting any blessing. This is the way the universe anticipates that you should get your prize.

You should live your day feeling appreciative and glad for what you as of now have. This is the ideal manner by which you can work on getting what you need even before the world offers it to you. These feelings can likewise be molded by how you decide to think. As needs are, it is suggested that you should live carefully by upgrading your mindfulness to stop yourself each time negative musings create in your brain.

From the start, it won't be a simple accomplishment to control your considerations and feelings. By and by, it is important that everything great calls for tolerance and practice. For the law of fascination in work for you, you must show restraint. You need to continue rehearsing the propensity for accepting. Above all, consistently recall that you can make your satisfaction.

CHAPTER 34:

How to Use Overthinking to Your Advantage?

Like I said, social media has become the most common distraction that people deal with. We all know that technology plays a huge part in people's' lives. We believe it is also a reason for the problem of procrastination. But ironically, it has the answers to your procrastination habits. Since there is technology, you don't have to worry about ending your habit of procrastination. Why? There are numerous ways to overcome the habit of procrastination. Yes, for example, through motivation, you can overcome procrastination, but apps and tools sound more practical than motivation. Don't they? So, if you are looking for the best anti-procrastination equipment, know that there are many.

Small Habits, Big Change

You already know small habits have a bigger impact on your life. For example, if you brush twice a day, you will not see the changes right away, but you will have a great set of teeth when you grow old. Just like that, when you practice simple habits, for now, there will be a massive impact on your life later. So, here are some of the tips that you should follow:

An Organized Individual

Do you think plans can't change your level of productivity? Well, try creating a plan, maybe for the work you have for next week or the work you have to complete tomorrow. And then, stick to the plan and see what happens. It might sound simple. You might even wonder if a simple plan can bring so much difference. Well, yes, it can! Through a plan, you organize the work that you have to do. When you organize the work, you understand the process clearly. For example, you have to complete a massive project, but if you just let the huge project be as massive as it is, you will not feel like doing it. You will not be able to see the amount of work you have to do in a day, and that will create boredom and ignorance. Thus, you have to organize the work that you have. Luckily, there are so many great tools and apps that you can find to organize work (more on this later).

Make It Simple

Another common reason for procrastination is due to having complex tasks. Of course, some tasks can be complicated, but it is not as if you can't simplify them. For that, you have to set simple, achievable goals. Instead of saying, "I'll complete the project," say, "I'll complete the first part of the project today." When you make it sound simple, it will be simple.

Have A Schedule

Once you have a goal, it is important to schedule it because scheduled work has a higher rate of achievement. Break your work into chunks and set a deadline. If you set your deadline, you will be able to achieve them before the actual deadline boggles your mind. Sometimes, you might come across unexpected situations in life; thus, completing the work before the deadline will help you stay in the safe zone.

Set Aside Distractions

You might already know the things that distract you. For example, if you are addicted to Snapchat, don't keep it your phone near you until you get the work done. Or, if you are a LinkedIn enthusiast like me, stay offline until you complete the work. Don't even add the Google chrome extension of LinkedIn because it is incredibly distracting. The moment you see the notification, you might want to check the messages even if you have so much to do. Thus, it is better to put all your distractions aside and focus on the work you have.

The Commodore Technique

If you don't know what this means, this approach promotes working for 25 minutes and taking a break for 5 minutes. Most people consider this as an effective and excellent solution for procrastination. Honestly, this is a fantastic technique, and you will be able to get a lot of things done if you follow this approach. Moreover, by following this technique, you can ensure the quality of your work as well. During the break, you must not get distracted. Thus do something like listening to music, walking, or even screaming to release stress. Whatever it may be, make sure it makes you feel relaxed and comfortable. Thus, the activity that you chose to do should be something that you like, but not will divert your focus!

Reward Yourself

I don't think anybody hates rewards, so it is highly recommended to reward yourself when you follow your plan. For example, if you set a goal to write 2500 words within 5 hours, you must treat yourself once you have achieved it! You can reward yourself with ice cream or an episode of your favorite show. However, make sure that you'll get back to your routine once you've rewarded yourself.

The Myth of Doing the Hard Thing

So far, you have probably heard that doing hard things first helps you get other things done sooner. REALLY? Let me ask that again? The rule of doing the hard things first don't work for me. If it works for you, then please ignore this point. But if you give it a thought, you will understand the underlying concept. When you do what's possible, you become motivated to do the hard things too. Besides, when you try to the hard tasks and if it looks harder than it seemed, you might even delay the work. Thus, it is usually better to do things that are manageable first.

These are the small habits and changes that you must incorporate to become a productive individual. But there are many more anti-procrastination tips that I want to share with you.

Getting Started Technique

If you want to do something, you must get started. People usually procrastinate at the beginning of a project, so it is important to understand the techniques to get started. How can you do it? Starting a project or a task will not be easy; in fact, it can be the reason for delayed submission. Whenever you plan to do a task, you need something to boost your mood. At first, getting started can be difficult, but when you move on with the task, it might seem possible. Thus, compare the way you feel when you start the work and the way you feel when you delay the work.

Even if you have done a little from the whole project, it's a good start. Starting the project is important, so it doesn't matter even if you do a very little portion of the whole project. There is a trick to make your mind like the work, and that is to start thinking about the work. When you keep your mind occupied with the task, you might somehow end up starting it. The reason is it is tiring to think, so you eventually start work.

For example, say you should edit an article. If you don't begin editing, you will never do it. Thus, just take the draft and change a few words. Eventually, you'll end up changing the major units where you wanted to change. You will do it even without forcing yourself to do it, which is amazing!

Or you can set a timer. What can you do with a timer if you really can't start the work? Simple, set the timer to 10 minutes or less, and then, once the timer starts working, you just remain seated. Even if you don't do the work, just sit there. Eventually, you'll start work, and you will not even feel that you have started. This is an easy trick because when you are within your workspace, you can't help but work.

Thus, these tricks and tips might help you get better at what you are doing. The simplest mantra is, "get started!"

Useful Tools and Apps

Now that you've learned almost all the possible trips and tricks, it is time to get a grip on the tools and apps available. Beating procrastination will not be easy until you get help from the technology that you blamed for your reluctance. You have so many great tools and apps to select from, yet we'll discuss a few beneficial tools that you can rely on. Here we go!

Procrustes

This is one of the procrastination-busting apps, but compatible only for iPad and iPhone. The app will support you throughout the procedures by providing the right answers and advice as to the option that you provide. For example, if you select the option "I don't know how to start," the app will suggest breaking the tasks into chunks. It provides not only ideas but also guidance to do the work. You'll find a rhythm to your work, and you can even check the statistics related to your productivity. The statistics will become a motivation to reach the goals.

Stand stand

Anecdotally, it is considered that changes in the working environment can cause positive changes to your productivity. Thus, the introduction of the portable standing working table has become a great piece of equipment to fight against procrastination. Sometimes, you might get bored by sitting for long hours. In such a case, you can consider the StandStand table. The StandStand table helps to increase productivity by allowing you to alternate between sitting and standing at your workstation. Once you change your posture, you'd be able to do focus and get a lot of things done. This is available for purchase on Amazon.

Focusrite

If you want to type something on the laptop or computer screen, you must make sure that you don't get distracted. It is easy to get distracted when you have the option to open as many as tabs as you want. While working on a screen, if you have too many tabs open, it will kill your productivity. So, for that, Focuswriter is a great tool. This is a program that works exactly like a Word document. It also has built-in timers, better ambiance, daily goals, and many other options. This program supports Windows, Mac, and Linux systems. By using this tool, you will be able to do your work on time with better productivity. Moreover, the time that you usually kill can be saved.

Freedom

This app provides peace of mind by helping you focus on the important things and avoiding distractions. Once the app does it for you, you will be able to focus on the work you do. People often procrastinate when they slowly shift from an important task to another entertaining activity. For example, say that you are working on a project, but meanwhile, you are scrolling through Facebook feeds, so do you think that you can give your best to work? I don't think so. When your attention is divided among other unimportant tasks, you will not be able to give the best to your MOST important project. So, the Freedom app will help you by blocking sites such as Twitter, Facebook, and so on. The Freedom app will block almost all the time-consuming sites. So, there's no reason why you must not consider it.

CHAPTER 35:

How to Stop Overthinking with Positive Self-Talk

Take a moment and reflect on some of the things that you have said to yourself today. Are you helping to build your self-esteem, or are you criticizing yourself for not being perfect? Self-talk is the inner discussion that you have with yourself. This discussion can either have a positive or negative impact on your life. If you continue to talk to yourself about negative things, then there is a certainty that your mind will be corrupted with negative thoughts. On the contrary, if you engage in positive self-talk, you will increase the likelihood of building your esteem.

What is Self-Talk?

Self-talk is the inner discussion that you have with yourself. Everybody engages in self-talk. However, the impact of self-talk is only evident when you are using it positively. The power of self-talk can lead to an overall boost in your self-esteem and confidence. Moreover, if you convince your inner-self that you are beyond certain emotions, then you will also find it easy to overcome emotions that seem to weigh you down. If you can master the art of positive self-talk, you will be more confident about yourself, and this can transform your life in amazing ways.

You can't be sure that you will always talk to yourself positively. Therefore, it is important to understand that self-talk can go in both directions. At times, you will find yourself reflecting on negative things. In other cases, you will think about the good things that you have achieved. Bearing this in mind, you must practice positive self-talk. This can be understood as pushing yourself to think positively even when you are going through challenges. When you do this, you will approach life more optimistically. As such, overcoming challenges will not be a daunting feat for you since you can see past the hurdles you are experiencing.

If your self-talk is always inclined to think negatively, it doesn't mean that there is nothing you can do about it. With regular practice, you can shift your negative thinking into positive thinking. In time, this will transform you into a more optimistic person that is full of life.

Importance of Positive Self-Talk

Research shows that positive self-talk can have a positive impact on your general wellbeing. The following are other benefits that you can get by regularly practicing positive self-talk.

Boosts Your Confidence

Do you often feel shy when talking to other people? Maybe you don't completely believe in your skills and abilities. Well, positive self-talk can transform the perceptions that you have about yourself and your abilities. Negative self-talk can hold you back from achieving things in life. It can even prevent you from even trying in the first place. Unfortunately, this can drive you to overthink about the things that you feel as though you should do. So, instead of acting, you end up wasting your time overthinking about them.

Positive self-talk lets you put aside any doubts that you could have about accomplishing a particular goal. Therefore, you will be motivated to act without worrying whether you will succeed or not. You're simply optimistic about life. There is nothing that can stop you from trying your best when attending any activity.

Saves You from Depression

Overthinking can make you more susceptible to depression because you garner the perception that you are incapable of performing well. Frankly, this affects your emotional and physical wellbeing. Some of the effects that you will experience when you're depressed include lack of sleep, lethargy, loss of appetite, nervousness, etc. Positive self-talk can change all of this. It will fill you with the optimism that you need to see past your challenges. As a result, instead of believing that you can't do it, you will begin to convince yourself that you can do it. Positive self-talk can transform how you feel. It's just a matter of changing how you perceive the world around you.

Eliminates Stress

There are many stressors that we have to overcome every day. The truth is that we all go through stress. The only difference is how we deal with stress. Some people allow stress to overwhelm them. Often, you will find such folks with a negative outlook on life. They will have all sorts of negative comments about life. "Life is hard," "I can't take it anymore," "I'm always tired," "Things never get easier," etc. We've heard such comments coming from our friends who have given up on life. The reality is that stress can get the best of you if you surrender. Practicing positive self-talk can help you realize that stress comes and goes. It is a common thing that everybody experiences. Therefore, there is no need to allow it to overwhelm you. When you begin to understand that you can change how you think and overcome stress, you will be less anxious and calmer. As such, this reduces the likelihood of overthinking.

Protects Your Heart

We all know that stress is not good for our health. Stress leads to many diseases, including cardiovascular diseases such as stroke. Therefore, by practicing positive self-talk, you will be protecting your heart.

Boost Your Performance

Positive self-talk can also help boost your performance in anything that you do. There are times when you find yourself feeling tired and discouraged. For instance, when you wake up in the morning feeling as though you ran several kilometers, this can be draining. It affects how you attend to your daily activities. With positive self-talk, you can tap into your energy reserves and boost your performance. It is surprising how you can quickly change how you feel by thinking positively.

How Positive Self-Talk Works

Before getting into detail about practicing self-talk, it is important to understand how negative thinking works. There are several ways in which you can think negatively, including:

- Personalizing

This form of negative thinking occurs when you blame yourself for anything bad that happens to you.

- Catastrophizing

If you expect the worst to happen to you, then you are simply catastrophizing everything. The issue here is that you don't allow logic to help you understand that some things are not the way you think.

- Magnifying

Here, you pay more attention to negative things. In most cases, you will block your mind from thinking positively about any situation that you might be going through.

- Polarizing

You look to extremes when it comes to judging the things that are happening around you. From the perceptions that you have developed in your mind, something is either good or bad.

The importance of identifying these forms of negative thinking is that you can work to transform them into positive thinking. Sure, on paper, this might sound to be an easy task. However, the truth is that it takes time for you to live an optimistic life. You need to practice positive self-talk every day and in everything that you do. This is the best way in which you will develop a habit of thinking positively regardless of the obstacles that you are facing.

CHAPTER 36:

Tips for Practicing Positive Self-Talk

Have a Purpose

There is a good reason why you will hear most people argue that it is important to live a purposeful life. Undeniably, when you strongly believe that you are here on this earth for a good reason, you will strive to be the best version of yourself. You will be constantly motivated to try to achieve your goals in life. The best part is that you will feel good about your accomplishments. This is because they are an indication that you are heading in the right direction towards your goals. Therefore, when practicing self-talk, always look to a higher purpose that you yearn to achieve. This will keep you on the move without worrying too much about the number of times you stumble.

Get Rid of Toxic People

It is common to have a bad day. We cannot deny the fact that there are times when life seems difficult. Usually, this happens when our emotions overwhelm us. Despite this fact, some people have these bad days every day. They never seem to stop talking about their worst experiences. Unfortunately, this can take a negative toll on your life, especially when interacting with other people. Picture a scenario where you are always told about how life is difficult. Your friend keeps mentioning to you that life has changed, and you can't realize your dreams. In time, this is the mindset that you will also develop. There is nothing good that you will see in your life since you can't think positively. The interesting thing is that you might be making positive changes, but you will unlikely notice.

Accordingly, positive self-talk will work best when you eliminate toxic people from your life. Sure, it might be daunting to let go of your supposed true friends, but if they can't change their mentality, then it's time to let them go. There is more to anticipate in life than ruminating on the worst that could happen.

Be Grateful

Positive self-talk can also take the form of showing gratitude to the little things that you have. Showing that you are grateful for what you have is a great way of changing your attitude towards life. For instance, if you were not happy about a certain achievement, being grateful that you achieved anything at all can help you develop a positive attitude towards what you achieved.

The best part is that it gives you a reason to approach life with optimism. Develop the habit of expressing your gratitude every day. Start your day by writing down some of the things that you are grateful for. It can be as small as being thankful that you woke up. The point here is to tune your mind to focus more on the positive side of life.

Never Compare Yourself to Others

It is easy to compare yourself to other people more so when you feel that you lack something. Sadly, such comparisons only push you to look down on yourself. The comparison game will blind you from seeing the valuable qualities that you have. You will develop a negative attitude towards your abilities as you assume that other people are better than you.

By expressing how you are thankful for what you have, you can identify the numerous things that make you different from other people. This is a great way of developing your personality and helping you believe in yourself.

Talk Positively with Other People

Talking positively with other people will have an impact on your self-talk. If you constantly talk about negative things with those around you, then there is a likelihood that you will also engage in negative self-talk. There are probably numerous times where you've heard people say that you become what and how you think. Therefore, if you keep focusing on the negative, expect negativity to flow through your mind. Stop this by trying your best to surround yourself with positivity, starting with the way you talk to other people.

Believe in Your Success

The best way of propelling yourself to succeed in your endeavors is by believing that you can do it. If you don't believe that you can do it, then this holds you back from trying anything. This should be applied to everything you do. For example, if you are working towards losing weight, you should convince yourself that you can do it. This is the first step that will give you the energy you need to overcome challenges on your way to success.

Overcome the Fear of Failure

Succeeding in life also demands that you should overcome the fear of failure. You should always bear in mind that your failures are learning lessons. Most people who have succeeded in life have failed at some point. When you overcome the fear of failure, you will be more than willing to try anything without hesitation. This opens doors to plenty of opportunities. The good news is that you will have learned a lot from the experience of failing.

Use Positive Affirmations

You can also give a positive boost to your self-talk by using positive affirmations. The best way to use these affirmations is by writing them down. Note them somewhere you can easily view them. For instance, you can stick them on your refrigerator or your vision board if you have one.

The importance of positioning them in a convenient place is to guarantee that you motivate yourself every day. Ideally, this is an effective strategy of training your mind always to think positively. Examples of positive affirmations that you can note down include:

- I am blessed.
- I am a successful person.
- I embrace what life offers me.
- I am happy today.
- I allow myself to be filled with joy.
- I value the people around me.
- I am proud of myself.
- I am kind to people around me.

Avoid Dwelling in the Past

When you think too much about the past, it will likely be difficult to focus on the present.

This will have an impact on your self-talk. If you keep regretting the mistakes that you have made, there is a good chance that you will think negatively. Your emotions will blind you from thinking clearly. As such, this can have an impact on the decisions you make.

You must find a balance between thinking about the future and the present. When thinking about your future, focus on the positive. If there is something that you want, think in that direction and convince yourself that you already have it.

Recommended Steps to Changing Your Self-Talk

If you have never practiced self-talk, you might be wondering how you will start or what you will do to ensure that you change your negative thoughts into positive ones. The following are steps that you should take to change how you converse with your inner self.

Step 1: Observe Your Self-Talk

The first thing that you should do is to take note of the kind of talk that you have with yourself regularly. Notice how your mind tends to think either negatively or positively or in both directions. While doing your best to understand how your mind thinks, recognize the effects of positive self-talk and negative self-talk. This step should help you recognize the self-talk that you frequently have with yourself without changing anything.

Step 2: Choose Positive Self-Talk

The next step is for you to choose positive self-talk. Certainly, you don't wish to talk to yourself negatively. When choosing positive self-talk, this will be influenced by how you perceive things around you. Therefore, it is strongly recommended that you should begin looking at the bright side of things. For any situation that you go through, focus on looking at the bright side. Sure, you might be going through a tough time, but consider the benefit that you are gaining. For instance, it will make a huge difference if you learn something beneficial from the bitter experience you are facing.

Step 3: Recognize When You Engage in Negative Self-Talk

You can't be certain that you will always practice positive self-talk. Some situations will drive you to think negatively and, therefore, will influence your self-talk. Accordingly, you must catch yourself when you notice your mind drifting and thinking negatively. Practicing this more often guarantees that you can shift your negative self-talk into positive self-talk and benefit from it.

Step 4: Manage Your Self-Talk

To transform how you think through self-talk, you will have to manage your self-talk every day. You should manage your self-talk since there are times when you might forget its importance. Make a habit of talking to yourself positively every morning. Prioritizing your self-talk will warrant that you start your day on a positive note.

Step 5: Choose Self-Talk that Suits You Best

After practicing positive self-talk for a while, you will notice that some methods have a bigger impact compared to others. Use these techniques to keep yourself motivated throughout the day. You need a lot of energy to overcome your daily stressors. As such, strive to find the perfect type of self-talk that brings out the best in you.

Step 6: Eliminate Self-Defeating Self-Talk

An important part of self-talk is eliminating self-defeating self-talk. Negative thinking will exaggerate situations, and you may have difficulty solving challenging situations. The worst thing is that engaging in negative self-talk will only cause unnecessary stress.

Generally, positive self-talk can help you avoid overthinking. The basic concept behind positive self-talk is that you will develop a habit of focusing on the bright side of life. Sure, challenges will always be there. You can't avoid stress completely, but you will improve how you deal with stress. Don't think that positive self-talk is something that you can master in a day. It takes time for you to develop the right mindset required to see life from a positive angle. Therefore, practice positive self-talk every day, and you will notice its benefits after some time. To conclude, you should always remember that practicing positive self-talk is the best way in which you can constantly remind yourself of all the great qualities about yourself and the great aspects of life.

CHAPTER 37:

What Happens When You Think Too Much?

Psychology of Overthinking

Overthinking is also often associated with psychological health issues like depression, anxiety, stressful events, or any future uncertainties. Family issues and personal issues could also be the reason for your overthinking. Overthinking too much can trigger us to focus on our failures and weaknesses and increase the risk of anxiety and depression. This will also lead the over-thinker to collapse into an exceedingly acidic loop of ruminating as their mental wellbeing decreases.

Overthinking may sometimes contribute to psychological distress. Any over thinkers turn to unhealthy ways of coping, such as liquor, drugs, or psychoactive substances, to self-treat the stress. This may also contribute to bad attitude, restlessness, and frustration. Overthink have physical effects and have sleep problems because sometimes the mind does not shut off except for sleep.

Depression

Depression is a widespread neurological condition, and as much as 50%–60% of individuals are expected to experience a depressive episode throughout their lifespan. Overthinking is amongst the most harmful behaviors and is not only a result of depression. Overthinking is a behavior, and a person who has this habit spends too much time overthinking each day. Regardless of which thoughts people overthink, they get nothing out of it as acts have been absent from overthinking the process, and that can be a result of long term depression.

When a person's brain overwhelms, he cannot afford thoughts. All the pessimistic feelings start flowing without a couple of seconds pause because it will be too frustrating that all those emotions are the ones he has got from the people he has ever seen in his experiences. People start feeling depressed when they feel out of control, and they are expecting bad things.

Stress and Anxiety

There are so much tension and fear in everyone's life. Any individuals have an anxiety condition. They feel worried about many things and get anxious. They sometimes think over little issues as well. Some people might have anxiety attacks. An anxiety seizure is a sudden sense of intense fear. Normal anxiety does not interact with day-to-day routines. But when it is constant, unavoidable, and interferes with living, anxiety starts overwhelming. Although everyone sometimes overthinks such circumstances, extreme over-thinkers waste much of their waking hours ruminating, placing strain on themselves. They convert their stress into overthinking.

Stress is triggered in many situations by your emotions, interactions, and all that occurs in you by specific individuals or outside circumstances. When you concentrate on ruminating constantly, you create it a routine. It is a cycle, so the more you do it, the tighter it becomes to avoid it. Overthinking is harmful and exhausting physically. It can place your fitness and overall safety at risk easily. Rumination allows you more vulnerable to fear and depression. It's a mental habit that could break you as an overactive mind can make life wretched.

Stressful Events and Embarrassing Moment

Past events that could be embarrassing in our lives can have a significant impact on our mood and how we feel about ourselves. Conditions of overthinking may arise from one or more traumatic life experiences. Specific events that affect are job pressures or work changes, life conditions adjustments, pregnancy duration, and childbirth, household and family problems, significant cognitive distress during a painful or unpleasant experience, assault or mental, psychological, financial, or emotional abuse and death or absence of one of those you love.

Stressful events have been associated with higher exposure to anxiety. Some kinds of traumatic events, specifically serious incidents and household conflict events, were preferentially predictors of increased sensitivity to overthinking. Exposure to overthinking has established the clinical association between traumatic life experiences and sleep problems.

Insecurities and Fears

Most people may experience insecurity, or a tendency to lack experience or surety in themselves, concerning some aspect of human life. For most, it is possible to resolve this insecurity before they have prolonged, adverse effects. However, when one struggles with systemic anxiety for an extended period, the fears and depressive emotions encountered can have a profound impact on life and cause overthinking.

Insecurity also contributes to pessimistic feelings regarding one's ability to mix in with friends, achieve targets, or gain encouragement and acceptance. The condition often includes anxiety: individuals start experiencing the feelings of insecurity, feel worried, and fears that try to frame anxiety and overthinking. If you are always frustrated and criticizing yourself for something less than ideal, you will begin feeling insecure and valueless. While trying your best and striving diligently can give you an edge, there are other unpleasant factors of perfectionism. Harassing to oneself and continually thinking about not being perfect enough may contribute to anxiety and depression, mental illnesses, or severe insomnia.

Future Uncertainties

Uncertainty is a useful source of conflict which may impact your current and future health. Change threatens our future planning capacity. Based on past experiences, our brains make decisions for the future. Whether the future is unpredictable when we learn anything different, we cannot depend on previous encounters to make our choices. Without that device, we might worry about what the future could bring, running around and thinking for potential possibilities.

It is the chaos-filled overthinking mind that keeps us off from going ahead in life. It is the overthinking mind that makes people feel depressed and obsessed with the future. It is the mind of the negative thoughts that we do need better regulation to quit stressing and keep living. If you are frequently alert and anxious for instability and the future, you may establish a persistent stress habit and are more vulnerable to panic and anxiety that can cause overthinking. Overall, acknowledging the uncertainty in existence is essential to your emotional wellbeing because there will often be things beyond your power.

Negativity

Due to any event or any situation that affected your mind leads you to think more and more about it. The more you think, the higher will be chances to be getting a mental illness. You start thinking negative thoughts about that event. You question yourself, and these questions gather up in your mind that causes negativity. This negativity leads to anxiety and depression that can ruin your daily life routine.

If you overthink, your decisions are unclear, and the discomfort intensified. Too much energy you waste on the negative that can cause you to get tough when you behave. When you happen to be depressive and overthink, it is no surprise that your dreams transform into negative feelings.

Lack of Inner Peace

Depression and anxiety become a deeply rooted pattern. Many individuals are struggling most from fear. Stress is an emotional or behavioral condition. Thoughts tinged anxiety, pessimistic emotions, and unrealistic perceptions are overthinking. These thoughts arrive uninvited, and they disrupt one's calm. They take hold of the behavior and mental processes of the human and intervene with acts.

This is a pattern of becoming nervous, slowly, and that when you do not do something about it, it can evolve and worsen. It may become a degrading and torturing condition that will trigger and disturb the peace that further develops overthinking.

CHAPTER 38:

Work Through Worry, Fear, and Anxiety

Fear lies at the center of anxiety. But it isn't the usual, healthy fear that everyone feels from time to time. Whereas healthy fear prompts us to make a decision and fades as we move forward, anxiety creates a fear that persists long after the trigger is behind us. It sticks to us like a shadow, and as time passes by, it worsens until it's well beyond our control and has come to dominate our lives.

What many people find when they begin CBT is that when they stop to analyze what triggers their anxiety and what thoughts said anxiety produces, the fear is often both irrational and out of proportion. It's a tremendous relief, to be sure, but how do we get to a point where we can even see the distortions that clearly?

Let's talk about the distortion first. Fear, to a degree, is a healthy response to the things and situations we perceive as dangerous. It may influence a decision we make for a brief period, but once that decision is made, it's gone. The prolonged fear through anxiety follows us around because our attention is fixated on the trigger moment; we never give ourselves the chance to put the fear behind us. As a result, we feel as though we need to be constantly on guard. And as we keep thinking about it, our brain scrambles to find more and more ways for what we fear to become a reality, limiting how we choose to live our lives. Even more damaging is how we eventually find other things to become afraid of: "Well, this may not happen, but what if this does?" Because of the cyclical nature of our thinking, we seek these things out even though they make us miserable.

It's almost like an addiction, and in short order, we have a list of reasons to be afraid that's increasingly more removed from where we started. To get on with living, we need to take a step back and put them in proper perspective. Generally, you can do this by breaking things down like this:

Articulate what you're afraid of. Is it a consequence of something you think you've done? Is it how you perceive others will feel about you? If it is, be honest with yourself and put it down on paper. The benefit of having your fear in a tangible form this way cannot be understated.

Now, write how you think this could come to reality. This part proves to be difficult for some, but it's also very revealing to how you think. How does what you fear to become a reality? Explain it in fine detail. Connect as many dots, as many as there may be.

Finally, how realistic is the result? Let's say that, for example, you have social anxiety and you avoid going out with friends. One of your fears could be that they'll learn of something embarrassing you did years ago and will think less of you because of it. How would they learn about it? You don't want to tell them, so it must be some other way. Does someone else know? Perhaps an old friend you haven't seen in a long time was present. They could resurface and decide to tell everyone. It's entirely possible, sure, but how possible is it? This is where the details of the preceding step come into play. How many dots did you have to connect to make your fear reality? This is a useful way to gauge how plausible your fears may or may not be. Of course, because of how anxiety distorts our thinking, you may find it difficult to see things as they are. This is when an outside perspective—like that of a professional therapist—can be useful.

Another method requires a change of perspective. It's a technique called fear-setting that was popularized by blogger and Guru Tim Ferris. After finding himself lost in uncertainty, he changed his focus from what he most wanted from his goals to what he was most afraid of.

They are two sides of the same coin when you stop and think about them: one is where we hope to find ourselves, and the other is what we want to avoid on the way. But Ferris has suggested that if we're better able to realize our fears and assess the risks, we'd be better suited to handle them and make a decision on how to move forward.

This is how he practices fear-setting. In the past, he has said he does this at least once a month, so that's a good place to start.

Narrow down a decision you're concerned about. A good example is "What if I quit my job?" A lot of us are unhappy with our work and want to pursue something else. But we find lots of reasons for not changing paths. On top of a piece of paper, write out this decision. Then divide the paper into three columns. The title "define" the first column and list ten negative consequences that could result from you making a choice you want to make. Be as detailed as possible. In our example, it's easy to think of such consequences.

Maybe you'd take a pay cut if you changed careers. Maybe you would be denied the job you wanted or couldn't find work right away. But if you have extreme fears, they need to be addressed too. What are the worst-case scenarios? Maybe you won't be able to pay your bills for some time, but in your nightmare, you think you'll end up homeless.

In the second column, write "prevent." For each item you wrote beforehand, write at least one way you can prevent or minimize the chances of these outcomes from happening should you choose to go forward.

Don't think of it solely regarding what you alone can do. If you have a support network of friends and family, can they help out in any of the scenarios? This isn't a list of what to do if your fears come to life: it's the things that you can do to mitigate the chances of that happening in the first place.

In the last column, write "repair." Here you'll write you can fix things, even if only slightly, should those things you fear come to pass. In our last example, where you leave your current job for other pursuits and some of those consequences, like being unable to pay your bills or (in the worst case) losing your home occurs, what can you do to fix the situation? When coming up with these ideas, be sure to include your support network too. Is there someone who can help pay your bills or give you a place to stay? Come up with a response to each item written thus far.

These columns are the analysis of your fears from every angle: what you're afraid of, what you think could result from these fears, and what you can do if they come to pass. Now for the following step: on a new piece of paper, write down what rewards you could gain from making the decision. As with the defined column, you'll want to include everything you can think of. However, you'll want to focus on all aspects of yourself. If you leave the job that's making you so miserable and get the job you want, will you be happier? How? Will there be other mental benefits? Physical? It's important at this phase to keep your "what-if" expectations within reason. You get the job, climb the ladder, and become the boss—it's not impossible—but there are other, more easily obtained goals you have a better chance of reaching.

Finally, on the final piece of paper, jot down what impact not taking this chance will have on your life moving forward.

Usually, this includes a continuation of whatever you're feeling right now ("If I stick around here, I'll continue to be unhappy"), but try to imagine how long that period will be. In most instances, you'll find that the circumstances last until you take action and make a change.

With everything you've written, try to determine if the rewards outweigh the risks and if you're equipped to combat the fears that should be realized.

The point of all this is to deflate your disproportionate thoughts. With everything out in the open and under a microscope in this way, you can re-balance your thoughts to make better decisions. Think of it as a push you're giving yourself to face your fears, only now you see them for what they are, not the monsters they pretend to be.

CHAPTER 39:

Cope With Worry and Anxiety

Here are some tips on how to cope with worry and anxiety.

Make Yourself Calm

Take a deep breath to make your mind calm. Breathing exercises is a quick but powerful procedure that can relax your brain, promote peacefulness, and immediately raise your condition. This reduces stress levels and activates the nervous system that helps calm the mind. In addition to being pain relief, relaxation techniques often promote focus and improve the immune system.

Breathing is the most potent strategy to reduce frustration and fear quickly. You seem to take short, shallow breaths when you are nervous or upset. A warning is sent to the brain, which would trigger a positive reinforcement cycle and enhance the reaction and fight or flight. That is why taking slow, profoundly soothing breath breaks the cycle and allows you to settle down. There are several relaxation methods designed to help calm you down. Three-part breathing needs to take one long sigh in and then completely exhale while the body is being cared about.

Cover your eyes while taking a few slow breaths and imagine yourself relaxed. See your exposure level, and assume yourself starting to work through a scenario that causes depression and anxiety by remaining focused and concentrated. Step out of the area as quickly as possible to head outside, even though it is only a few minutes. The new weather does not only help settle you calm, but the change in the scene will also disrupt the nervous or angry cycle in thinking.

Gather Your Thoughts and Write

Writing is the one easy way of decluttering the mind and soul. This is not the only place to continue, but it is a decent spot. When you go to sleep, try this correctly. Note down about what is happening with the mind and emotions. It may be something good, something unhappy, something hard, or something beautiful. Write everything down and plan to look through the list again when you awaken. The commitment helps the brain to let it go quickly. Create a new chart every night. You could be remembering much from the day before, and it is beautiful. Keep writing down all, make a plan to check in the meantime, and then respond once you wake up. With time, you will find that most of your fears start to disappear, whether you are taking steps to fix them, whether entirely forgot about them. You will continue to realize the items you have no power over are also not worth sticking onto.

Although you cannot see the outcomes as quickly and you will see a tidy cutting board, you only can realize it. You will experience discomfort within so that you can change the outward view. If you have stuff in your head, it allows you to move them out of your head and then on sheet. That is one of the simple practices of writing down your activities and thoughts. It stops your mind from being full of what you need to update and consider.

Writing reduces repetitive harmful case thinking and enhances functioning performance. Scientists agree that such changes will, in turn, free up your brain energy for specific behavioral tasks, including the opportunity to control stress more efficiently.

Working in a regular newspaper will also help relieve anxiety and deal with stress. It is a safe medium for expressing suppressed feelings. To start with a journal article, you do not have to be a famous blogger. Writing your thoughts down is the simplest method for beginners to play with.

Set Your Priorities

Prioritization is a smart opportunity to efficiently and effectively take accountability for your success. The first move is to find out what matters most to you, what are your life expectations and your long-term aims? Compile a set of your key goals to ensure the objectives you set to match your behavior and the choices you make.

You have already got an infinite list of to-does, and your brain is full of thoughts. Yet you have limited resources and a small period which must be taken into account while preparing your plans. Prevent cognitive overload and tiredness when knowing you cannot accomplish anything. This would better help you to concentrate on performing a few items well. Choose those fields, like partnerships, creative ideas, or fitness, and entirely devote to them, avoiding anything else that stressed you. This will allow you to keep focused and to level stuff up.

The next move is to develop an action plan to achieve the defined targets and concentrate on how to break the time and reflect on each element in the checklist. It is important to remember that your list of goals shifts when you get older, and that is perfectly fine as long as you check in frequently with yourself to make sure that those objectives always represent you.

Identify that which is essential to you in all your written stuff. Whether you choose to reduce or declutter, the first move is to define what is crucial. Classify in this context like what is significant about your career and what is most necessary to work on right now. Create a brief list of those items for each. Now that the importance has been defined, you will recognize what is not essential. What items are not needed or valuable to you during your life? Get the trash out of your mind by removing as many of those thoughts as you can.

Get Enough Sleep

The correlation between sleep patterns has long been the focus of research. One inference, there is a connection between sleep deprivation and anxiety. One research explores suicide mortality rates over six months. It concludes that sleep deprivation is a factor that contributes to most of these fatalities. Another study shows that individuals with sleep problems, including insomnia, are likely to exhibit depression symptoms. Sleep is a critical part, sometimes overlooked, of the overall health and welfare of an individual. Rest is necessary as it helps the mind to regenerate and to be prepared for another day. Sufficient sleep can also reduce unnecessary weight issues, cardiac failure, and decreased time of sickness. Having an adequate rest at the right period will help relieve stress, wellbeing, physical health, quality of living, and protection.

When you sleep, the mind is getting ready for the next day. To help you understand and recall knowledge, it is creating new pathways. A better night's sleep helps you learn more. Sleep helps to develop thinking habits and problem-solving capabilities. Sleep always allows you to be alert, make choices, and be imaginative. In some regions of the brain, insufficient sleep affects behavior. If you are deficient in sleep, you can find it challenging to make choices, solve problems, regulate your thoughts and practice, and respond to change. Deficiency in sleep was often linked with anxiety, depression, and risk-taking behavior.

The way you act when sleeping partly depends on what happened when you rest. Your body works throughout the night to help balanced brain activity and sustain fitness levels. Sleep also tends to promote the development of children and adolescents. For your wellbeing, a good night's rest is crucial. It is just as important actually as having a healthy diet. Often you do not get that much sleep, so your cures are not exceptional. You are indeed going to alter your sleep when you want to declutter your mind, and it will occasionally do well. But if you are not giving it any consideration, you are not going to know how much your sleep impacts you.

Take Regular Walks

Going up and performing some sort of physical workout is a perfect way to keep things out of your mind. You can enjoy cycling or garden work, so it does not affect what you are doing. Spending any personal exercise makes the mind focused. Walk, sweat, and stay healthy and cut the stress hormones off. Exercise lets you change your health, avoid infection, raise stamina, and enhance your attitude. It allows you to sleep well, look better, and focus more.

Choose an exercise that heightens your pulse rate, movement, meditation, martial arts, biking, cycling, etc. If you continue any regular exercise frequency, it will benefit your fitness and wellness for the coming years. While removing the challenges that overtake us, we should be following activities that will help us relax and cool. Popular events such as cognitive ability and relaxation are deemed successful in obtaining this goal.

CHAPTER 40:

More Tips on How to Cope with Worry and Anxiety

Focus on Single Tasking and Avoid Multitasking

Take the chart of to-do and add half the stuff on it. Only pick specific things to do every day, and concentrate on all those. Let the others go. When you practice less, you would have less in consideration. In the most time, multitasking is a pleasant way to clutter the head with a variety of actions, even without resulting in increased levels of success or satisfaction. Alternatively, seek single-task and only focus on one job at a time. Take away everything else before the job is finished. Focus instead on the next step, and so forth.

You will systematize tasks that you do daily, such as planning menus, running, washing up, or bringing your animal on the run, by developing a standard daily and yearly to-do plan. Stop multitasking by actually devoting yourself with one essential job at a time and putting the others to the side.

Swapping from duty to job, you feel that at the same moment, you consider anything and everything surrounding you. Yet, in truth, you are not; you cannot pay more attention to one or two items at a time. This assumes multitasking is not merely detrimental. This even allows the subconscious to fail as it seeks to know what is happening. The best choice is to concentrate on one issue at a time.

Although there is no benefit in periodic multitasking, frequent switching between activities reduces your attention and concentration, increases discomfort, and generates unnecessary noise in your mind. Extreme multitasking reduces productivity and can affect working memory.

Single-task as far as practicable is the remedy. Create a list of items you intend to do every day. Keep a quick and practical to-do chart. Start with whatever is essential, then move down the list to accomplish one job at a time.

Focus on Positivity

Self-doubt appears to be too deep-embedded to react to ground platitudes you do not believe in. Telling yourself that you are amazing cannot block the negative away. The most straightforward approach to cope with self-doubt is by actively questioning the pessimistic thoughts. Your subconscious will continue to look at you differently each time you show to your mind that you are more intelligent than you assume. You are loading your brain with trash, thinking, "no one cares," or "sad stuff still happening to me," warnings. If you analyze yourself feeling like this, take a deep breath and recognize one course in practice that you can take directly to change your life.

Discussing your self-doubt regularly can alter the way you perceive. You ought to take steps with bravery but no doubt. Be able to verify yourself incorrectly and admit that your brain may not always be accurate. If you declutter your thoughts, you will have more resources and attention to dedicate to meaningful and efficient issues. This should help you develop peace of mind; you have to be the most reliable reflection.

Learn to Let Go of Things

Something to think about? Angry with someone? Feeling frustrated? Should bear a grudge?

Though these are all normal emotions and feelings, they are not necessary. See if you should just get them out. It is more difficult than it seems, but the dedication is good enough to justify it. Notify your suffering and get yourself to believe that your issues are more than someone's self-pity and disappointment, but if you are not cautious, then it is going to leave you trapped.

Support yourself, value yourself, and start to move ahead. You ought to give up on what holds you back if you want to climb. It is necessary to get over the painful feelings and thoughts that leave you feeling trapped. Eliminating unwanted emotions, worries, and doubts helps alleviate tension, improve self-esteem, and open up space in the mind. Regularly track your feelings, and seek to swap the pessimistic feelings with positive emotions.

Productive problem-solving is beneficial, whether you are trying to cover your expenses or have a rough time coping with a colleague. Imprisoning the same stuff again and again, predicting disastrous consequences, and your choices do not lead you far. This is a difficulty that can be corrected, then focus on modifying the setting. When you can do more to remedy the issue, then focus on improving your attitude and let go of negative stuff from your mind. When getting stuff out from your brain will be a feature of daily routine, you will find that you are less stressed and have much more energy to work on activities. You will see an improvement in your efficiency and the reliability of your performance.

Listen to Motivational Talks

Recognizing motivation presents one with other useful perspectives about human existence. It describes why we set targets, aim for success and strength, why we crave mental comfort and physical identity, why we feel emotions such as anxiety, rage, and sympathy. Learning regarding motivation is essential as it allows one to determine where inspiration comes from, why it happens, what improves and reduces it, what parts of it can and cannot be recovered. It enables one to address the query as to whether other forms of motivation are more effective.

Motivation represents something special for everyone and helps us attain cherished goals such as increased success, strengthened wellbeing, personal development, or a sense of mission. Motivation is a mechanism to improve the way we perceive, act, and respond. The effects of inspiration on the way we lead our lives are apparent. In the face of rapidly changing situations, when we are continuously reacting to shifts in our climate, we require encouragement to take appropriate action.

Motivation is a crucial factor that helps people to adjust, work productively, and sustain wellbeing in the face of a series of prospects and challenges that are continuously evolving. The improved drive has many health benefits. Motivation as a condition of mind is connected to our anatomy. Our health and fitness suffer as our energy becomes exhausted. Listening or reading to motivational stuff can help you to declutter your mind from harmful and unnecessary thoughts.

CHAPTER 41:

Aromatherapy Against Depression and Anxiety

Depression can influence how you feel, how you think, and how you act. Even though it's a state of mind issue, despondency can cause both physical and passionate side effects. These can shift contingent upon the individual. However, they frequently include:

- nervousness
- eagerness
- trouble
- despair
- trouble concentrating
- trouble resting

Individuals utilize basic oils as corresponding medicines for some conditions, including gloom. Note that fundamental oils aren't a solution for wretchedness. They're a medication-free choice that may help soothe a portion of your manifestations and assist you with dealing with the condition. By and large, fundamental oils are sheltered and liberated from reactions.

What The Examination Says

Albeit many basic oils are available, research on the expected advantages, dangers, and viability is regularly restricted.

Lavender

The botanical yet hearty fragrance of lavender oil is regularly esteemed for its quieting impacts. Exploration proposes that fragrant lavender healing may help:

- mitigate nervousness
- decline pressure
- improve disposition
- advance unwinding

The spice itself may likewise support despondency. Analysts in a recent report thought about the viability of lavender color to the stimulant imipramine. Color is unique about a fundamental oil. Colors are produced using new spices and grain liquor, for example, vodka. Analysts inferred that lavender color might be a useful adjuvant treatment to get mellow moderate melancholy.

Wild ginger

As indicated by a 2014 creature study, wild ginger may have upper characteristics. Scientists found that pressure tested mice that breathed in wild ginger oil experienced less pressure. They likewise displayed less misery like practices. It's the idea that the oil may initiate the serotonergic framework, which is an arrangement of cerebrum transmitters related to discouragement. This may slow the arrival of stress hormones.

Bergamot

The citrus aroma of bergamot oil is known for being both elevating and quieting. As per a 2013 study Trusted Source, bergamot oil fragrance based treatment essentially diminished nervousness in patients anticipating outpatient medical procedure. Even though downturn and tension are various issues, they frequently occur simultaneously. Uneasiness is likewise a potential intricacy of despondency. It's indistinct how bergamot facilitates worry. It might help decrease the arrival of stress hormones during unpleasant circumstances.

Different oils

Studies have demonstrated that both ylang-ylang oil and rose oil Trusted Source have to quiet and loosening up impacts. The oils can likewise diminish what are classified "autonomic capacities, for example, your breathing rate, pulse, and circulatory strain.

Albeit other basic oils are thought to ease indications of gloom, supporting proof is, for the most part, recounted. A portion of these oils are:

- chamomile
- sweet orange
- grapefruit
- enroll
- frankincense
- jasmine
- sandalwood

Instructions to Utilize Basic Oils for Depression and Anxiety

These basic oils are essentially perceived for their fragrant consequences for gloom and its indications. Regardless of whether you decide to breathe in the aroma legitimately or permit it to scatter into the territory is up to you. You should even now have the option to profit by its belongings in any case.

Here are the most widely recognized strategies for fragrance inward breath:

Breathe in the fragrance legitimately from the oil's container or inhaler tube.

Spot a couple of drops of the fundamental oil onto a cotton ball and breathe in legitimately.

Include a few drops of the oil to a diffuser and breathe in a roundabout way.

Make a fragrant healing shower by including a few drops of fundamental oil weakened with nectar, milk, or a transporter oil to your bathwater.

Appreciate a fragrant healing back rub by including a couple of drops of the fundamental oil to your preferred back rub oil.

Consolidating fundamental oils may likewise support discouragement, as per a recent report. Fifty-eight hospitalized hospice patients with terminal disease got either a hand knead with general back rub oil or fragrance based treatment rub oil on seven successive days. The fragrant healing oil was made with frankincense, lavender, and bergamot basic oils. Individuals who got the fragrance based treatment knead experienced fundamentally less agony and wretchedness.

Individuals with respiratory issues, pregnant ladies, and youngsters shouldn't utilize fundamental oils except if under the management of a specialist or prepared aromatherapist.

Every single fundamental oil may cause an unfavorably susceptible response, so you ought to never apply them to your skin undiluted. If you do plan to apply a basic oil mix to your skin, you should include 1 ounce of a transporter oil to each 3 to 6 drops of fundamental oil. Basic transporter oils include:

- sweet almond oil
- olive oil

- coconut oil
- jojoba oil

You ought to likewise do a skin fix test before huge applications. Touch a limited quantity of your fundamental and transporter oil blend to a little fix of skin in any event 24 hours before to your arranged application. This will permit you to check whether the blend will make your skin have a response.

Try not to ingest basic oils.

The U.S. Food and Drug Administration doesn't direct basic oils. Just purchase oils from a legitimate producer. On the off chance that conceivable, approach a prepared aromatherapist for a suggestion.

Different Medicines for Depression

You shouldn't substitute your present treatment plan for gloom with fundamental oils without your primary care physician's endorsement. Fundamental oils are just intended to fill in as a correlative treatment, notwithstanding your present routine.

Ordinary medicines for melancholy include:

solution antidepressants

psychotherapy, remembering one-for-one and gathering meetings

inpatient mental treatment for serious instances of sorrow

electroconvulsive treatment for individuals who aren't reacting to the drug, can't take antidepressants, or are at great danger of self-destruction

transcranial attractive incitement for individuals who don't react to antidepressants

Untreated or blundered melancholy may prompt:

- physical agony

- tension issues
- self-destructive contemplations
- substance misuse

In case you're encountering despondency, talk with your primary care physician about your indications. They can work with you to build up the best treatment plan for you. When your treatment plan has been set, you should adhere to it as the most ideal. Missing arrangements or drugs may make your side effects return or cause side effects like withdrawal.

In case you're keen on utilizing basic oils, talk with your primary care physician or a prepared aromatherapist. They can assist you in deciding the ideal approach to consolidate fundamental oils into your present treatment plan.

CHAPTER 42:

Effective Steps to Stop Thinking Too Much with Conscious Meditation

One of the essential aspects of conscious meditation is usually to center your mind. Focusing your thoughts helps release your brain from the abundance of stimuli that bring tension and concern. You may concentrate your mind on items like a single object, an illustration, a mantra, or even a breath. If you are seated, lying face down, walking, or in certain places or tasks, you should practice meditation. Seek to feel relaxed enough to get the best out of the meditation. Goal to achieve a healthy pose during meditation. Without a decision, let the thoughts pass via your mind.

The method requires fast, even-paced respiration utilizing the muscle of the diaphragm to enlarge the lungs. The goal is to slow your breathing, draw in more air, and use the back muscles, chest, and neck area when exercising so that you breathe more effectively. When you are a beginner, it might be better to perform meditation if you are in a peaceful place with little disturbances with no screens, radios, or mobile phones. If you learn more mediation skills, you will be able to do it everywhere, particularly in circumstances of high tension when you profit more from mediation, such as a traffic accident, a challenging job conference, or a large group at the supermarket.

Do not let the idea of the "right" form of meditating contribute to the tension. You may visit specific centers of yoga or community courses taught by qualified teachers if you ever want to. But you can quickly perform meditation on your own, too. And you can do the exercise as structured or casual as you wish. However, it does fit your behavior and circumstance. Some people develop meditation through their everyday routines. They might begin or end every day, for example, with an afternoon of conscious meditation. But what you will need for meditation would be a few moments available.

1. Take a deep breath. For learners, this strategy is excellent since breathing is a standard feature. Limit your total concentration on breathing. Concentrate on sensing and responding as you breathe in and exhale from your nose. Breathe in gradually and thoroughly. When your mind wanders, turn your concentration back softly on your movement.
2. Body screening. Shift focus to various areas of the body while utilizing the technique. Be mindful of the different feelings in your body, whether discomfort, stress, comfort or relaxation. Merge body screening with breathing techniques and visualize various areas of the body experiencing heat or relief.
3. A mantra to say. You may build your very own slogan, Christian or Muslim. Instances of spiritual mantras include the Christian practice of Jesus Worship, the sacred name of God in Judaism, or the Hindu Om mantra, Buddhism, among other Eastern faiths.
4. Walk outside, and meditate. The combination of walking and meditation is an easy and safe place to calm. You can use this strategy everywhere you travel, whether in peaceful woodland, on a town street or in the store. Using this form, hold back to your walking speed to concentrate on any step of your hands or arms. Do not settle on one specific destination. Concentrate on your limbs and feet, repeated terms of practice in your head such as "raising", "rolling," and "placing" as you raise each foot, push your body forward and position your feet on the table.
5. Focus on prayer. Prayer is the best-recognized form of meditation and the most commonly performed. In most rituals of the church, the spoken handwritten prayers are

included. You may pray for words of your own, or hear prayers recorded by others. Speak regarding potential options through the priest, preacher, minister, or some other religious guide.

6. Write and remember. Most people experience loving reading poetry or holy books, and spending a few minutes to focus silently about their significance. You should also listen to the sacred songs, spoken words, or other songs that you consider soothing or encouraging. You may like to compose your thoughts in a book or talk to a relative or spiritual leader.
7. Concentrate on your affection and gratitude. You center your mind on a holy picture by being in some kind of relaxation, incorporating experiences of devotion, kindness, and appreciation into your thinking. You may even shut your head, using your creativity, or look at image representations.

Conscious Meditation

Conscious meditation on attention requires trying to focus on one level. It may include observing the air, chanting a single sentence or mantra, staring at a burning candle, responding to a repeated guitar, or counting on mala crystals. Since it is difficult to concentrate the mind, a practitioner may meditate for only a few moments and then practice on more extended periods. In this type of meditation, when you find your brain going, you start focusing your thoughts on the desired topic of interest. You should get them out instead of chasing the wild ideas. Through this method, your concentrating capacity improves.

Mindful Meditation

Meditation of mindfulness helps practices to track roaming emotions as they pass through the subconscious. The aim is not to indulge in or evaluate the thinking but merely to be mindful of each cognitive notice when it occurs. Via mindfulness practice, you can see how different habits appear to shift your emotions and feelings. Over time, the human ability to quickly evaluate an encounter as excellent or poor, fun, or negative may become more conscious. With practice, internal equilibrium forms. Students learn a mixture of focus and mindfulness in several mediation centers.

CHAPTER 43:

How to Support an Anxiety Partner?

Those who experience fear can become overwhelmed, angry or scared. But as far as their partners are concerned, it is difficult to support a partner with fear. Likewise, partners are confused, frustrated and may feel out of control and feel helpless to see how their loved ones suffer. It is difficult to know how you can help or participate in their healing and ultimate healing. “It can be a real struggle when someone experiences anxiety, and it can also be very difficult for their partner. There is a lack of understanding of fear because it can create fear in the person who supports it. They may feel out of control or feel experiencing despair.

Living with someone with a fear

Living with someone with fear is complicated. Anxiety is scary because it makes the person suffering from anxiety feel isolated and alone, and nobody understands what they are going through. How can they do it if fear and pain have also not suffered from fear? Decision-making becomes a daunting task as trust in itself, and the ability to control is lost. Sufferers feel they wander in the dark forever. They lack the confidence and strength to find their way back to normal. They just want them to feel calm and peaceful.

Unpleasant thoughts constantly bombard the mind and leave little room for "clear thinking", that is, without fear of "what happens" what happens. Even though they are distracted, such as watching a movie or chatting with friends, fear is hidden in the background as a virus that constantly evokes negative thoughts and scenarios. Nervous people may feel sorry for themselves; it sees itself as a complete failure; or feelings of resistance to others.

- **4 Helpful Tips to Support an Anxious Partner**

For people whose partner is suffering from anxiety, you will find the following help in understanding anxiety.

1. Mindful consciousness

In good practice, attentive attention opens the way to understanding and compassion. This happens when you identify your concerns and feelings about your partner's fear. Being aware of your answers, the level of tolerance, patience, and the language you use to talk to your anxious partner can help alleviate potential emotional conflict.

2. Let your partner be afraid

As a partner with someone who is suffering from anxiety, you must step back and understand that anxiety is not your problem.

You should allow your partner to possess fear and thus support them. It is important that you also be aware of your thoughts and feelings and how your partner's anxiety can affect them. In this way, you can avoid excessive reactions or be too emotional when a partner's anxiety is triggered.

3. Let your partner talk about fear

You should let the patient talk about the fear that someone else is worried, but do not go beyond it. Illness is a fear of feelings, and talking about such feelings is part of the healing process. But they talk too much about their feelings - i.e., where they get into the story - creating imbalances and increasing fear.

It is necessary to strengthen how much you love and care for your anxiety partner. You need to make sure that they know you are there and that they are ready to work together as a couple through your fear. You must be gentle. Be guided by your compassion and understanding.

4. Refrain from your judgment

Do not judge or try to draw any conclusions about how your partner feels or thinks. Do not call them "how to think" or "how to feel". It will be difficult for your partner to understand his fear and its effects. They will not necessarily understand why this is happening. When they hear things like "suck it" or "click on it," it will only put more unwanted and unnecessary pressure on them and make things worse. So try to judge it. What Your Fear Can Tell You - And Why It's Important to Listen

Is your fear of trying to tell you something important?

What if fear is not always a symptom to be treated, but a "health-seeking signal" that invites us to reconnect with the real parts that we have neglected or suppressed? Below is a recent example of my work as a licensed psychotherapist who illustrates how anxiety sometimes functions as an important messenger that calls us to treat psycho-disease wounds that occur in childhood and adolescence when we can stand alone and willing to vote and listen.

What is the fear?

Fear is often considered an automatic "built-in" response to perceived threats and is often referred to as a "combat or flight response" or "combat or flight response" as a species. Therefore, children who grew up in a chaotic, potentially traumatic home environment, where their combat or escape (excitement) reactions were often triggered, are prone to develop various types of anxiety disorders before adulthood. Therefore, there is concern that doctors and psychiatrists whose patients report anxiety that interferes with their daily functioning and quality of life usually prescribe anxiety medications but do not always recommend that their patients consult a qualified mental health professional. Investigate possible causes (causes) of anxiety, such as early childhood trauma that has been subconsciously suppressed (focus in this book), and identify possible additional or alternative (i.e., prescribed) treatments.

Psychotherapy as a means of successful treatment of chronic anxiety

What if fear was not always something we had to avoid and get rid of, but instead something that would help us be curious? One way to encourage your clients to explore this opportunity is to ask them to ask their thoughts, feelings and bodily feelings the next time they are afraid. What's going on now? Was there a "trigger" that caused fear? As the case study below illustrates, this simple exercise itself can provide valuable information about what anxiety signals it is trying to convey.

Anxiety and psychotropic drugs

While it is a personal choice to use anti-anxiety medications to minimize anxiety symptoms, and in some cases medically appropriate, there are other effective interventions that an anxiety person can take, such as cognitive behavioral therapy; maintain an information diary as part of ongoing work on the psychotherapeutic intrapsychic / family system (as outlined in the case study above); participate in deep breathing exercises; yoga; daily exercise; and homeopathic medicinal products as directed by a naturopathic physician.

As the short discussion above shows, there may be much more fear than meeting the eye. While it is understandable why someone suffering from anxiety wants to alleviate these extremely unpleasant symptoms, the symptoms themselves may point to possible solutions for those seeking to investigate their anxiety by carefully cultivating an attitude of acceptance, curiosity, and patience. Journalization, painting and other forms of creative expression, as well as psychotherapy and sharing in a support group, can be a means of revealing the fear of wisdom.

Almost every individual in the world has to deal with the stress and anxiety of everyday life. Unfortunately, there are many difficult situations when dealing with anxiety and stress. This leads to physical and mental problems.

While there are people who deal with anxiety addictions such as alcohol, others are looking for a list of anxiety disorders and various strategies such as therapy.

There are many different ways to deal with anxiety and stress without addiction. The list of anxiety disorders is your best bet to find help if your anxiety is high, and the usual methods and techniques to overcome anxiety do not work.

However, you can also get rid of your problem with some anxiety management strategies and techniques.

Here are some common symptoms of anxiety:

- The heartbeat faster
- panic attacks
- Cold and hot redness
- Snowball worried
- violent behavior
- Obsessive thinking
- Tightening of the chest

There are some of the most common symptoms of anxiety. If you notice any of these symptoms, contact your doctor immediately. Some forms of anxiety cannot be treated without medications.

Other forms require some natural techniques to relieve stress. You should discuss several options with a psychiatrist or practitioner.

CHAPTER 44:

Meditation and Mindfulness Techniques to Overcome the Worry and Anxiety

Meditation and mindfulness is a growing discipline and will help you handle the various aspects of depression that disturb your life. Meditation and mindfulness are not a solution, but it is everybody's best option. Once you can develop a little space among what you are witnessing and yourself, your discomfort can lighten. But if you get so attached to the familiar sound of tension that is still there, it may intensify slowly, forming a "daily ritual" of stress that is harmful to your safety and welfare. Therefore, as you get engaged in reactivity habits, you generate more pain in your life. That is why it would be essential to thoroughly recognize the difference between reacting unknowingly and trying to respond with mindfulness.

Meditation helps you understand to stay with uncomfortable experiences without evaluating them, preventing them, or motivating them. It also makes them disperse as you encourage yourself to experience and accept your fears, annoyances, traumatic experiences, and other troublesome feelings and behaviors. Mindfulness helps you consider the fundamental factors of tension and anxiety in a healthy way. You provide the ability to obtain insight into what causes your problems by sticking through what is occurring rather than expending time battling or moving away from it.

Meditation allows you to create storage inside your issues, so they do not make you miserable. When you start to comprehend the possible factors of your uneasiness, there naturally appears liberty and a feeling of lightness. Practicing mindfulness is essentially a part of understanding to believe and to stay with unpleasant sensations rather than attempting to flee or analyze them.

The very first step in handling fear is awareness. You will get a more robust understanding of causing circumstances through recognizing their unpredictable existence and how your concern continues to operate, and that is where mindfulness falls through. Anxiety is a neurological disorder related to being unable to control feelings. Yet work suggests that a daily practice of mindfulness reprograms synaptic processes within the brain and thereby enhances our capacity to control emotions.

Via mediation, we become associated with the emotions and plots that cause anxiety. We try and look at them, settle down with them, and get them out. We discover two crucial aspects: thoughts do not describe us, and opinions are not real. Through this renewed insight, we will slowly shift our interaction with fear, separating an emotional event and what is real. The advantage of this ability is gaining knowledge of the body, which teaches us to center our mind on the bodily stimuli that are occurring at the moment. This method includes a visual inspection of the body slowly, allowing us more sensitive to what is being visually felt. You relax with the perceptions, in the same manner, you relax in the mind while experiencing certain stimuli. This go-to strategy will have a secure space that can be reached regularly if fear threatens to set in.

What Is Mindfulness?

Anxiety will stress you emotionally, which can affect the human body. But remember that studies have said that you will reduce your levels of stress and anxiety with an essential practice of mindfulness until you get depressed about becoming nervous.

Mindfulness is mostly about paying enough attention to the day-to-day life and the usual things we hurry about. It is about pulling back the pressure of your head by returning to the body. Do not stress; you will not have to waste an hour's salary on a lecture and twist your body to avoid uncomfortable positions. You already still have all the resources that you need to be conscious of work. To relieve your fear and relax your mind, use specific techniques to incorporate small moments of relaxation during the day.

Mindfulness is about becoming wholly dedicated to the current moment. It means trying to reconnect with the personal reality the body's natural perceptions, the noises, feelings, colors, flavors, and emotions of the environment around you. As you move, it may be as easy as recognizing the earth under you or the water's sensation on your body as you wash. Awareness education is an effective means of achieving this. Repeated carefulness has been shown to alter the brain's composition and, especially, a stressed mind.

Mindfulness is a phase that contributes to a mental condition marked by non - directive knowledge of current events, such as feelings, emotions, body states, and the atmosphere. It helps us separate ourselves from our beliefs and opinions without getting them branded as positive or evil. Awareness counteracts procrastination and anxiety by concentrating our energies on the current moment. Dreaming more about coming and obsessing over the past are ill-adapted thought patterns.

After all, learning from the experience and looking forward to the future is essential; but, when people invest so much energy outside of the current moment, they become discouraged and nervous. For these situations, being conscious may be a useful resource to make one concentrate more on the present moment. Mindfulness operates through a variety of various means. It prompts us to open and also to recognize our feelings. Consequently, we are abler to acknowledge our feelings, feel them, and regulate them. Even mindfulness helps one to look at issues from various viewpoints.

Anxious feelings are hard-headed thinking. Thus the more you ask them for being kind, the more they will make you worried. Trying to fight fear feels like battering about in a flood once you are in the middle. It can make matters harder. Anxiety is a reaction to flight or fight, recall, and the more you struggle off your nervous emotions, the better your brain can pump you ready for the battle. Healthy thoughts may continue for a bit, but usually, it is only before your nervous mind determines that sufficient is finished.

Awareness postpones the fight. This improves your spirit to examine your emotions and opinions without battling or altering them. The brain knows from everyday practice in mindfulness that it is safe to have ideas and feelings enter or just go. Also, there would be moments where you decide to cling on to a concept or emotion as long as humanly possible, and that is not going to avoid being mindful. So it stimulates the mind to be more conscious of who is remaining and who is gone.

Basics of Mindfulness Practices

Mindfulness is just having the scene as it would be. That is where you allow emotions and opinions to come, and then you have to let them go, to find out what they assume. Caution will take time to become used to. You will typically notice that your mind wanders off in all manner of areas when you start. That is natural. Your subconscious has accomplished what it does all of your life, so it is going to require some motivation to stay quiet. Mindfulness may not reduce stress or other troubles; instead, by being conscious of the uncomfortable feelings and ideas that arise due to tough circumstances, people get more options to manage them. Exercising mindfulness does not mean that we would never get depressed; instead, it enables us to be more considerate about what we would like to react, whether peaceful and compassionate or, sometimes, with observed frustration.

Feel your Breath

Try only at the beginning for ten to fifteen minutes, and when you are prepared, moving up to more. Place yourself, so you will feel relaxed and appreciated. Cover your eyes once you are comfortable, and concentrate on breathing. Imagine the blood circulating through the body and out. If your brain goes, as it would undoubtedly do, you softly return to your breath. Allow your emotions to come and go, your perceptions, and your experiences. You have nothing to do with them. Let them out of your mind. Within a stressful environment, it is hard to calm down and appreciate anything. Continue to take a moment for all the sensations and feel the touch, tone, hearing, scent, and taste. When you consume a favorite meal, take the time to smell, feel, and appreciate it.

Seek to consciously add accessible knowledge, approval, and discernment to anything you do. Find happiness in simple pleasures. Recognize it the way a real friend should be served. Consider lying back while you are getting bad feelings, taking a deep breath and shut your head. Reflect on the air as it travels within and outside the mouth. Only just a minute of sitting and relaxing will improve your thoughts. Simple controlled breathing is like yoga. This would also recover the oxygen and carbon dioxide balance that was knocked away by fast, shallow respiration. It is also a way to trigger the calming reflex, which changes the reaction to fight or flight and alleviates the awful levels of anxiety

Be Mindful to your Thoughts

Easing fear is more than just being rid of stress and not even being afraid of it. Whenever you see this for what it is, the less influence it can affect you. Picture monitoring your thoughts and emotions on a residential street in the same way you would observe the flow. See them enter, and see them leave. You do not have to grasp the stream, and you do not have to adjust it. Instead of treating fear as being in your way, mindfulness allows you to consider it as being in the way. When you stop battling your subconscious, it stops fighting back against you.

A nervous mind is a brain packed with electricity. The explanation of why there is fear is to energize your body. Be conscious of your strength and why it is there. Projecting the strength into any form of action can help consume the power the brain provides to the body to make it stronger, quicker, and better. Since your body's strength as it will be there in your racy head, your weak muscles, and help it figure a way out. For a simple walk, jog, go up and downstairs, which consumes your extra strength and can help relax your nervous body and stressed mind.

Shift the thoughts outside of your mind. Interrupt the fear by stepping, only a bit, beyond yourself. Who are you watching? What is it you can detect? How are you feeling? Note the ground below you. And the breeze on the face. Which is it you may hear? Pause, react, and respond to the universe around you. It will support your ground up and steer your nervous mind to something that is not too frustrating.

Practice Regularly

Mindfulness could happen at any moment, so getting a daily routine can make this good exercise more likely to transform into a method. Shock the brain out of its standard form of listening to the universe by doing something else. It only has to be one small item outside of what you usually will do. Seek or listen to new songs, alternative walking paths, or relax in a bath instead of rushing around the tub. Choose a moment to be conscious of the universe inside and outside you. When you are eating your breakfast, on the route to work, as you get up, even before trying to sleep, it may be on your regular run. The mindfulness influence on the brain is not permanent, but it is durable and long term. The impact is coming from a robust and routine activity. Speak of something more like a nutrition-pill exercise. Be careful and diligent, and there would be positive stuff.

CHAPTER 45:

Meditation: Effective Steps in 15 Minutes

Life today is loaded up with bunches of commotion. Consistently you need to beat numerous types of interruptions from the advanced space and your general surroundings. Before racing to work, you should put in almost no time browsing your messages to see whether your undertaking proposition was acknowledged. While heading to work, you need to control yourself from getting frantic over foolish drivers while in rush hour gridlock. That is not all; in the wake of plunking down in the workplace and preparing to start working, your notices go off, and you're enticed to check your internet based life pages and take a gander at what your companions are stating.

Sadly, the messiness that we need to manage consistently keeps us from intuition unmistakably. It keeps us from settling on significant choices that could assist with liberating our psyches from stress. For example, rather than taking a shot at your task, you may be occupied and burn through a ton of time via web-based networking media. Toward the day's end, you will start focusing on yourself over the way that you may wind up presenting your undertaking late. Before you know it, you're overthinking about something that you should have managed before on.

Care contemplation proves to be useful as it can assist you with discovering harmony inside yourself and quiet your psyche. This area will talk about how to care contemplation can assist you with halting overthinking. On the whole, how about we comprehend what care reflection is and how you can rehearse it.

Care contemplation alludes to mental preparing practices that train your brain to focus on your encounters right now. These encounters are the sentiments and feelings that you are confronting now. The thought here is that you divert off your psyche from concentrating on regular babble and focus on the present. Rehearsing care reflection quiets down the brain as you possibly center around the present during the period when you're pondering.

The Most Effective Method to Meditate

First of all, locate a tranquil spot where there are negligible interruptions. You can decide to do this inside or outside as long as there is little commotion originating from the general condition. On the off chance that you incline toward doing this inside, guarantee that the lights don't occupy you. During the day, close your draperies and utilize characteristic light. Another significant thought that you should remember is that you ought to decide to think when you're free. Try not to do this when you are busy, something significant. This can wind up being an interruption, and it could keep you from accomplishing the complete center that it requires.

Instructions to Sit

The stance is an indispensable part of reflection. If you don't expect the correct stance, you will think that it's hard to reflect. You should expect a decent stance that is agreeable to you.

- Sitting Down. You can either decide to plunk down on a bed on a seat. Whatever alternative you pick, guarantee that you are agreeable. When plunking down, affirm that your back is in an upstanding position. You can utilize a cushion to balance out yourself and guarantee that you're not stressing to keep up an upstanding position.
- Positioning Your Legs. When sitting on the floor, fold your legs. People who are unsuitable to do this ought to think about utilizing a seat. If you decide to utilize a seat, ensure that feet contact the ground. Try not to utilize a seat that will leave your feet hanging.

- Positioning Your Arms. Your arms ought to be easily situated on the head of your legs. The point here is that there ought to be no firmness in the sitting position that you pick.

In the wake of accepting the correct position, the time has come to loosen up your body and brain. This is accomplished by concentrating on your breath. Focus on how you are relaxing. Follow how you are breathing in and breathing out. Notice how your chest moves because of the air coming all through your nose. Focus on every single substantial development that you are encountering. Feel the development of your chest. The swelling and emptying of your stomach. To effortlessly accomplish the center, you can decide every breath that you take. A total breath in and breath out can be considered one full breath. Keep doing this until you arrive at a tally of ten.

With time, your brain can become accustomed to your checking procedure. Along these lines, you could have a go at changing how you check, for instance, rather than tallying from 1 to 10, the tally from 10 to 1. This makes more spotlight on how you're breathing and could, consequently, assist you with quieting your brain. You will understand that your brain will, in general, meander to different musings. This is ordinary, so don't stress over it. Perceive that your brain is meandering and figure out how to welcome it back to concentrate on how you are relaxing. Keep in mind, and this ought to be done tenderly. Try not to attempt to constrain things when you are reflecting. You ought to be more mindful of yourself and the musings and feelings moving around your psyche. Subsequently, you can oversee how you think and what you think about it. Your spotlight here ought to be on a certain something; your relaxing. Give close consideration to what exactly is going on in your brain. Notice the contemplations and impressions that travel every which way. Try not to oppose them in any capacity. Simply notice them. You are not required to respond to these considerations and sensations in your psyche. Certainly, they may be oppressive, yet attempt to delicately come back to your place of center without deciding what you're feeling or encountering. As you finish, delicately carry your consideration regarding where you are currently. Sit for a second without taking any kind of action. Take in delicately and inhale out while permitting your body and brain to stream with it. Take another full breath while tenderly opening your eyes as you complete your reflection. Stop for a second, and choose what you need to accomplish inside your day. Care reflection practice is as straightforward as it sounds. In any case, it is quite difficult. Try not to expect quick outcomes when first beginning your contemplation works out. It requires some investment to ace how to do it. Along these lines, you should expect to make it a propensity as it fortifies your mindfulness. In time you will see the effect it has on you.

How Mindfulness Can Help You Deal with Anxiety

At the point when you figure out how to live carefully, you will be more attentive about what goes on around you. This implies your psyche will have the option to think obviously and recognize circumstances or triggers that can make you on edge. Coming up next are manners by which care can assist you with overseeing tension.

Associating You to the Present

Care will cause you to notice the present. There is an immense advantage that you increase here because overthinking will, in general, harp more on how we see our future and our past. Since you will ace more command over how you consider your future and past, you will be in a superior situation to keep yourself from stress.

Care Retrains the Brain

One of the most significant advantages that you gain by rehearsing care is that you can rework the cerebrum to think emphatically. There are new idea designs that you will make through the activities that you will receive. You will start to acknowledge what you have now and appreciate existence without laments. In doing as such, you develop contemplations and convictions that rouse you.

Care Helps Regulate Your Emotions

Care can likewise assist you in dealing with your feelings successfully. As you upgrade your mindfulness, you gather a more profound understanding of the best way to control your feelings when managing regular stressors. Your passionate knowledge will be given a tremendous lift. As needs are, there are insignificant possibilities that you will permit your brain to be overcome with nervousness.

Care Shifts Our Self-Perception

For the most part, the convictions that we hold about ourselves drives us to overthink about the things that we can do and those that we can't. For example, on the off chance that you accept that individuals don't care for you, at that point, you will fill your psyche with self-question. You will never be sufficiently gallant to contend unquestionably amid others. Luckily, care moves your observation as you will have confidence in your capacities. Rehearsing care all the more regularly will positively affect your life since you will be more empathetic about yourself and your general surroundings. The negligible actuality that you acknowledge what life brings to the table suggests that you will offer thanks for what you have. This is an incredible method to see life from a positive point.

Care and Healthy Sleeping

Other than helping you adapt to tension, care can likewise guarantee that you rest soundly around evening time. By living carefully, your brain gets the chance to unwind. You will avoid overthinking things and encounters that you have no influence over. There is no way around your past and certain parts of your future. In that capacity, it has neither rhyme nor reason that you should squander your vitality ruminating over them. Care will, in this manner, urge your psyche to live in the current, which will lastingly affect your rest designs.

Clutching negative convictions about ourselves can keep us from being cheerful. This reflective practice is intended to carry you closer to the accounts that you have created in your psyche about yourself and your life. Take a couple of moments during the day to acquaint yourself with the suppositions that you have created about yourself. These are the recognitions that are keeping you from being genuinely glad. Ordinarily, they will dissuade you from using your maximum capacity. In the wake of seeing how you feel about these suppositions, consider a situation where you didn't need to accept every one of these things about yourself. OK, be more joyful? Perceive the significance of relinquishing these recognitions that you have as it will permit you to carry on with your existence without restrictions.

Care contemplation is an extraordinary method of overseeing how you think. By upgrading your mindfulness, you are in a superior situation to see when your psyche is pondering something. From the data given, there is nothing that you profit by overthinking.

CHAPTER 46:

Anxiety Medication

Are anti-anxiety drugs right for you? Discover typical negative effects, dangers, and how to take them properly.

Lady holds the tablet in between thumb and forefinger, pondering it, a glass of water on the other hand

The function of medication in stress and anxiety treatment

When you're overwhelmed by heart-pounding panic, disabled by worry, or tired from yet another sleep-deprived night invested fretting, you'll do practically anything to get relief. And there's no question that when stress and anxiety are disabling, medication might assist.

Various kinds of medications are utilized in the treatment of stress and anxiety conditions, consisting of standard anti-anxiety drugs such as benzodiazepines (generally recommended for short-term usage) and more recent alternatives like SSRI antidepressants (frequently advised as long-lasting stress and anxiety service). These drugs can offer short-lived relief. However, they likewise feature negative effects and security issues-- some substantial.

According to the American Academy of Family Physicians, benzodiazepines lose their therapeutic anti-anxiety result after 4 to 6 months of routine usage. And a current analysis reported in JAMA Psychiatry discovered that the efficiency of SSRIs in dealing with stress and anxiety had been overstated, and in some cases, is no much better than placebo.

What's more, it can be challenging to leave stress and anxiety medications without challenging withdrawals, consisting of rebound stress and anxiety that can be even worse than your initial issue.

Where does that leave you if you're suffering? Even when stress and anxiety relief features adverse effects and threats, that can still seem like a fair trade when panic and worry are ruling your life.

The bottom line is that there are a time and location for stress and anxiety medication. Medication might be practical, particularly as a short-term treatment if you have extreme stress and anxiety that's interfering with your capability to operate. Numerous individuals utilize anti-anxiety medication when treatment, workout, or other self-help techniques would work simply as well or much better, minus the downsides.

Stress and anxiety medications can alleviate signs. However, they're wrong for everybody, and they're not the only response. It's up to you to assess your alternatives and choose what's finest for you.

- Benzodiazepines for stress and anxiety
- Kinds of benzodiazepines
- Xanax (alprazolam).
- Klonopin (clonazepam).
- Valium (diazepam).
- Ativan (lorazepam).

Benzodiazepines (likewise called tranquilizers) are the most commonly recommended kind of medication for stress and anxiety.

When taken throughout a panic attack or another frustrating stress and anxiety episode, because they work rapidly, generally bringing relief within 30 minutes to an hour, they're efficient.

They are physically addicting and not advised for long-lasting treatment.

Benzodiazepines work by slowing down the worried system, assisting you to unwind both physically and psychologically. The medication hangover can last into the next day.

Typical negative effects of benzodiazepines consist of.

- ☐ Sleepiness.
- ☐ Lightheadedness.
- ☐ Poor balance or coordination.
- ☐ Slurred speech.
- ☐ Difficulty focusing.
- ☐ Memory issues.
- ☐ Confusion.
- ☐ Indigestion.
- ☐ Headache.
- ☐ Blurred vision.
- ☐ Benzodiazepines can make anxiety even worse.

According to the FDA, benzodiazepines can aggravate cases of pre-existing anxiety, and more current research studies recommend that they might result in treatment-resistant anxiety. Benzodiazepines can trigger psychological blunting or feeling numb and boost self-destructive ideas and sensations.

- ☐ Benzodiazepine security issues.
- ☐ Substance abuse and withdrawal.

When taken frequently, benzodiazepines result in physical reliance and tolerance, with progressively bigger dosages required to get the same stress and anxiety relief as in the past. This occurs rapidly, typically within several months, however, in some cases in as low as a couple of weeks. If you quickly stop taking your medication, you might experience serious withdrawal signs such as.

□ Increased stress and anxiety, uneasiness, shaking.

□ Sleeping disorders, confusion, stomach discomfort.

□ Anxiety, confusion, anxiety attack.

□ Pounding heart, sweating, and in extreme cases, seizure.

Many individual's error withdrawal signs for a return of their initial stress and anxiety condition, making them believe they require to reboot the medication. Slowly lessening the drug will assist decrease the withdrawal response.

Drug interactions and overdose.

While benzodiazepines are reasonably safe when taken just sometimes and in little dosages, they can be even lethal and harmful when integrated with other main nerve system depressants. Constantly speak to your medical professional or pharmacist before integrating medications.

Do not consume on benzodiazepines. When combined with alcohol, benzodiazepines can cause a deadly overdose.

Do not combine with pain relievers or sleeping tablets. Taking benzodiazepines with prescription discomfort or sleeping tablets can likewise result in a deadly overdose.

Antihistamines magnify their impacts. Antihistamines-- discovered in numerous over the counter sleep, cold, and allergic reaction medications-- are sedating by themselves. When blending with benzodiazepines to prevent over-sedation, be mindful.

When integrating with antidepressants, be careful. SSRIs such as Prozac and Zoloft can increase benzodiazepine toxicity. You might require to change your dosage appropriately.

Paradoxical results of benzodiazepines.

The benzodiazepines work since they slow down the worried system. In some cases, for factors that aren't well comprehended, they have the opposite impact.

- Increased stress and anxiety, irritation, hostility, rage, and agitation.
- Mania, spontaneous habits, and hallucinations.
- Unique benzodiazepine threat aspects.

Anybody who takes benzodiazepines can experience unsafe or undesirable negative effects. Specific people are at greater danger.

Older grownups are more delicate to the sedating impacts of benzodiazepines. Long-lasting benzodiazepine usage likewise increases the threat of Alzheimer's illness and dementia.

Individuals with a history of substance abuse. Because they are potentially addictive and dangerous and own when mixed with alcohol and other substances, benzodiazepines need to be treated with extreme caution by someone with a current or prior substance addiction problem.

Benzodiazepine usage throughout pregnancy can lead to reliance on the establishing child, with withdrawal following birth. Benzodiazepines are likewise excreted in breast milk.

The connection between mishaps and benzodiazepines.

Benzodiazepines trigger sleepiness and bad coordination, which increases your danger for mishaps in the house, at work, and on the road. When on benzodiazepines, be mindful when driving, running equipment, or doing anything else that needs physical coordination.

- SSRI antidepressants for stress and anxiety.
- Girl resting on the sofa, clutching herself with a distant gaze.

Numerous medicinal products initially authorized for anxiety treatment are also recommended for stress and anxiety. Their use is limited to issues of persistent stress and anxiety, which require continuous treatment.

The most frequently recommended antidepressants for stress and anxiety are SSRIs like Prozac, Zoloft, Paxil, Lexapro, and Celera. SSRIs have been utilized to deal with generalized stress and anxiety condition (GAD), obsessive-compulsive condition (OCD), panic attack, social stress and anxiety condition, and trauma.

Typical negative effects of SSRIs consist of.

- Tiredness.
- Queasiness.
- Agitation.
- Sleepiness.
- Weight gain.
- Diarrhea.
- Sleeping disorders.
- Sexual dysfunction.
- Anxiety.
- Headaches.
- Dry mouth.
- Increased sweating.
- SSRI withdrawal.

Physical reliance is not as fast to establish with antidepressants, and withdrawal can still be a problem. If ceased too rapidly, antidepressant withdrawal can set off signs such as severe anxiety and tiredness, irritation, stress and anxiety, flu-like signs, and sleeping disorders.

Antidepressant medication and suicide threat.

Antidepressants can make anxiety even worse instead of much better for some individuals, causing an increased danger of suicide, hostility, and even bloodthirsty habits. While this is especially real of kids and young people, anybody taking antidepressants must be carefully enjoyed. Tracking is particularly crucial if this is the individual's very first time on anxiety medication or if the dosage has just recently been altered.

Indications that medication is making things even worse consist of stress and anxiety, anxiety attack, sleeping disorders, hostility, uneasiness, and severe agitation-- especially if the signs appear all of a sudden or quickly degrade. Get in touch with a physician or therapist right away if you identify the caution indications in yourself or an enjoyed one.

See Suicide Prevention if you are worried that a buddy or household member is considering suicide. The suicide danger is biggest throughout the very first two months of antidepressant treatment.

- Other kinds of medication for stress and anxiety.
- Buspirone (BuSpar).

Buspirone eases stress and anxiety by increasing serotonin in the brain as the SSRIs do and reducing dopamine. Compared to benzodiazepines, buspirone is sluggish performing-- taking about two weeks to begin working.

Because the danger of reliance is low, and it has no major drug interactions, buspirone is a much better choice for older people and individuals with a history of substance abuse. Its efficiency is restricted. It works for generalized stress and anxiety condition (GAD). However, it does not appear to assist other kinds of stress and anxiety conditions.

Typical negative effects of buspirone consist of.

- Queasiness.
- Headaches.
- Lightheadedness.
- Sleepiness.
- Weight gain.
- Upset stomach.
- Irregularity.
- Anxiety.
- Diarrhea.
- Dry mouth.
- Beta-blockers.

They are likewise recommended off-label for stress and anxiety. These assist the physical signs of stress and anxiety such as quick heart rate, a shivering voice, sweating, lightheadedness, and unsteady hands.

Because beta-blockers do not impact the psychological signs of stress and anxiety, such as concern, they're most useful for fears, especially social fear and efficiency stress and anxiety. If you're expecting a particular anxiety-producing circumstance (such as offering a speech), taking a beta-blocker ahead of time can help in reducing your "nerves.".

Typical adverse effects of beta-blockers consist of.

- Lightheadedness.
- Drowsiness.
- Weak point.
- Tiredness.

- Queasiness.
- Headache.
- Irregularity.
- Diarrhea.

Medication isn't your only alternative for stress and anxiety relief.

Young couple running along a narrow dirt path in nature location, the female smiling as she takes a look at her smiling partner.

Stress and anxiety medication will not resolve your issues if you're distressed because of installing expenses, a propensity to leap to "worst-case circumstances," or an unhealthy relationship. That's where self-help, treatment, and another way of life modifications can be found in. These non-drug treatments can produce long-lasting modifications and long-lasting relief.

Workout-- Exercise is effective stress and anxiety treatment. Research studies reveal that routine exercises can alleviate signs just as efficiently as medication.

Worry busting methods-- You can train your brain to stop stressing and take a look at life from a more well balanced and calm viewpoint.

Treatment-- Cognitive behavior modification can teach you how to manage your stress and anxiety levels, stop uneasy ideas and dominate your worries.

Yoga and tai chi Yoga and tai chi are mind-body interventions that engage you mentally and spiritually. The information has revealed their effectiveness for various medical conditions, consisting of stress and anxiety.

Mindfulness and meditation-- Mindfulness is a mindset where you find out to observe your ideas, sensations, and habits in the present, caring, and non-judgmental method. It typically brings a sense of calm and relaxation.

If stress and anxiety medication are ideal for you, choosing.

It's likewise essential to find out about the typical side impacts of the stress and anxiety medication you are thinking about. Side impacts of stress and anxiety medication vary from moderate problems such as dry mouth to more serious issues such as severe nausea or noticeable weight gain.

CHAPTER 47:

Ways and Habits to Attract and Keep Good Energy During the Day

Focusing on disposing of the negative behavior patterns isn't sufficient. Presently, we have to bring great propensities into our new life.

I trust you've kept on taking a shot at the idea interference for every one of those unneeded or negative contemplations. You've even thought of another action to attempt or return to from when you were a youngster. It's an ideal opportunity to concentrate on you as an individual and what it is you deeply desire. There are a few things you can normally do to prime your psyche, body, and soul for the enhanced you. How about we take a gander at a couple of them.

Organize Associations with Individuals—Not Things

In the present data immersed society, it has gotten normal to pull out our telephones at whatever point there is any measure of vacation. At the point when you see individuals holding up in lines everywhere throughout the city, in stores, standing by to be situated in eateries, they are continually sitting or remaining with their heads down, noses in their telephones. Furthermore, why not? On our telephones, we can play fun games, visit with companions, read blog entries or reports, stay aware of our big names… Wait for a second, didn't we simply discuss disposing of data over-burden? The truth is out. Presently, the time has come to organize your connections and associations with others. How about we set a little test for you to attempt this week.

At the point when you go to the supermarket or the bank, or perhaps when you take your family out to supper next time, leave your telephone in the vehicle. What?! Indeed, that is the thing that I said. Leave your telephone in the vehicle. At the point when you are remaining in line—and this might be a little nerve-wracking—have a go at saying something pleasant to the individual behind or before you. I know… you may get a bewildered look, possibly they will be too soaked in their telephones to see you, or maybe it will have been for such a long time since they've gotten genuine human correspondence, they won't realize what to do. Be that as it may, I challenge you. Converse with somebody in line; at that point, see what occurs. Chances are, you will have positive cooperation that will remain with you for the remainder of the day. The vast majority truly appreciate chattering with outsiders at the store. Many individuals get a major lift in disposition from even the humblest collaborations like that.

Keep Up That Diary and Keep Tabs On Your Development

Journaling is an incredible method to keep your psyche centered and record your advancement. It's incredible to have a composed source to return to whenever you sense that you need a little support. Attempt to compose a smidgen in your diary every day. Expound on how you feel, what difficulties you've prevailing with as of late, and your assurance to continue onward. This is additionally an extraordinary method to keep yourself responsible. Record what you've provoked yourself to do this week and record it immediately when you've finished it. Continue onward, and soon you will have pages of extraordinary work to think back on when you believe you are losing steam or need a jolt of energy. Since we as a whole have those days, and that is alright! Like I said before, this is a major endeavor.

What's more, it's significant that your difficulties not go into errands and wellsprings of overthinking simply like the ones you've endeavored to oust! Challenge yourself, however, don't over-trouble yourself. Try not to attempt to take on every tip in this unit on the double! I'm spreading out a few choices in the expectations that there will be a not many that truly stand apart to you as something you might suspect would radically improve your everyday life and your points of view. Furthermore, recollect, nobody changes their carries on with for the time being!

Eat More Advantageous

Eat more advantageous—not "practice good eating habits." I put it like this because there is no surer method to crash your advancement than to over-burden you with a test like totally changing how you eat right away. On the off chance that you are now a sound eater, that is extraordinary! In any case, I would alert you and others the same to not get excessively got up to speed in any sustenance publicity or prevailing fashion that is by all accounts gobbling up your Facebook and internet-based life channels. This is another incredible case of letting something proposed to improve your life become a wellspring of fixation, overthinking, stress, and sentiments of disappointment. Nourishment plans and supplement advertising is similarly as large as some other type of showcasing, and you ought to never embrace an eating routine or sustenance plan as the last day of nourishment. Utilize some presence of mind, don't indulge, and attempt to eat more sound stuff than undesirable stuff. That is all you have to stress over this moment. Try not to go on an amazingly low-carb diet at present. You're managing something undeniably more significant than that.

Exercise

It's the ideal opportunity for the universally adored sound brain/body tip—work out! Presently, don't moan. No, you don't need to begin preparing for a long-distance race or purchase a total arrangement of free weights for your new offhand home exercise center. I've said it beforehand, and I'll state it for about each tip on this rundown—each little advance in turn. At the point when you separate things and approach them slowly and carefully, you will be unmistakably more fruitful with your objectives than if you attempt to take on a lot at once.

It is additionally so significant, particularly with work out, to assess your condition and capacity. Try not to contrast yourself and the YouTube wellness stars doing insane exercises every day and chugging protein shakes. This is about you and your improvement, and nobody else's arrangement is going to coordinate your impeccably.

Much the same as with smart dieting propensity, the initial step is to investigate what you are as of now doing and climb a crosspiece on the movement level. That's it in a nutshell. In case your somebody who appreciates turning out to be, however, doesn't appear to discover an opportunity to do it, at that point, I'm getting you out! Exercise isn't about the period, it's about how hard you work, and I'm just talking a couple of moments every day to begin. If you are beginning at zero movements, at that point, your objective is just to consider the potentials for success to walk or have rather than sit. On the off chance that you can, fit in a stroll around the square or go to a recreation center and walk a smidgen. In case you're at home, get up from your work area, and accomplish something physical consistently or two to get your blood siphoning somewhat more. It's tied in with rolling out little improvements. Transform those little changes into propensities at that point. Focus on climbing another bar.

Set Aside a Few Minutes for You Normally

This is another that can mean heaps of various things to various individuals. Setting aside a few minutes for you essentially means putting aside time each day to take part in a movement that causes you to feel great and quiet you. The exemption I will propose here is that you don't make this chocolate or lousy nourishment time. Indeed, chocolate causes you to feel great… for a couple of moments… however, generally speaking, it would be a horrible plan to shape the propensity for eating severely for the sake of "you" time. I'm certain there are other, more solid other options!

Do you appreciate kneads? A great many people won't get a back rub every day, except possibly once per month, you treat yourself to an expert back rub. Every day, discover something that loosens up you and put aside thirty minutes or all the more only for that. Regardless of whether it's simply sleeping! Peruse a book, light a flame, accomplish something that clears and loosens up your brain, and doesn't stir you up. This is tied in with loosening up, yet as opposed to supplanting the pressure of your day with something loud and diverting for the remainder of the night, the objective is to quiet and calm your brain and body. Extending is an incredible method to do this, particularly on the off chance that you've been stuck in an office seat throughout the day.

Plans for The Day

A significant number of us like to have each day sorted out, and that is incredible. The issue of overthinking creeps in when we begin to fixate on getting each and everything on the rundown done, even those things that are not fundamental. Some portion of shaping great propensities is realizing when to disapprove of something that you simply don't have the psychological vitality to do if it is unnecessary, on the off chance that you feel pushed. However, you're gaining ground toward transforming you and propensities; at that point, it's alright if you need to pass on that work excursion or that birthday celebration for a companion of a companion you don't know quite well. If you figure your time would be better spent at home unwinding or accomplishing something, you appreciate, at that point, pick yourself. You don't generally need to decide to give your important time and vitality to others since they request it. The commitment is amazing power in many individuals' lives, and numerous individuals wind up feeling remorseful if they don't generally say yes to solicitations or solicitations.

Request Help When You Need It

This can be a major one for those over thinkers who are additionally overachievers and sticklers! In some cases, we focus on something over the top, at that point, feel committed to propelling ourselves too difficult to even think about fulfilling what we've focused on. Try not to attempt to be superman or superwoman. There will be times when you need assistance, particularly on the off chance that you are shuffling duties with work and a family. Have a discussion about it with your friends and family and companions, and you will locate that more often than not, they are happy to get you out. Significantly, you do not feel like a disappointment for requesting help. Nobody gets past life alone. Leave this alone a chance to bond and structure new associations while figuring out how to cooperate.

Be Thankful

Freeing your psyche from the mess is additionally about purifying your feelings. At the point when you start to dispose of messiness and negative musings in your brain, just as the mess in your condition that is associated with unsafe feelings, it is significant that you begin supplanting those negative emotions with positive ones. From the start, it might require exertion and a composed suggestion to make yourself go, yet in the long run, the objective is to make these considerations programmed.

Appreciation is an amazing thing for the psyche. It can right away turn an unpleasant, awful day into something positive and confident. Rather than focusing on the difficulties you are confronting and the things you don't have, consider the entirety of the great things throughout your life.

CHAPTER 48:

Face The Negative Mindsets That Surround You

If there is an element that all humans have, in particular, it is our deep need to achieve pleasure. While the road to satisfaction is filled with different forms of hurdles, there is something that always gets in our way: anxiety. A vast number of people identify themselves as "worried". Additionally, many confess they do not know how to avoid planning for the future.

Constant stresses will quickly contribute to fear and can lead to mental disorders in effect. Increased mental wellbeing issues seem miserable because this behavior will also impact your physical safety. It is not enjoyable to worry, and you feel scared while those thoughts stay. Anxiety has been found to have a detrimental health effect, affecting a variety of symptoms from depression and lung to cardiac problems to stomach disorders.

Recognize the stimuli and cultivate the consciousness of your psychological thinking and methods. You cannot learn how and when to stop thinking about the future when you do not know what will trigger the problematic process. Consider whose restrictive values lead you to think about it. Feeling such as worrying, for example, when you do not want to be taken off track, to avoid frustration, or that it is a way to prove you appreciate the things. Such proposals have a useful purpose, but instead of empowering, they are often restricting.

Excessive anxiety about future events produces "disturbance" and can be overcome by the methods of brain-cleaning. There is an extensive range of items to do, so draw focus to the simple one to bring you into the routine of regularly clearing your mind.

Self-comparisons may be useful at times. They will send you the transformation plan and encourage you to progress. Other occasions, they may be a way of pulling yourself out and identifying all you believe is false. Comparing ourselves against others, we reflect on both their qualities and successes and neglect ours. It is typically an unequal reference point for contrast. Consequently, if you look at somebody's conditions versus your flaws, you can still come out wrong. Also, if you equate power to intensity, there will always be stronger ones and weaker ones.

Even if you perform well compared to others, this comparison can cause you to pump up unnecessarily. You end up resenting people for doing better, without understanding the right individual quite well. You might realize that if you have ever resented anyone for seeing them first, and later realizing that you have had the wrong concept. You might wind up thinking more than you should about your achievements.

Much of the time, we perform such psychological associations without knowing why we do. It is a human act, and as a consequence, it is something performed without awareness. And the answer is to become aware of putting these feelings to the center of the awareness by being searched for them. It becomes much better with time if you concentrate on these feelings for a couple of days, and then it will be impossible not to notice. When you understand that you have such comparisons, allow yourself a rest. Do not panic or feel guilty and just accept your feelings and shift your concentration gently.

No one here is perfect. Mentally, we all realize that, however socially, when we do not achieve excellence, we tend to feel terrible. You are not perfect, and you will never be perfect. Try to change, but do not believe you can ever be the "ideal person." If you take a closer look at something, the imperfection allows you who you are, and you are always good.

If you ever desire what some have, you are never going to get it. You are just going to want more. That is an infinite loop, and it is never going to deliver happiness. Regardless of how many dresses you obtain, irrespective of how many homes you buy, no matter how many luxury vehicles you obtain, you can never get plenty. Rather, learn to know what you already have would be enough. You are lucky because you have a roof above your head, meal on the plate, clothes on the body, and people that love you. Everything you have above and beyond that, and let's say that we have more than enough. Be great with that, and you will be satisfied with that.

CHAPTER 49:

Secrets to Successfully Tame Your Thoughts

Occasionally, everyone has negative thoughts; however, persistent negative thoughts might point to a more serious problem that might require therapy and medication. The causes of negative thoughts could be chronic worrying, personality problems, drugs, life events, illness, or a combination of these factors.

Therefore, treating this problem or stopping negative thoughts depends on the frequency of the symptoms or the results of a psychological test. The correlation between worrying and negative thinking, however, is a bit tricky because people experiencing this problem cannot determine whether worrying is making them have negative thoughts, or if those thoughts are making them experience depression. Even things like allergies, sleep deprivation, hunger, stress, tiredness, or even a common cold can make people worry, which can ultimately lead to negative thoughts. Often, negative thoughts can lead to depression, just as depression can lead to negative thoughts. This cycle can make people lose control of their lives.

People's emotions stem from what they are thinking, and dwelling on their negative thoughts can make them spiral down into stress and depression. This is the guiding principle behind CBT, which is the cognitive behavioral therapy developed by Dr. Aaron Beck in the 1960s at the University of Pennsylvania. Negative thinking slows down the achievement of people's personal and professional goals. Therefore, it is important to get rid of such thoughts. Fortunately, there are several ways to do this; however, one's choice is a personal decision. The most important thing to remember is that one is in control.

Some of the most effective ways to stop negative thinking include:

Speaking to the Negative Thought

When negative thoughts pop up, it is important to have an awareness of them; for example, when one is feeling stressed, disappointed, hungry, tired, or something else. Ignoring these thoughts does not make them go away; actually, they tend to latch on to one's mind.

To overcome them, one needs to identify them and recognize them, which will help one squash them, which involves telling oneself that they are not true. According to the Intentional Coaching LLC by Frances McIntosh, this will get rid of the negativity quite fast.

Spend Time with Positive People

The fastest way to catch a cold is to spend time with someone with a cold. Therefore, people who frequently experience negative thoughts will worsen their situation by spending time with negative people. People plagued by negative thoughts tend to gravitate towards like-minded people.

This is unfortunate because negative people are pessimistic, and their pessimism rubs off on others. Therefore, people who tend to have negative thoughts should physically avoid negative individuals and seek more positive people and situations.

Understand that no one and Nothing is Perfect

Expecting perfection in everything can be debilitating and prevent one from finding contentment and happiness. Therefore, it is important to understand that no one is perfect and base one's success in reality.

For example, if one is expecting a promotion this year, one should envision it next year. After all, one year will not make a huge difference. It is important to work towards certain goals with a sort of detachment to the results because imperfection can be liberating in a certain way. Essentially, it allows people to live on their terms.

Have an Active Mindset

There is no substitution for having a healthy and active mindset, apart from practicing what works. Based on their experience dealing with different kinds of people, psychologists understand that everyone is different. However, one thing stands true. There is no standard practice to treat or prevent negative thinking.

People's mindsets are completely personalized and depend on the limiting beliefs they are trying to remove, as well as the positive mindsets and beliefs they want to instill in their lives. Therefore, people should find a positive routine and stick to it to the point where it will start to work.

Positive Morning Routine

Sometimes, to counter negative thinking, one might need to start controlling one's thinking early in the morning. When people control their thinking, they tend to gain the ability to control their lives. Negative thinking tends to slow down people. Leaders, for example, should take their thoughts captive and replace negative thoughts with belief and hope.

Many leaders and successful individuals employ a strategy of using an effective morning routine, which includes reading something positive and encouraging as soon as they wake up. As simple as this might sound, it is extremely effective when it comes to facing and ending negative thoughts.

Slow Everything Down and Breathe

To overcome negative thinking, people need to learn how to identify or recognize their unhealthy thoughts, breathe deeply, and slow things down. A good way to do this is to incorporate literal time blocks, reminders, or alarms into one's schedule. Essentially, breathing will help one stay more at ease, in addition to helping one maintain self-awareness.

Intentionally Assume a Positive Attitude

Contrary to what many people seem to believe, it is possible to assume a positive attitude as soon as one wakes up in the morning.

To do this, one needs to be aware of when negative thoughts start to dominate one's mind. People can choose their focus and attitude. Therefore, they should try to let go of things that do not serve their goals and well-being.

Displacement Theory

It is impossible to stop thinking; therefore, people who want to break their negative thought patterns should replace them with positive thoughts.

Since people think about one thing at a time, it makes sense to pick positive things and think about them, which will eventually displace negative thoughts and feelings.

Focus on the Possibility of a Positive Outcome

Despite one's current situation at home, work, or anywhere else, there will always be instances where negative thoughts will intrude into one's mind. When this happens, focus on where one is heading and why it is important to get there.

Essentially, life is about one's goals, but not the negative parts or aspects of meeting one's objectives.

Identify the Root Cause

Negative thinking, for most people, stems from issues hiding below the surface. Essentially, they stem from the root. Pun intended! Anyway, most people have ingrained negative thought patterns.

They need to discover the roots of such thought patterns and find ways of preventing them. Only then will they be able to face their problems.

Make a Conscious Choice

It is difficult to deal with negative thinking patterns unless one understands what they are. Therefore, it is important to identify one's negative thoughts and things that trigger such thoughts.

Essentially, one needs to be self-aware to identify negative thoughts and shift one's way of thinking and perspective at the moment.

CHAPTER 50:

The Psychology Behind Rethinking and Thinking Too Much

Worries, anxieties, and doubts are normal human emotions. It is normal to stress over a first date, a conference, a meeting, unpaid bills, and the sky is the limit from there. Typical concern, be that as it may, can get over the top when it is wild and persevering. A few people will participate in general stress each day over most pessimistic scenario situations, and can't keep adverse contemplations off of their mind, which meddles with their everyday life Steady stressing, continually anticipating the most exceedingly terrible result, and thinking contrarily can negatively affect one's physical and enthusiastic wellbeing. It can deplete one's sure vitality, leaving one inclination anxious and eager. Over the top concern can likewise cause muscle pressure, a sleeping disorder, stomach issues, and cerebral pains making it hard to concentrate on work or school. Tragically, a few people who experience the ill effects of constant stressing will, in general, take out their contrary sentiments on those nearest to them or attempt to divert themselves by observing an excess of TV or self-sedate with drugs as well as liquor. This issue can likewise be a side effect of summed up uneasiness issue, which includes an overall sentiment of disquiet, apprehension, and pressure that hues as long as one can remember. Individuals tormented by constant stressing can complete a few things to turn off their negative contemplations. They can bring an end to this psychological propensity and train themselves to remain quiet and take a gander at life from a not so much frightful but rather more adjusted point of view. For most constant worriers, negative deduction originates from their convictions, both positive and negative, about stress.

Positive Beliefs about Worry

A few people accept their stressing encourages them to get ready for the most exceedingly awful, evade awful circumstances, or discover an answer. As per their perspective, by stressing over an issue sufficiently long, they will make sense of the issue and, in the end, think of a decent arrangement.

Others think stressing is an approach to keep them from ignoring something or is the mindful activity. Lamentably, it is amazingly hard to get out from under this propensity on the off chance that one trusts it fills a positive need. On the off chance that one is managing this issue, one ought to comprehend that stressing is neither a positive thing nor an answer. In this way, it is essential to recover control of one's psyche.

Negative Beliefs About Worrying

A few people experiencing interminable or over the top stressing accept that their stressing is hurtful and may make them crazy or lead to other physical medical issues. They dread that their stressing will assume control over their lives and constantly control them. They wind up agonizing overstressing, which builds their tension and props their concern up securely.

Chronic Worriers

While the vast majority stress over something when it occurs, others will, in general, anticipate that negative things should occur. They have negative reasoning that makes them inclined to "imagine a scenario where malady." The head of the American Institute for Cognitive Therapy, Robert L. Leahy, Ph.D., recommends there is a hereditary factor to incessant stressing, including non-sustain and support factors.

Individuals whose guardians separated, for instance, have an improved probability of building up a summed up uneasiness issue. Individuals with this issue experience pressure, overstated concern, and interminable tension. Moreover, overprotective guardians and guardians who power their children to deal with them because of their brokenness may wind up raising inordinate worriers too.

Basically, as per a few psychoanalysts and clinicians, there might be ecological and organic segments to ceaseless stressing. Guardians ought to guarantee that their children have a sentiment of wellbeing. They should believe that their folks would protect them. They ought to grow up with this sentiment of wellbeing, which will help make them secure grown-ups.

For instance, individuals who grew up with conflicting and inaccessible guardians frequently come to accept that the world is a dangerous sport. Overprotection and separation can likewise consume children's sentiments of security and inward wellbeing, which will prompt negative reasoning and incessant stressing in adulthood.

Why People Worry

Individuals with this issue regularly stress since they figure something awful could or will occur. Like this, they trigger an excessively touchy concern procedure dependent on the conviction that their stressing will keep awful things from occurring or give answers for apparent or anticipated issues. They feel that things will turn crazy if they keep themselves from stress.

At the end of the day, as indicated by them, on the off chance that one can envision a negative circumstance, it is one's duty to stress over it to locate a preemptive arrangement. Shockingly, this can negatively affect their psychological and physical wellbeing, which is the reason ceaseless worriers will, in general, be perhaps the biggest weight on medicinal services frameworks.

- How to Prevent Yourself from Worrying

Exorbitant or ceaseless stressing can keep individuals from having a profitable existence. Stress and nervousness will, in general, command their considerations, which occupy them from their home, work, or school life. A few people propose it merits allowing oneself to stress for a brief timeframe, however postponing harping on it until some other time. To do this, they suggest the accompanying:

1. Creating a period and spot for stressing

2. Writing down one's concerns

3. Going over this rundown of stresses during the concern time frame

Specialists, in any case, propose more approaches to decrease constant stressing that can have negative physical and mental impacts. Nobody needs to be an apprehensive Nellie or wet blanket and continually stress over everything without exception, which is proportional to stressing one's life away. Different approaches to managing ceaseless concern include:

Identify and Embrace Uncertainty

Having investigated one's unwarranted concerns, one ought to recognize what one needs to acknowledge and proceed onward. This may mean tolerating one's impediments or tolerating a specific degree of vulnerability. For instance, individuals stressed over getting disease ought to acknowledge that it is a chance, as nobody comprehends what will occur later on.

As opposed to what most interminable worriers accept, vulnerability isn't an affirmation of an awful result; rather, it is unbiased. At the point when individuals acknowledge vulnerability, they don't have to continue stressing any longer.

Acknowledgment, for this situation, implies an understanding that vulnerability exists in all parts of life, relinquishing one's concerns and focusing on things one can acknowledge, appreciate, and control.

Bore Oneself Calm

In specific circumstances, continually rehashing a dreaded idea can cause it to appear to be exhausting, which will cause it to disappear. Confronting the mirror and expressing one's concern ordinarily will cause it to lose its capacity.

Making Oneself Uncomfortable

Most constant worriers will, in general, feel that they can't be discomforted. In any case, on the off chance that they make themselves awkward, they will go to the acknowledgment that they can deal with it and achieve more than they suspected conceivable. The objective of doing this is to have the option to do what they would prefer not to do or to do things that make them awkward.

It is regular for ceaseless worriers to evade new circumstances, things, and individuals that make them awkward. They utilize a preemptive concern system planned for helping them stay away from distress. Be that as it may, by doing things that make them awkward, they will figure out how to depend more on adapting procedures, as opposed to stressing.

Stop The Clock

It is normal for interminable worriers to have an urgent need to keep moving. They will, in general, think they need a quick arrangement; in any case, something awful will occur. Be that as it may, it is critical to think about the upsides and downsides of such desperation. Rather than concentrating on the need to keep moving, it is smarter to concentrate on the current happenings.

They ought to think about their alternatives and figure out what they can do in the present to make their carries on with important and wonderful. They can either concentrate their brains on finding the correct arrangement or spotlight on improving their current second.

As per most specialists, the last is the best procedure to conquer extreme stressing. They ought to likewise take full breaths, tune in to music, or read to stop the clock and assume responsibility for their tension.

It Is Never as Bad as People Think It Will Be

Incessant stressing or tension is all over 'what uncertainties' and expectations. Be that as it may, what a great many people foresee or stress over is regularly more regrettable than what occurs. Constant worriers will, in general, stress over things they can deal with and control. As per a few specialists, exorbitant worriers are, in reality, acceptable with regards to managing genuine issues.

Cry for All to Hear

The amygdala, which is the enthusiastic piece of the human cerebrum, encounters such concealment when individuals stress. A short time later, this feeling storms in and accompanies indications, for example, an expanded pulse, exhaustion, and gastrointestinal issues. In this way, it is smarter to shout for all to hear or communicate one's feelings in a proper way, as opposed to attempting to dodge them. At the point when one is angry or crying, one will stop being concerned. This is the reason specialists and analysts frequently request that individuals cry or let out their annoyance and feelings.

Share One's Worries

Notwithstanding crying out loud, which is an intellectual treatment procedure, individuals should discuss their concerns to an advisor or a friend or family member. This will make them stress less, as it will assist them with getting to the underlying driver of their concerns. As a rule, psychological conduct treatment and talk treatment can cooperate to convey astonishing outcomes. Individuals need to recognize what triggers their concerns. By profoundly dissecting their concerns and glancing back at the early bases, they will distinguish the underlying foundations of their uneasiness and discover approaches to conquer their concerns.

CHAPTER 51:

React by Counteracting Excessive Thinking and Changing Habits

To gather how to change your habits and actions towards positivity, we first need to appreciate what positive thinking is. Positivity entails the mental process that refocuses our emotions towards optimism. These people are often inclined to always think the best out of any situation. They would not focus mostly on the negative side, even if the results are not the way they expected. A positive mindset is a behavior or a state of mind that keeps recurring. Many people trying to achieve a positive mindset are often drawn to the fact that this is not a onetime thing and that its nature is cyclical. Whenever you find yourself in a situation that is demanding, a positive mindset will always seek the best out of every situation. To achieve positive thinking, you ought to embrace positivity at its finest. Most of us forget that positivity sits at the helm of a positive mindset.

Positivity goes beyond the personal attributes of having a face that suggests that you are cheerful and happy. It encompasses the overall outlook that we give to life. The way we generally look at life will work a great deal in determining our focus in life. A positive minded individual will have less focus on what tends to annoy them in life. Their energies are redirected to what works for them, what makes them happy. To embrace positivity, one needs to understand that challenges are part of life, and learning to embrace them and live with them is one step towards achieving your goal. A mindset, from the wording of it, refers to the situation that you decide to put your mind in. For one to be referred to as to have embraced positivity, several pointers manifest themselves. Some of these pointers are:

Optimistic Mindset

Embracing optimism acts as the first stage towards achieving positivity. Optimism can be said to a synonym of positivity in that they both have the same effect. Optimism is the act of expecting every outcome to impress you, and when this does not happen, you will find happiness in every situation you may be in. The opposite of optimism is pessimistic. A pessimist is an individual who only visualizes the bad in every situation. They are often inclined to negativity. Time and again, it has been a bone of contention that people with a negative mindset are always prone to accept results the way they are because if it is not a loss, then it is a win, and surely it still works for you. However, this is not the case as when an optimist is faced with a loss. They will find a way to make it work for themselves even in that time of need. Positive minded people tend, giving it a shot rather than dismissing it entirely.

Maturity

Maturity here exhibits itself in a manner that suggests that you are accepting that things may not always happen in a manner that you want them to. Accepting the factors and consequences of the occurrence is what helps a great deal in making your move on. To achieve this, we need to appreciate that we are prone to making mistakes, and when this happens, we should build upon our mistakes rather than letting them affect us entirely.

The Heart of Resilience

You will be pulled down emotionally almost whenever your plans do not go through, or maybe the outcome was undesired. People who let this kind of energy burn them from the inside will always have an effect of feeling low. Having maturity as your guide is one way to ensure that you get through this peacefully.

To get back from an occurrence that was painful to you, one needs to understand the fact that letting this feeling pin you down will not work best for you.

You have to come off your comfort zone and make sure that you are getting rid of all the negative energy that needs to be channeled in the way that makes you happy.

Gratefulness

When something happens, that is in your favor, and you ought to find it in your heart to say thank you. Good things have a sense of rarity, which makes them gold. Whenever a good thing happens to you, no matter how frequent, it is good practice and behavior always to say thank you. This should be done continuously.

Having a Present Mind

Absent-minded people tend to build castles in the air; thus, they will stumble upon any thought just to pass the time. These are the type of people who tend to think negatively. When you are present in mind, this means that you will be conscious of any arising factor and that whenever this happens, you will be ready in your mind to face it.

Being Integral

An integral person is an individual who embraces integrity. Integrity refers to the act of doing what is right and sticking to it. You are honorable about your failure, no matter how low you fall. Some people tend to lie to themselves about a pre-determined facet in life. This is often not advisable as the reality will always dawn one way or another.

However, there are ways in which an individual may achieve a positive mindset. This is through embracing positivity every step of the way. The transformative steps towards being positive minded include:

Commence Your Day with a Positive Attitude

The way you commence your day will determine how best you will operate throughout the day. When your day is distorted at its onset, this is a pointer that the rest of the day is going to follow suit. It is not dependent on the fact that you have bad luck but rather dependent on a much bigger perspective. People who have a rough start at the onset of the day tend to link this to the rest of the day. Thus, in turn, the day looks to be a wholesome boring spree, yet it is your perception that effects this. To make sure that you are devoid of this, you ought to start your day by having a monologue with yourself. To achieve this, the mirror is your biggest tool. Take a moment and look into the mirror. Tell yourself that the day will be successful and, in turn, channel this energy throughout the rest of the day.

Radar on Positivity No Matter How Remote

As part of life, obstacles come in to create challenges that make our lives more adventurous. These obstacles will always manifest themselves, whether we like it or not. Getting through an obstacle will be the biggest determinant of how you move on. You are encountered with a problem; for instance, what you need to do is focus on the minute facets that bring about happiness and joy to you. For instance, you can make a joke out of the worst scenario possible. This brings in a light mood.

Your Mistakes Should Be Your Biggest Lessons

When you make a mistake, you should not be overwhelmed by the consequences of the mistake, no matter how dire they might see. Falling will necessarily determine how you rise. Focus on your mistakes; let them form part of you. Tell yourself that you are not going to fall for these mistakes ever again. With this in place, device new methods of getting through the day.

Be Transformative

Transformation here involves two facets, one is the positive thoughts, and the other one is the negated one. This transformation should assume a manner that it should be from the negative to the positive. Never should you adopt the vice versa because it is a downhill road. Avoid putting yourself down in that you are telling yourself you cannot do something because you are not good at it.

A rather different approach to this would be challenging yourself to be the best in what may seem subtle to you at the moment.

Take Note of the Present

Negativity draws its base from sentiments that might have been expressed by an individual, for instance, of higher hierarchy than you. It may be that your boss has scolded you for the occurrence of something particular. You ought not to think deeply about these sentiments because the deeper you go, the severe the effects are. Take a point of not holding on to the bad energy. Do this by releasing yourself from the shackles of bad thoughts. You can decide to erase this from your mind as soon as it is done and move on to be mindful of what is in the present.

Believe you can, and you're halfway there.

CHAPTER 52:

Towards A Better Life

Pierre Corneille Self-love feels narcissistic in the ground; after all, who would only love himself would become in vain.' Self-love is the cause of every other enjoy.' But self-love varies from narcissism because it concerns self-respect and private well-being rather than a mere need to be careful and loved to cope with our insecurities and questions about self-esteem. Self-love, which only saves your energy without any benefit, is crucial to combating anxiety, concern, and rethinking in adverse topics.

It is essential to have a fundamental knowledge of who you are in a group if you want to manage your lives. We are the key to our lives by our self-image, which our image is the one we tend to keep in mind. This shaped picture is consistent with all of our deeds, emotions, or behavior, and even our skills. We are nearly the type of person we tend to think we are a unit. What we want to remember is that as soon as we keep this picture, no effort, determination, engagement, or self-command will lead to the United States being an alternative technique, because we are always inclined to use our technique. To be different, we tend to look at our self-image first, however.

You have Created Yourself

You have developed yourself, whether you realize it or not, in a significant manner. You borrowed, imitated, or created your personality characteristics, mannerisms, forms of speech, facial expressions, gestures, and even methods of thinking and thinking. It could have been of an excessively large novel or a film by a parent or other parents, a favorite teacher, a friend, or an artist.

Perhaps you borrowed from someone you didn't want. Maybe it was from some UN organization that you felt uneasy or scared. It could have been some way to construct that individual to imitate you feel less frightened and discouraged.

Never Reject Yourselves in Any Way

You need to check the temperament you developed. Maybe because you were an associate imitator, you can find one of the reasons you can keep from doing so. It is not unusual to cause hunger. It can help to understand that no one is from scratch producing a Self. All have a comparable problem to do. Of what is out there, everyone decides. Even if you have imitated your temperament, you are not a fake. No one else has ever combined the same as you did. Don't miss that there are twelve notes alone in the region, and yet several thousand distinctive and magnificent blending areas have been produced. All of it is a question of how you find your location.

It doesn't construct you to be a new individual taken from others. The great problem with this is that you can change it at any moment since you put it together from scratch. You never stuck. You are never stuck. It's not a catastrophe to discover that you're not the person you believed you were. It's the start of the catastrophe, on the contrary.

To change the experiences you experience, it is essential, to begin with, a transparent knowledge that you can never help yourself by dismissing a quarter of yourself. We get hated because we tend to create a picture. Still, we expect that our families, peer groups, companionship, religion, and the community where we live always have to take the foundation. The sad half could be that we cannot live up to the pictures, pictures, designs, norms, or ideas of how we are always inclined to assume that we must be. It's a dead-end for psychology.

Freedom Begins with Self-Acceptation

We permitted our ego to manipulate us into an insufficient, insufficient, uncertain, silly, poor, evil, or undignified method of cognition. All this can be described as bad inefficiency and bad self-image. We will still have bad self-esteem and a bad self-image until we construct a conscious call to change our thought habits. In your lives, the first and most important thing is to accept yourself, to enjoy who you are. You can only start to like others once you are interested in yourself. Many people believe they must first state that they have to tell themselves and love others. Well, that method doesn't operate. The reality is that you first have to face all the errors-all, the presumed crimes that have been committed, every time you felt like a jerk, and the whole day you acted inadequately. You must be prepared to stand in front of the entire globe without an apology. If you do that, you come back from an unconditional love situation.

How you see yourself produces your conduct, creating your workplace or outcomes. This behavior You determine dissatisfaction when you link your self-worth to your achievements and conduct. No matter how difficult you attempt, someone will believe you're not all right. Remember this: in somebody's sight, and you could always be a failure. You won't all over, not even a majority sometimes. Consider how much your existence is about obtaining approval and understand that needed reality: you might not receive the approval you are seeking! You just can't help everyone, therefore learn to please yourself, and you're a UN organization.

You Can't Fail as an Individual

It's worthwhile continuing that the UN organization in your region is spiritually outstanding, but it's not always outstanding what you do. What you do may be successful or failing, but by the fundamental mental method, you will be detached from the outcomes and will never be a success or inability to support what you have and do. There is no way you will only fail as a person in existence. That's not how you came about.

You're in suffering when you hate yourself for all that you did or did not, or when you hate other people for what they haven't given you. Suffering is a form of self-determination. It's a way to get irritated. It is disappointing for you not to live up to expectations that we are inclined to have something about ourselves or that someone else has people if you get to know that anger and pain and an absence of happiness in our lives are right there.

I have discovered that the main reason for their conduct is self-hatred while collaborating with people from United Nations agencies. Their self-hatred was the reality that they had not met the standards of someone else. Most of us decide to think of what we have or don't have and what we did or what we didn't have.

We think we've let others down once we tend to unite a mistake in the region.

We do not say that we are not intelligent if we tend not to live up to our people's expectations, employers, religion, family, or partner. This is called the judgment of oneself.

If you judge yourself, you are going to decide that you are wrong. And once you're placing yourself for one thing you did, or something that wasn't accomplished, or a position wherever you're foiling someone else, you feel bad. But the only way to sculpt the very small lack of judgment is to sculpt it. It's not going to be nice; it's just destruction.

Everyone in the U.S. indeed has stuff in our life that we tend to regret, but we tend to avoid moving and regretting houses for some reason.

The class must be found, and the skill is thrown away.

So long we tend to area unit against each other, we won't be for anyone. Being against others means being against us. This is a psychological and spiritual truth. The most corrupt thing we can do is judge someone. One of the main adverse and self-destructive behavior a person can have is to remove another person and remove the life of another.

Release with Yourself Everyone What if you had no regrets from the past? Try to figure out what would occur when, whatever they did with you, you completely forgave everyone in your life? Hoped you would make sure that you preserve unhappiness, poverty, illness, lack, and limitation in your life in the way that you cannot excuse, be it yourselves or anybody else.

Many people don't want other people to apologize. You tell stuff like: "Why when they came to me, should I let them go off the hook? The enemy always is someone we imagine that is going to harm America or take away something from us, but it is not true that anyone is going to harm America. Through ourselves, people harm the US. They don't harm the U.S.A. to the least. However, we give instructions for treating the United States, and they obey it.

Call instantly to ransom you, as you will eventually be destroyed in the end. "Yes," you say, "I agree with you, but you don't recognize my circumstances. You hurt Maine. Maybe I'm going to give up the rancor at some point, but right now, I can't give up." Understand that this mentality is dangerous and dangerous to you more than it is to the individuals whom you are feeling. "The pupils and followers replied." It was an exemplary life you lived. You brought the U.S.A. out from the Moses area. He softly answered, "When I met my Maker, he will not lift:' You were like Moses and Salomon? You judged us demonstrating wisdom like Salomon. "It's going to wonder,' Were you? The tale demonstrates that people lived to be their own throughout time. Why do we still have a desire to fight?

It's because we have to help others that fight.

You're sure that you will be angry with somebody-your boss, wife, relatives, and kids by assuming your destiny. At first, it will be a solo technique to assume your destiny, and it can seem that everybody is against you. But your picture is the only one you have to stick to. The views of those who agree or disagree with the region are irrelevant.

Your obligation is the choice to assess your lives. You are responsible for the results of your own lives. Your accountability is your decision or your inaction. Alternative people can often have values and convictions that clash with you. And when you see yourself living against their principles and convictions, it can be appalling because, in some manner, it is threatening their basis.

When you face your faith, the inner fight of nursing is waged by an Associate, and the fight is, "Could they potentially be correct? And if they are, then I can be incorrect." The UN organization knows who they are, is not susceptible because of people's views. Those who are insecure in their region and do not understand the United Nations organization can always be scared by any UN organization which threatens their scheme of faith, directly or indirectly.

CHAPTER 53:

Positive Energies and Positive Thinking

The power of positive thinking and positive positives is an astonishing thing that could change your life altogether. As shown in the following review: recently in The School of Greatness, I had the chance to meet Steve Weatherford, former NFL Super Bowl Champion, who was one of every most efficient punter in the globe, about his keys to achievement, which can create the distinction. A favorable mentality will help you conquer all the barriers you face. One of the greatest factors that entrepreneurs and prospective customers can extract from the sport and athletic world is that visual picture efficacy and positive thought are effective. Studies after studies indicate the strength, but life changes follow this simple one.

Most individuals don't realize that the people in this region are always learning at the top of their successful field match, feeding their hearts and minds to keep growing. One of Steve's favorite stuff is that he is lucky, in his place of the highest 1%, one of the most successful in the globe, yet in his future career (enterprise), he is moving to another problem and still learns and grows.

"Take a look at your day in detail. Make a 30-day commitment, and your lives can change. "Steve's main hack of existence is favorable self-representation. He advises you to "speak about yourself daily. Say in the afternoon,' Today will be a great day, and it may be because.'. Visualize your day in detail. Commit yourself to the practice over thirty days, and your lives may be amended. It completely visualizes every problem. Does that? Does that? No, not so. But it can make a huge difference with the sight. "You must see in your life stuff occur before you can figure it out. It is viewing. I have discovered that as an athlete, I have never been to the business industry, asking what you do not want. Because once you have these negative thinking, they can express a lot.

"If you are in an excessively good position, your percentage of chance to punch the excellent punt is much higher, because your unit of area wonder what you want to do and what you don't want to do or do. Steve claims his strategy to life is to imagine and schedule and take benefit from Monday's impetus. "Your brain is the strongest muscle that you have."

"Every Sunday, I write down the objectives I want to achieve in every aspect of my life; I have prepared to achieve some incredible stuff. For the week, I map out and split it into days, my vision, and my achievements. I assault the most on Monday. On Monday, as a consequence, my mentality, my productivity, my effectiveness lays the tone for the rest of the week.

Self-Scouting

Steve speaks about the significance of "self-scouting" or thinking about your day to look at videos of observation when a match is like a sportsman. Identify what you were likely doing well and identify objects that you did not lock and freight for the following day.

Propensity Formulae

Steve's prosperity recipe identifies what you see, keeps your concentrate, trusts your strategy, works every day and is diligent, coherent, and good because you've got your view wherever you want to be. A consistent job is the most important part of building your vision.

"For me, prosperity is a mixture of health, prosperity, appreciation, and charity, and that I gravitate towards individuals like this because I want {I want need} to do so, and I want it to share that further with other individuals." Daily Investment Steve suggests that there are everywhere dreamers. You plan to take these daily decisions, which will complement your interest over time, to get closer to your objective. People have a true disadvantage in choosing what they need ten years from what they need now.

Make it a matter of importance to encourage and educate you about beneficial people from the UN organization.

Impact Steve claims he now wants to make a difference to people. "You will live." You will die. But what region of your lives will you do together? You cannot bring it with you. You cannot take it with you. What will you do between now and when you leave? Will it affect individuals? "That's Steve's thinking:" I plan, or I am going to know, I plan to fight, I will learn. Any error you make, any weakness you have. I don't believe in failure; I just believe in ways to develop. Your zone unit does not plan to lose, and else you plan to gain. You plan to know. We can all learn something from Steve's positive winner's thinking; below is a guideline for a positive attitude:

1. Question yourself as to whether you think positively

The fear of defeat cannot stop you. Ask yourself,' Do I think so?' Don't you know if you are a negative man or not? Take this wellness questionnaire, which will not only give you a positive rating but help you to determine the reverse abilities, which can best enhance your happiness and well-being. If you want to operate on performance, continue reading if somebody is a UN organization.

2. Build up your favorable data memory

Were you aware that you could be willing to improve your performance by merely storing favorable word lists? When you push your brain to use beneficial phrases often, you create these terms (and their fundamental significance) much more available, linked, and easier to use in your brain. Once you get a phrase or schedule from your memory, beneficial people can simply go back to the primary. Don't you know the word positive area unit? Thousands of phrases have been tightly evaluated by psychologists to work out how good and negative they are.

In the useful term manual for adolescents and the positive textbook for kids, I only collected useful phrases. Try this original approach if you are disturbed to think it is good. It can help your brain grow in the respect that may promote the implementation of reverse positive thinking.

3. Enhance the capacity of your brain to function with positive data.

When your brain is building strong neural networks, attempt extending these networks by requesting your brain to use beneficial information in fresh respects; for instance, you may use favorable phrases AND warn you, in inverse order, an hour ago, to recall these phrases.

Or you can write those phrases on cards, and you can split them into two pieces, stack them all, and then you can realize the game for each card. For your brain to fit the term parts, you have to look for innumerable beneficial data to look for the data you are looking for. For instance, "laughter" would be diverted into "laugh" and "her" This beneficial reminder job can help you generate it when you attempt to assume that it is good.

4. Reinforce the capacity of your brain to take care of beneficial issues

Do you think of yourself as one of those people who notice the dangerous thing — such as once someone has cut you off or your food doesn't fit as much as you wanted? Then you taught your brain for the negative, and your brain became very sensitive. This coaching can be incredibly hard to reverse. Train your brain to focus on the good even better. Just concentrate on useful information and turn your attention away from the negative. Do you need to make the benefits more attentive? Examine those matches of quality.

5. Make sure you have random value moments.

Did you know that you are simply going to make a difference? You most probably learned about the research of Pavlov's dog if you ever took AN intro to the science class. This is a fast update: Pavlov had a dog. To inform his dog that it was nearly feeding time, Pavlov would ring a clock. As with most animals, once he was about to get fed, Pavlov's dog would be thrilled. So, he would drool throughout the site. What occurred? What occurred? Well, suddenly, even though food isn't a donation, Pavlov's dog just began to be thrilled. Food and the bell's noise were combined in the brain of the dog. The dog was presently thrilled by something so nonsensical as a bell.

This impact is referred to as conditioning. The first reaction evoked by the second stimulus (Food) is now evoked by the First Stimulation alone (Bell) in the idea that two stimuli zone units are constantly combined. This is without the U.S.A. knowing it all the time. Some items that we eat as a child with our parents are, for instance, the favorite meals for several people. The favorable emotions of being with the family and the particular meals in our brain have probably occurred. As a consequence, we tend to have at present the warm, lukewarm feeling that we get with family from the disbursement of our meals alone, even if our family is not present when we consume it.

Though you acquire a certain atmosphere, you can always respond in a way that allows you to use classical purchases to enhance your performance if you understand what you are doing, what Pavlov did you do precisely. You just connect dull stuff again and again (such as a ringing bell) with beneficial ideas and emotions. Very quickly, this boring stuff will mechanically produce value. This is the traditional purchase at the job. This can help you think you have these little beneficial times that maintain you energetically and reasonable as a consequence of once your region has been moving about, perhaps even feeling tumbled by pressures or difficulties.

6. Think good, but not too abundant, and once you should believe negative.

Naturally, positive thinking has its advantages. But positive thinking is not always the most efficient answer. Edges are sometimes also negative ideas.

If we tend to be dissatisfied or grieving in the region, having adverse ideas and displaying the feelings created by those thinking enables America to transmit to others that they tend to be supportive and kind. Our thoughts can inspire the US to act, change our lives, and change the planets if we're treated under the belt and become angry. If these negatives are casually pushed aside without serious consideration of their roots, these effects will be negative. Then question yourself; once you are specialized in the negative, does this adverse emotion lead to an intervention that will improve your lives? If so, then maintain it. If not, operate to modify it.

7. Practice Feeling

There is a unit AN of endless things about which to be upset, sad, or worried. But the reality is that there is also a range of items that make you feel passionate, happy, and enthusiastic. We must decide to specialize in this. It is up to us.

One way to train your brain to get a beneficial impression. Gratitude, for the people, stuff, and feelings, is once we tend to feel or special gratitude. If we tend to feel particular at the job, we tend to earn the consideration and friendship of those with whom we operate. If we tend to thank our associates or colleagues for our region, they are friendly and helpful to the United States. When we tend to thank the unit for the restricted stuff that is going on in our everyday life, we realize many of that in our life. Do you have to create a routine of sensation? Try to follow the sensation of these five directions.

CHAPTER 54:

Minimalism for Anxiety Relief

Never allow little things to occupy your mind; live a simple life.

Exploring Decluttering for Relieving Anxiety

Minimalism: What does it mean?

Minimalism refers to a conscious situation or condition of having lesser belongings. A minimalist sees minimalism as a conscious condition, which is often driven by purpose, intentionality, and clarity.

Minimalism helps us to reduce our needs only to things that are of the greatest value to us. Additionally, it helps to rid ourselves of those things which are of less importance and less value to us. This type of life helps us to make decisions intentionally.

How to Become a Minimalist?

Does minimalism help us to relieve anxiety? What gives birth to anxiety? Some of the things that contribute to this condition include overthinking, excessive worrying, and stressful life situations or conditions. When you feel anxious, you also feel overwhelmed, and your mind may feel cluttered. And when your space is now clustered, filled with numerous unorganized tasks and projects, focusing on your life and relaxing becomes impossible. By definition, minimalism is a way of life that focuses on intentional decisions and is designed to eliminate clutter and distractions.

An anxious life is filled with anxiety, worries, lack focus and purpose, and delivers excessive stress. Anxiety will negatively impact your life, mood, and choices.

Attaining Anxiety Relief by Keeping Clutter to a Minimum

When you are cluttered physically, it affects your mental life. Our mental space is filled with records of belongings we have, visible or not. The organization of our space and possession is so important to a human's mental health that we dedicate time to decorate, organize, and clean at every possible opportunity. With more belongings, you are filling your life with extra tasks and decisions. If your wardrobe is getting filled with clothes, you start to have issues deciding which clothes to wear when going out, and you waste time on making such decisions. Do you keep gifts that you don't need? You may think that getting rid of such a gift is bad, but if you never end up using it, it will just add to the clutter in your life. With such clutter around you, your mind has more information to process and store. As individuals, we have different tolerance levels for clutter and cleanliness. But no one feels bad when their space is clean, clear, and organized. Try cleaning the clutter in your life, and you'll find some relief from your anxiety.

How to Reduce Physical Clutters

The first step to reducing clutter is by making a list of all your belongings and going through them all. This might be a huge task, but it is worth doing. It is not a must to complete this task in a day, but make sure you build that list. Your goal is to identify whether you love that item or not. When you're done with the list, it is best to hold on only to items that you treasure: the ones you love to own, not the ones you dislike or never use. Use these four categories to assist you with your decluttering goals;

- Keep
- Donate
- Sell
- Discard

You can donate some of your used items to thrift stores or organizations around you that will give them to people who are really in need. You will be happy, and the receiver will also be happy. You don't need to push them all out at once, do this bit by bit, and see the joy that comes from giving.

When you clear clutter from your life, you will eliminate some stress and anxiety, and when belongings around you include only those items which bring you the greatest value, you will feel more joyful, fulfilled, and happy.

Decluttering Digitally

9. How to Reduce Digital Clutter

Have you ever heard of digital clutter? This type of clutter often includes your email and the files on your hard drive or your Google drive.

They can be just as overwhelming as physical clutter. When you have numerous emails to respond to, you might also feel overworked.

When you go through them and get rid of the ones that are not useful, and when you organize the ones that you need, you will feel a measure of joy and satisfaction.

You don't need to address everything at once; you can just take 15 minutes out each day, week, or month to delete them bit by bit.

This will result in getting rid of all your digital clutter.

You will spend a healthy amount of time to get rid of this clutter; it is not an easy job.

But how do you prevent clutter from building up again?

Here are some tips that will help you do this effectively; they are simple and straight to the point.

Tips to Maintain a Clutter-Free Space to Relieve Anxiety

10. The Number One Rule

Here is a 1-in-1-out rule you should follow. For every new item that you bring to your home, try to take one old item out. What makes up the clutter you once battle with? Clothes? Electronics? Whatever it is, always get rid of one when you buy another. You don't need to waste it; you can donate old items.

11. Build Space for All Items

Try to be organized. If you are organized, and you have an efficient way of storing items, they won't turn into junk. You can find shelves or boxes and keep your essentials there. But if your belongings are left in random places, even the important ones will become junk.

12. Ensure That Your Surfaces Are Clear

Another thing you must do is always to keep your surface clean. If you put one item on the table, it can become a natural place to keep other items. Instead of doing this, it is best to put all items in proper places to keep your surfaces clear. If you keep your surface clear, you easily get rid of clutter in your home.

13. Regularly Organize Your Paperwork

Is your clutter filled with mail and other paperwork? One thing you can do is to file your mail and paperwork in folders (digital or paper ones). Put them in places so that you can easily access them when you need to. This will help prevent important files from becoming junk.

14. Build an Outbox

This is not your email outbox. This should be a box or carton that should be in your garage or entrance where you keep items you will be donating or ones you will be getting rid of.

If you are not sure whether you would like to give something away or not, you can keep it in this outbox. If you don't miss it while it is in the box, it is junk; get rid of the item.

15. Buy Intentionally

Are you struggling to get rid of the clutter in your life? One other important thing that you can do is to become intentional with the things you buy for your house. Don't just buy things because you have the money.

Always ask yourself the following question before buying anything:

- Do I need this item?
- Do I love this item?
- What do I need this item for?

If you can't find answers to any of these questions, or if you realize that you don't like the item, don't buy it. You should buy only the things you need and love. This will prevent your home and life from becoming cluttered with items that you don't use.

Do you have roommates or a romantic partner that you purchase things with? If you make purchases together, talk to them about things you will be buying and the reasons you have for wanting to buy them. You should be the one to make the final decisions about the belongings that you choose to have in your personal space, though; no one should make this decision for you.

Some people may think that a minimalist home is just a white-walled box containing furniture and no decorations. That's far from the truth. Being a minimalist does not mean you get rid of all the things you need in your life. Even if you need to stay happy and reduce overthinking by getting rid of things which others see as important, do it.

Create a healthy space for yourself: a space filled with essential things that you love to see, things that make you happy. This will reduce your anxiety dramatically.

CHAPTER 55:

Ways to Attract Good Energy

Never expend your energy on worry; rather, use it to develop yourself and live a great life.

Positive energy can improve how we feel and communicate with the people around us. In our daily dealings with other people, we receive the kind of energy we send out. This energy is usually within our entire body, spirit, and mind, and, when it vibrates out, it's usually felt by others around us.

The way we feel about the people around us is a result of the kind of energy we carry around and the energy that we pick up on from them. We may feel free and cheerful being around some people and feel awkward and cold when we're around other people. Maintaining positive energy will improve our total well-being and help us to communicate more positively with people.

On the other hand, negative energy negatively affects our entire well-being due to the feelings of resentment, discord, and unhappiness that accompany it. So, your ultimate goal should be to resist negative energy and embrace positive energy.

You can achieve this by increasing your energy level and surrounding yourself with positivity. Here are nine daily ways to help you boost your inner vibration and help you to feel the energy flow around you.

Pay Attention to the Energy You Release

If you're releasing lots of negative energy, there's no way you'll attract positive energy. How others feel when they are with you tells a lot about the kind of energy you discharge. Do people feel calm and happy or gloomy and sad when they're with you? Your answer to this will help you know if you have to work on boosting your energy or not.

Negative energy will always impact your relationships negatively, and your attitude towards others is a reflection of who you are. Ask yourself: What kind of impression do I make on people?

If you're the type that always reaches out to people and creates great relationships, you may be releasing positive energy. If you're the type that people avoid, you may be releasing negative energy. Therefore, you must focus on emitting positive energy.

Change the Way You Think

If you spend most of your time thinking about negativity, you'll become a pessimist in no time. But if you spend most of your time thinking about the positive aspects of your life, no matter how hard that can sometimes be, you'll easily attract good things. Always ensure that the positive thoughts guide you in all you do.

If you're battling a bad situation, resist the urge to slip into pessimism. Instead, tell yourself that it's only a phase, and it will soon pass. Always engage in positive affirmations, especially when things go wrong. When you receive bad news, try not to dwell on it or catastrophize. Replace negative thoughts with positive ones. Let the inspiration for your actions come from positive and realistic thoughts about yourself.

Discard Negative Influences

Quit surrounding yourself with negative people, things, or places will take away your happiness and total well-being. Some people are toxic, and you should be far away from them. These are people that always try to discourage you from everything that you do and look for every means to bring you down constantly. If you're not observant enough, you may begin to pick up bad habits from these toxic places, people, or things.

When you disengage yourself from these negative influences, you'll be able to design the kind of life you want for yourself. At times, disentangling from these influences may seem difficult because they're a part of your daily life. If this is the case, avoid them at all costs and prepare yourself mentally if you cannot avoid running into them.

Increase Your Circle

As you discard negative influences, increase your circle of positive influences. Surround yourself with people of like minds that can influence you positively and inspire you to be the best you can be. Ensure that you hold these relationships in high esteem and nurture them.

These people should be able to be honest and authentic with you, but it shouldn't be done to spite you or make you feel less sure of yourself. The positive energy that radiates from this group will help you live a happier life.

Be Kind and Compassionate

Some little acts of kindness can have a significant impact on the receiver and the giver. Being kind and compassionate towards others has been proven to attract lots of positivity and good relationships. So, the more you give and show compassion to others, the better your physical and mental well-being will be.

Being kind is also a great way to motivate the people around you and inspire them to be kind to others. Smiling to people around you, serving someone a cup of tea, or doing anything that makes people around your happy sends loads of positive energy to you, and this boosts your inner happiness.

Be Grateful

Each day, if you dwell too much on negative thoughts, you'll find it hard to see the things you ought to be grateful for. Devote most of your quiet time to thinking about the little things in your life and be thankful for them. Doing this will help you let go of harmful and toxic emotions.

Think of the good things and people in your life and why you are grateful for them. Doing this for a few minutes every day will help you a great deal. If you can't think of any right now, you can begin by keeping a gratitude journal and jotting down a list of things that make you happy and feel contented. Being grateful will help you reflect on the bad times you've had and how you overcame them all.

Discover Your Inner Strength

Taking your focus away from all the negative thinking that may erode your confidence level and cause feelings of insecurity and self-doubt and shifting to positive thinking that boosts your self-esteem and confidence is essential to developing inner strength. Inner strength is what will make you resilient in the face of stressful situations and help boost your energy level so that you can handle whatever comes your way.

Align Your Current Self with Your Future Self

The things you spend your time and money on will determine how far you'll go in life. These choices you make today will shape your life tomorrow. Ask yourself: What do I desire most in the world? Work towards being the person that your future self will be proud of by building healthy relationships and a healthy lifestyle.

Develop a picture in your mind of who you would like to be in the future and start taking steps to make it a reality. Doing this will help you exert more control over your life, and the more positive actions you take, the more positive the reality you will create for your future self.

Actin Good Faith

There's a general belief in business that both parties act in good faith as they work together. We all benefit from treating one another fairly, but only a few people understand that this principle should be followed as we interact daily personally or professionally.

Endeavor to be nice to everyone you meet and treat them with respect, and in most cases, you'll receive the same gesture. Even if someone wrongs to you, avoid retaliation as it won't make you feel better. They may be having a bad day and react negatively to you for this reason. So, when you act nicely to everyone, even when they react harshly to you, you can be sure to attract positive energy, and this will help a lot

CHAPTER 56:

Applying Positivity

The thing about positive thinking is that it is contagious, just like negative thinking. Meaning, when you are around a positive person, you can take on this "vibe" or energy and become positive yourself. Positivity affects more than just you; it affects the people and environment around you. For example, if you were to go to a job interview and you showed up with a confident, positive attitude, then the employer would be more inclined to hire you. If you showed up tired, hungry, or exhausted, then this would show in your attitude, and you wouldn't be able to put your best foot forward. The employer would most likely turn a blind eye to you and hire the next positive person that came for an interview. It's simple - positive attracts positive, and negative attracts negative.

It has been proven that our brains change shape and form, depending on how we think and live our lives. With that being said, what's more, interesting is that when we repeat habits, thoughts, and behaviors, we are training our brain. We can train our brain to act and behave in any way we want because when we repeat things, our brains connect synapses that weren't formerly there and then associate these thoughts with behaviors, turning them into habits. So it makes sense to say that when we think negatively, we are repeating bad thoughts to ourselves. While our brain associates the negative thoughts with the behaviors that we act on, we then continue to repeat bad habits. We can do this with positive thinking, as well. Ever heard the saying: "Life is what you make it?" This is true because when we implement negative thoughts, we act, see, feel, and apply negative habits. However, when we repeat positive things to ourselves (even when we don't believe them), we start to see, hear, think and apply positive behaviors.

The reason negativity is mostly seen in this generation or society is because negativity is addictive. It's hard to escape, and once we think negatively, we cannot stop as it acts like a drug. We do these things because we don't like to take the blame; instead, we want to blame our negative thoughts on why we are depressed. We blame our worrying about why we are anxious.

We blame our overthinking for our actions. It's a hard truth to accept, but the only person to blame for your negative thinking is YOU. The thing is, change isn't easy. What's easy is what we continue to do, which is familiar to us. So it's no wonder we don't just wake up one day and say, "Hey, I am going to be positive today." But that is the answer, it's your choice to wake up and be positive, and it is that easy. However, what is not easy is continuing to do something new and different. This is why changing and rewiring your brain to be positive takes commitment and dedication if you truly want to escape the nightmare of negativity you have been living in.

How to Think Positive

When you develop and improve positive thinking, it goes beyond just what you think about that makes you smile. It becomes your environment. It becomes who you are as an individual. Positivity - just like negativity - consumes us. It can be very difficult to think positively when having a rough day or when everyone or everything around you seems depressing or worrisome.

But the truth is when you think positively, your aura and mind stop looking for the bad in every situation, and you become grateful for these hard days and extreme failures because they shape your destiny. In every horrible scenario, there is good that you can take from it.

At first, seeing the positive in situations can be very difficult, but over time, it will become so easy that you won't even need to think.

The positive will just be there.

So, how do we do it? Here are four ways you can develop positivity in your life:

1. Focus On Three (Or More) Positive Things Daily

Before going to sleep at night, rehash your day in your mind. Think about everything that happened and take three positive perceptions away from the day. It could be anything. Was the sun shining? Did you reconnect with an old friend? Maybe your boss or coworker wasn't that grumpy today, which made it a less stressful day. The more you start seeing the small positive effects, the more your perception of positivity will develop, and the quicker happiness and success will come.

2. Do Something Nice For Someone

It may not seem like it, but acts of kindness can not only lift your spirits but lift someone else's spirits as well. When we do nice for others, we are feeding our soul with positivity because those chemical endorphins shoot off in our brains as a reward response. These acts could be anything, such as smiling at a stranger, waving to a coworker, or pausing to do something thoughtful for someone you know. When you make someone else smile, your heart smiles, which makes you feel better about yourself and develop confidence.

3. Be In The Present

If I haven't said this enough, then let me say it again: Be mindful! When we stay in the present moment, it creates balance and structure in our awareness of what is going on around us. When we become aware of our surroundings while staying in the present moment, we will be able to better pick up on the positive things that happen, and negativity will seem like a distant friend.

4. Practice Self-Love And Gratitude

The thing about positivity is that when you love yourself, it becomes easier to help others and give back to the universe. Just think about it - if you don't love yourself, then your relationships fall apart faster, your job never seems to feel satisfying. You constantly second-guess your ability to handle stressful situations. However, when you do love yourself, you can be thankful for what you have because you have it. You won't be asking for more or for things that you don't have, and envy or jealousy won't seem like important things to worry about anymore. Being grateful for the human, you have required self-acceptance and a deeper understanding of what you want in life. So, any chance you get, be grateful for what you have rather than envy what you don't have. The grass is rarely greener on the other side.

Changing Your Mood

Most of the time, we get stuck in negative thought-patterns because our moods are dark. It's a cycle - negative or worrisome thoughts bring on bad moods, which bring on perceptions of more negative outcomes, which then makes it hard to make important decisions because our minds are crowded, which then leads to overthinking (or negative thoughts), and so on. Some days we don't want to get out of bed, and other days we are motivated, producing "feel good" chemicals that result in getting more done. On the days you feel down, stressed out, anxious, or depressed, think about the productive days and try to draw from that energy. Also, sometimes it is okay to give in to your dark mood. Just try not to sulk or make it a daily habit. Here are ways you can change your dark mood to a lighter one when you feel stuck in the mud:

1. Get Exercise

We talked about this already as well. When you work out, those "feel good" chemicals released in your brain and can change your mood instantly. Also, it is a good distraction from your bad mood because instead of focusing on what got you so upset, you can focus on other things, like the scenery or your breathing. Make sure to drink water while you work out as being dehydrated can make you feel worse.

2. Listen To Or Watch Motivational Material

When you don't feel like moving or getting out of bed, and it's just one of those days, then watch an inspiring movie or listen to an uplifting podcast. Even though we tend to listen to music that matches what we feel in our down moments, ignore this urge and do the opposite - crank some happy, upbeat tunes. Who knows, it may even want to make you dance or sing. It will lift your spirit 60% faster by listening to or watching motivational material over listening to or watching negative, depressing material. Interestingly, when we listen to what suits our mood at the moment, we are training our brains that these attitudes are okay, and then we find ourselves falling deeper into the negative cycle.

3. Change Your Body Language

This means that you should act and behave the way you want to feel. So, if you want to feel confident, then prance around the house in the sexiest or wackiest thing you have and pose in front of a mirror with your chest puffed out and your back straight. If you want to feel relaxed, then throw on your comfy clothes and lounge around, but be mindful of what you tell yourself. Force yourself to smile for 60 seconds, and I guarantee your mood will lift, even if only slightly. Don't let negativity consume you; break free by being you. Be funny, laugh, tickle yourself, talk to someone about your aspirations and dreams, or do whatever you need to get out of the funk you are in and into the mood you want to have.

4. Be Grateful Or Have An Appreciation For Everything

Here is a weird, funny fact: We find it normal when someone goes around and complains about everything. We listen to our friend's vent, our parents bicker, our bosses complain, and even strangers are arguing with themselves sometimes. It is "normal" to listen to someone complain and bicker about things, but wouldn't it be weird if we heard someone going off about how grateful and appreciative they are about everything? How often do you hear someone say, "It's raining outside, and I am so grateful for the rain," or, "Food is often taken for granted, so I just wanted to take a moment to feel blessed for this food." Have you ever heard someone say, "I appreciate that my kids scream and give me attitude because it means they are growing human beings?" No, you probably haven't. Imagine if you said out loud everything you were grateful for today, everything you appreciated today and even yesterday. Imagine how you would feel and how you would make others feel. You may even have a good laugh, but isn't that the point? Practice this.

5. Force Positivity, Even When You Don't Feel Like It

The truth about your thoughts is that they do not control you. This is the same with moods; they don't control you. So when you have a hard time practicing or enforcing the previous techniques, just do it anyway. Force yourself to smile, force yourself to get out of bed and dance, force yourself to feel grateful. Once you get up and force positivity into your day, you are taking control of your surroundings and your behavior. This teaches your brain that, even in down times and dark moods, you are in control of how you react to them, creating positivity and healthy habits.

CHAPTER 57:

The Pitfalls of Overthinking

The process of rewiring your mind requires that you avoid negative influences, to keep your mind fresh and relaxed for optimal rewiring results. One of the biggest negative influences that can obstruct your progress is overthinking.

Overthinking is different from thinking things through. Some certain decisions or actions need to be thought through very carefully – in fact, not thinking them through could have hazardous outcomes. For example, say your company offers you the position of heading a new regional office, which would require you and your family to relocate overseas for three years. That is the kind of decision that requires hours, if not days of pondering, and this is perfectly healthy and desirable.

But spending 10 minutes trying to decide what to have for breakfast, which ties to wear, or spending hours comparing goods before making a purchase is, in the larger scheme of things, pretty trivial and a waste of time. Overthinking also includes running past conversations or encounters in our heads over and over again, and looking for hidden signs or meanings, or thinking that we should have said this rather than that. It also includes envisioning future scenarios, all the things that could go wrong, what you should say in case of this or that… in short, overloading your mind with useless thoughts that have no; logical basis. You can't go back and redo what has happened, nor will worrying or overanalyzing gives you the ability to control what happens in the future.

Keep in mind this simple equation: Overthinking leads to worry. Worry leads to stress. Stress causes a stress response and panic attacks. Moreover, overthinking can lead to crippling uncertainty, where you are unable to decide on even the simplest thing, and you lose control over your life.

You are overthinking leads to two types of destructive thought patterns: ruminating and worrying.

1) Ruminating involves rehashing things over and over again, thinking:

"I shouldn't have said that; that made me look really bad..."

"Maybe If I'd dressed more conservatively, I'd have gotten the job."

"I should have done this instead. Now I've blown my chances..."

"I wasn't clear enough. He/ she probably misunderstood what I meant..."

"That email wasn't phrased well. I should have mentioned..."

No further explanation is needed here. This thought pattern is extremely detrimental to anyone who gets caught up in it. Also, it's usually not accompanied by positive self-talk like, "We all make mistakes, I'll make sure not to do that again in a similar situation."

2) Worrying is a form of overthinking where your mind leaps ahead and imagines failures or disastrous outcomes to things that haven't happened yet, such as:

"I haven't studied well enough. What if I fail the exam? I won't be able to graduate this year..."

"The weather's bad, and it's getting dark. I'm worried my spouse will have an accident driving home..."

"My doctor says the surgery is simple and safe, but what if I have complications? What if they get worse and I die? "

Worrying and ruminating are two thought patterns that bombard your brain with a constant stream of unhealthy and unproductive thoughts.

If left unchecked, you will be dragged into a vicious cycle that will eat away at your very sanity.

Some pitfalls of overthinking

- Impairs judgment because it keeps your brain. in a constant fog of fear and indecision.
- Elevates stress levels.
- Creates chronic indecision, which can destroy self-confidence over time and cause depression.
- Burns up mental and physical energy to the point where you can either get physically sick or develop a mental disorder.
- Leads you to mistrust feeling happy.

Stress Response and panic attacks are a direct outcome of overthinking

The stress response is also known as the "fear and flight" response. It is vital for our safety in times of real danger, such as a shooting, God forbid, or some other violent situation. The stress response releases certain hormones in our body (including the well-known adrenalin rush) that trigger a heightened pulse and heart rate, dilated pupils, shaking, and intense fear and alertness. Having a stress response when there is no real danger that can wreak havoc on your mind and body.

A panic attack triggers the same response when constant anxiety builds up.

Stress response or panic attack is followed by what is called the "relaxation response" when your body calms down, and its functions return to normal. This commonly takes between 20 to 30 minutes. Every time you have a stress response or panic attack, you are pushing your mind and body to extremes that, over time, will take their toll on your health. It's no surprise that stress is the number one cause of high blood pressure, stroke, heart disease, diabetes, and many other serious illnesses. This alone should be a huge motivator for rewiring the mind.

12 Tips for Rewiring your mind to stop overthinking

Overthinking is a habit, and as we have learned, habits are hard to break because you have to train your mind to abandon the old neural pathways and create new ones. Again, constant practice is the key here. Use the following tips help rewire your mind for overthinking:

1. Catch yourself doing it:

Be on the lookout! When you find yourself overthinking, stop immediately. Go and do something else that will divert your focus; solve a crossword puzzle, meditate, do yoga, or read a few pages of a book. Choose an activity that will allow your brain to stay focused on the task at hand and not start to wander again to that phone call you had last week or the meeting with your boss next week.

2. Schedule time for thinking and decision-making:

Write down any problems you need to solve or the decisions you need to make. Schedule a specific time to sit down and think calmly and reasonably. Stay alert. If you catch yourself overthinking, stop yourself immediately.

3. Set deadlines for making decisions and stick to them:

Depending on the issue, you can allow yourself 10 minutes, half an hour or an hour. Be firm with yourself and stick to the deadline. For example, you've been asked to coordinate a bake sale for your local church. Realistically, you would need no more than 5 minutes to review your schedule, decide if you would enjoy doing it and reply with a yes or no. On the other hand, if your boss offers you a promotion which would mean more money but longer working hours, you may need more to weigh all the options and decide if it worth it. You may even need to sleep on and see if you still feel the same way the next day.

4. Practice mindfulness:

We will discuss mindfulness skills in another episode because they are crucial tools to keep your mind focused on the present, rather than overthinking abbot the past or the future

5. Accept that you are doing your best:

Affirm to yourself that you have always done your best in past situations and will do your best in the future. Tell yourself that yes, you have made mistakes, but on the whole, you always manage to do pretty well.

6. Don't expect perfection:

Perfection in anything is very rare, so it's unreasonable to expect things to go exactly the way we want. It's good to be ambitious and strives for the best but not to fret or worry over outcomes that are less than perfect.

7. Don't personalize:

Not everything that happens in your life is necessarily about you. If your boss returns your "good morning" with a grumpy "Hi," don't run to your office and start pulling your hair out, thinking he's planning to fire you. Maybe he has family issues or is going through a personal crisis that has nothing to do with you at all. The clerk at the store who is a little short with you may not be feeling well. Taking everything personally leads to feelings of victimhood, which are very unhealthy. It also makes you insensitive to other people's feelings and circumstances so that you appear selfish and unkind. This is why empath is an important quality to develop through meditation, as we will see later.

8. Challenge your fears:

Just because something didn't work out in the past doesn't mean you're doomed to failure forever. Every new beginning is a new chance to succeed and to shine. Train your brain to see every new beginning as an opportunity, a new adventure that you are excited about and eager to experience.

9. Think of what can go right:

Overthinking is triggered by fear. If you do need to think something out and find yourself going off on a negative tangent, rewire your mind by challenging the thought and then thinking of all the things that could go right. Visualize yourself in that situation and how you would feel. Start to create a new and positive neural pathway.

Put things into perspective: Shut down overthinking by asking yourself how much a particular issue will matter in two years, six months, a month, and so on. Changing time frames puts the issue into perspective and helps you realize that its overall impact on your life will not be devastating. This will shift your outlook and make you feel more positive

11. Get enough sleep:

A good night's sleep is the best "brain balancer" there is. A well-rested mind and body is the magic charm that helps us think calmly, weight options realistically, and see things in perspective. A jittery and sleep-starved brain is more prone to overthinking, anxiety, and worry.

12. Be grateful:

Each night before you go to bed, make a mental list of all the positive things that have happened throughout the day and the things and people you are grateful for. Stop yourself if you start dwelling on the negative and think pf someone you love, your partner or your child, and how grateful you are to have them in your life. Be grateful that they are happy and healthy and that they give you so much joy. You will find that next to that, anything else is pretty insignificant. This exercise will help you drift off to sleep with nothing but positive, loving thoughts in your head.

What Science Tells Us About Overthinking?

Overthinking is a result of growing older and becoming more cynical and disillusioned about life. Scientists have found that when we lose our childhood curiosity and wonder, we become more prone to worrying and overthinking. To a child, everything is wondrous and new; life is a great adventure waiting to be explored, and each discovery is more wonderful than the last. A child's brain makes billions of positive connections when that child is at the peak of curiosity.

We can regain some of those childhood qualities by observing children at play, playing with our children or grandchildren, and relearning to see the world through their eyes. What a fun and easy way to rewire our brain with optimism and joy!

CHAPTER 58:

Self Confidence

Now that you have accepted the fact that self-confidence can be developed, let's look at practical ways you can go about the business of building your self-confidence. I will give you a step by step approach that will give you the results you want if you follow it diligently. Are you ready? Let's go!

Step 1. Determination

How many times have you decided to do something only to go back on your words and your old ways after just a few attempts? It is the same with building self-confidence. Anybody can decide to do something, but only a determined person can keep to their words. This is usually because, as the saying goes, nothing good comes easy. Anything worthwhile you want to do in life will cost you something.

Becoming a confident person will cost you some time, effort, strength and courage, and if you are not very determined, you will go back to your former ways even before you have started. It is a determination that will make you intentionally blind to all the reasons why you should abandon your journey of self-confidence.

Step 2: Learning from Confident People

There is hardly anything you will want to do in life that someone else has not done already, but maybe differently. Even when it comes to learning how to develop self-confidence, you have to watch people too. This is not very difficult, though, and I will show you how to do it. I believe you have role models and people you admire and wish to be like. Good. You can start observing their lifestyles within a closer range now. Study their biography, read their stories, follow them on social media, listen to them. Their lifestyle will gradually start rubbing off on you, in the sense that you will begin to learn to face life with the same level of confidence they exude. This does not mean that you should copy other people's way of life. A self-confident person does not imitate other people. The difference here is that you are trying to get started, and you need some kind of direction. While you observe the lifestyles of the people you admire and become used to how confident they are, your subconscious mind will get the message and start exhibiting confidence as well. So, do not imitate; just watch. Learn, and grow your style.

Step 3. Attacking your Fears

Anybody who lacks self-confidence misses many opportunities in life because they do not believe that they have what it takes to go out there and do stuff. The only way to face your fear to do it head-on, and until you start confronting those fears, you will remain at the same level that you have always been. To grow, you have to act, and you have to act now. What are the things that make your heart skip anytime you remember them? What are the things you should do but have not done because you think you are not good enough? You have to get up right now and start doing them. You may have heard that courage is not the absence of fear, but the ability to do what you are supposed to do even in fear. You do not have to wait till you have zero fears because that time will never come. Nobody has zero fears at any point in their lives. Even your role models and great public speakers have their fears, but you hear their names today because they do not allow their fears to determine their actions. For you to become a confident person, you need to get up and act on your fears. They will always be there, anyway, so do not let them stop you.

Step 4. Constant Practice

Anything you repeat over and over again becomes a habit with time, and you can start doing it without having to think about it. The same goes for developing self-confidence. When you have acted on your fears, you should repeat your actions over and over again until they become a habit. Generally, if you repeat the same thing for at least 21 days, it will become a habit. Make a list of the major areas of your life that you are lagging because of the lack of self-confidence. Pick the items on the list one after the other and practice each of them for at least 21 days. You'd be amazed at the result you will get. Repetition makes for emphasis, and when you emphasize your abilities over your fears, you will push the fears down and be in charge of your life. Conquering a particular fear once is never enough. You need to conquer it again and again and on different occasions until you believe beyond doubt that you are capable. If, for instance, you are scared of speaking in public, there are things you can do to bring yourself out of that fear and build the kind of confidence you need to make things work for you. You could start by standing in front of the mirror in your room and speaking to yourself alone. When you know that no one else is watching and you are not scared of being judged, you can freely try your best. When you speak, turn on the voice recorder on your phone, and record your voice so that when you hear the sound of your voice, the fact that you were able to speak successfully will be a push for you to continue. fter perfecting the act of speaking in front of your mirror, you can use your video recorder to record the performance, which you can then post on your social media channels. You should feel free and remember that you are not participating in a competition or anything; it's all for the fun of it. When you get feedback from your online friends, you will know your areas of strengths and weaknesses and improve accordingly. If you keep doing this, you will no longer find public speaking as terrifying as it used to be for you. And after you have made your first actual speaking engagement fearlessly in the presence of an audience, do not relax and think that you have won the battle. You should search for other speaking engagements so that you will have more opportunities to keep the fire burning.

Step 5. Persistence

As I am showing you things you could do to build your self-confidence, they may sound very interesting to you, and you might want to jump right in and start trying them out. That is fine, but there is something else you should know, too. It is never easy to change from one lifestyle to another. After becoming determined to be a confident person, learning from other confident people, attacking your fears, and repeating the process through constant practice, you should be persistent in all these and know that even when you feel like throwing in the towel and giving up, you can continue in your practice. It is not easy, and there are times when you will feel low. But you should be persistent enough to continue and make it work. Every successful person you know had to pass through a period of persistence in little actions, which yielded the big success you see now.

Step 6. Positive Affirmations

In order not to give up on your journey of self-confidence halfway, you need to have constant reminders in the form of positive affirmations. Keep telling yourself that you can do it. The words we speak to ourselves are powerful, and they get registered in our subconscious mind with time. Write down some things you want to see manifest in your life and speak them to yourself every day while taking the necessary actions to bring them to reality. You can write these positive words and place them at strategic points in your room or in your office, where you will see them every day to remind yourself what your goal is.

Confident Affirmation

I believe in myself and in my ability to do whatever I put my mind to. I will keep working on myself until I have no fear left in me. I am the master of my own life. I have confidence in myself.

CHAPTER 59:

Self-Esteem

We have a higher risk of depression if our self-esteem is small, which is characteristic of codependency. Maladaptive behavior is acquired, and the values and self-esteem and habits that cause low self-esteem and codependency are therefore learned. Self-esteem is what we feel of ourselves. We think of ourselves. Positive and negative self-assessments are included. Good self-esteem is a real, optimistic perception of oneself. This represents respect for oneself and implies a sense of value, which is not defined by contrast with others or approval. Self-acceptance (included by some writers in mutual-esteem) is even more profound. It feels good enough, not ideal, not incomplete. They believe that we are worth and loved, not just due to appearance, talent, accomplishment, intellect, rank, or popularity. It's a feeling of deep fulfillment. All of us have intrinsic value, not dependent on our success or what we do or offer. Just as every baby and race is exceptional and caring, so are we. Regrettably, many of us as codependents grew up in families with no affection, conditional love, or won. We figured we had to gain or win a parent's attention. As a consequence, we hate being honest because of fear that we may be unhappy. We must follow people who can't love or deny lovers. In interactions and at work, we "over-do" and "over-give," and finally feel resentful, manipulated, or abused. If you think that self-esteem is important for living and having healthy relationships, enjoying and enduring, you are wrong! The following bad habits, common to codependents, can make you feel insecure, embarrassed, depressed, sad, and hopeless:

- Negative comparing yourself with others.
- You may find fault with yourself

- Tyrannizing yourself in "Should."
- Project self-criticism with others and believe they judge you.
- Feeling "better than" is a sure way to offset the underlying guilt and
- Low self-confidence.
- The lift we're getting is wrong. It would be better to ask why we have to equate ourselves to someone else.
- It's self-shaming when we compare ourselves negatively.
- We feel lower, we lose trust, and we like less.
- It depresses and discourages our mood.

An active "inner critic" assassinates us with what we ought to and ought not to do and assaults what we have already done. Usual discovery of faults can lead us to believe that others see us as we see ourselves. This way, we project our criticism towards others and expect and feel the effect of criticism or judgment, even if no objection exists.

Reduced self-confidence makes us afraid to make mistakes, to look stupid, or to fail. Our self-esteem is always on the line, and it is, therefore, safer to try nothing new to avoid incompetence or failure. This is another reason why we continue to create unique and strenuous activities or experiences. At the same time, we criticize ourselves for not achieving our objectives. Instead of taking a chance, we are wrong not to try, which means "failure" and low self-esteem.

Adjusting others from a young age leaves us uncertain about our values and beliefs and promotes trust in others. It becomes challenging to make choices, even paralyzing. Low self-esteem and guilt raise our concern that mistakes lead to self-doubt, weakness, and indecision. Slightly, we procrastinate and look for affirmation, feedback, and reactions that further weaken our self-confidence and self-esteem.

Adapting to others also alienates us from our needs and desires. They tell ourselves and others that they are not essential by not understanding, communicating, and satisfying our needs and wants. On the other hand, it creates self-esteem by taking responsibility to meet our needs and demanding (such as requesting an increase). If not, we feel helpless, a victim of circumstances and others.

The respect and love that we deny ourselves makes us vulnerable to abuse and exploitation. We do not feel worthy of well-being handled and refuse to be humiliated or disregarded, justify, or rationalize. Based on the acceptance of others, we are wary of setting limits so that we do not alienate those we love or need. We quickly blame ourselves and accept blame from others because we're guilty of shame. Although we forgive others ' mistakes, mainly when we are excused, we are not as kind to ourselves. Excuses for us don't count. In reality, due to past mistakes, we can punish ourselves and blame ourselves for years.

How Low Self Esteem Affects Our Wellbeing

Self Esteem is defined as a psychological state based on an individual's assessment of its importance. The definition describes a person's point of view as to whether or not they are worthy of praise. Briefly, your self-esteem defines the metric you want.

Low self-esteem is a term that we use to describe people who don't think very much about themselves. People with low self-esteem have a little self-worth measure. You don't feel good about who and what you are. This can create any number of problems for the person experiencing the condition.

Low self-esteem also may be defined simply as "feeling insecure." Low self-esteem can include the following common indicators: relying too heavily on others for decisions and guidance and frequently feeling overwhelmed by normal life pressures, feeling more depressed than others physically: appearance, age, height, weight, etc. Such patterns are superficial, and we appear to associate them automatically with more optimistic or over-trustful personalities. Besides, these measures reflect a kind of over-compensation designed to hide feelings of failure or inferiority. These might include the following: general frustration and a need for a Quick vengeance fuse, prone to regular irritations that quickly escalate to outbursts. Conflictive, immediately confronted, and violently reply, Blaming issue with others. Argumentation over trivial or insignificant issues. Although it is a common condition, and we are all familiar with it, many would identify it as a disorder or disease. This has been addressed, but low self-esteem fulfills many requirements describing conditions. For example, low self-esteem can be uncontrollable; for many, it shows similar symptoms and seems to be related to similar past experiences such as violence in many situations. People with low self-esteem may be more vulnerable to other health problems because of the added pressures which this disorder frequently accompanies.

Something good is that we can boost our self-esteem through rigorous training and disciplined practice. As we regain our self-esteem, we begin to move back into prosperous and more successful, enjoyable lives. We can do many things every day to improve our self-esteem. There are some things to do next.

- Avoid being attacked. Punishing yourself for perceived failures only over time can decrease your self-esteem.
- Work to achieve your goals. Each day, strive to advance to at least one of the goals of your life. Be patient, and devote yourself to doing so every day. It strengthens your confidence and takes you closer to your dreams.
- Please fill your mind with positive data. Read books on self-help. Participate in trust courses. Research and return to the things that inspire you.
- Make an effort. Make efforts. Do try to accomplish assignments and challenges. When you excel, it will strengthen your self-esteem and confidence. It won't matter if you fail, because you know you've done your hardest.

- Emphasize your strengths and note them as often as possible. Use them to deal with difficult circumstances.
- Identify areas of difficulty and bad habits. If you're frustrated, learn to identify it and stop until it gets out of hand. It takes practice to change bad habits but immediate positive results. Think of all the negative situations you can prevent by not negatively affecting the climate.
- Visualize every day, your performance. When you wake up, imagine yourself fulfilling your goals and becoming the person you want to be. This is your inspiration to make an honest effort every day to come closer to this potential reality.
- Affirmations. Comments. Mind your positive characteristics as often as you can. Speak openly to them. Be brave. Be bold.
- Concentrate on growth. Never look back. The past has nothing to do with it. Seek and follow your wishes every day.
- Know how you can quit by taking yourself seriously. Everything doesn't have to be so awful sometimes. Relax, and have fun. Learn to be humble and smile at yourself. We're all wrong. We're humorous sometimes.
- Be yourself. You can do whatever you want, but you can't be somebody else. Learn to accept all your faults and yourself. You can improve them, and you can refine them. You don't describe your shortcomings.
- When you find that you need to boost your self-esteem, hold specific suggestions close at hand. Self-esteem is a vital part of life. This influences our emotions, our way of functioning, and our way of interacting with the world around us. If you acquire low self-esteem, it can affect anything you think about and make you feel negative. It can influence all you do in all aspects of your life. Going through life with low self-esteem is like swimming with the attached links.
- Free yourself from low self-respect, and your life will be better in innumerable ways.

The Mystery Surrounding Self-Esteem

Every person creates self-esteem in whatever fields he is commanding.

If you kill love, you kill the person.

Low self-esteem denies an individual the opportunity to be themselves. Psychologists mostly work with the person affected, trying to mend their damaged trust, but no one tells us that there is more to the cycle of managing this mess than just the people changed.

Low self-esteem frustrates a human being and can lead to depression or, even worse, suicide. Unfortunately, it is not well understood, and more of a personal problem is considered during the development phases. It only becomes a family problem if it has been discouraged.

Without everyone taking personal responsibility for this rapidly growing "low self-esteem" family, devastation can get out of hand more than it does, right from home to other institutions.

Who Is Responsible For Low Self-Esteem?

Looking at yourself, you can feel so confident that you have never contributed to lowering the confidence of someone else! As long as you relate to people every day, you may be part of somebody else's appreciation.

Your closest family members are most concerned since you feel the pain if you lose. Nonetheless, note that what affects your child may be an example of what you did to an alien.

Some acts that reduce self-esteem. As teachers, if you humiliate a poor student at any point, you can trigger a withdrawal syndrome in him and affect him after that. You don't think an adolescent knows how his teacher mocked him for poor performance. And worse, he knows the name of the instructor. You might not be able to persuade him if he is an average student that he can do better.

Conclusion

Decisions must be made every day, consciously or unconsciously. Some we take without realizing it, others we think about many times, and others we feel we never get to take. But we are wrong. Even if you think you don't decide whether black or white, right or left, someone will do it for you. Now you will think: and who will come to resolve for me? It is not that there is someone physically next to you using your words, but the fact of not making a decision is to make it in another sense. I'm going to explain myself.

When we have doubts because we do not know what to decide, we take our time; some people need minutes, others a few hours, other days, other months, and others are just postponing it forever. It depends on many factors, the type of a decision, its importance, the situation, what we risk, the price we have to pay. In short, it depends on each moment and each circumstance. It costs a bit then to generalize. What is clear is that when we postpone something important, we hope that the situation is ideal, that we are prepared, that the time is right, that there is an adequate, total environment, that we continue and continue in a loop that goes around and around. Turns and the situation we have does not change; it is more, many times it worsens because we are still evaluating whether I dare or not,

It has happened to all of us; I think that in a very high percentage at some point, you go through a situation like that. And what result do you get? That you feel frustrated, anxious, and you are lowering the level of trust that you have in yourself precisely because of that same situation. I'm not saying that you have to decide things to the tune, or that nobody squeezes you to do something when you don't feel ready. Still, between taking the time and preferring to stay like these indefinitely (which is not supposed to be the ideal situation and that's why you want to change), there is away.

Has it ever happened to you that you have an idea, something occurs to you, and after a while, you see that someone has had the same or a very similar one and it got ahead of you? Me. When I saw that someone had implemented what had occurred to me, I was very angry with myself for not having decided at that time. It seemed that someone had told him my thoughts. Perhaps not in the same way, but worryingly similar. People who move, who are proactive, who often succeed are because they implement, they act, while others are still thinking about it. They just got down to business and started looking for how they could do what they wanted instead of going round and round without making the decision. I concluded that sometimes the speed in making a decision, apart from gaining time and momentum.

After a few times, I decided that it couldn't happen again, so I don't make decisions lightly, but I don't take long to assess whether something interests me or not. Let's see how we can make decision-making something that flows more and is not a massive slab that we carry for a long time.

Value What You Are Deciding

You have to see if the decision you have to make depends on you, that is, if you have to have full responsibility for it. So, we will have to consider what affects the result and how. What degree of commitment do we have with what we are deciding, are we very involved, do we care much, little, or do we not care? It is not the same to determine what we are going to wear to go to a party or decide how much we are going to invest in our company during the following year. How important is the result of that decision for you? That is an important question. And afterward set yourself most last time, in that period you need to settle on a choice.

List of Pros and Cons

Everything has a risk, and many people do not make decisions for fear of not getting it right and making the wrong decision. So we always run risks, big or small, but what you have to be very clear about is that not taking any it is already a risk in itself, staying as you are or even worse. I recommend that you make a list of the advantages and disadvantages that may be involved in deciding one thing or another, in the case of having two options, that is valid for as many options as we have.

Check Your Feeling

Once you have made the decision, close your eyes, and feel it. It is essential to know how we think about the choice we have made. That does not mean it is a confirmation like: You have made the right decision! You will feel a little uneasy, uncomfortable, and even afraid of not knowing if it is correct, but it does not matter; it is reasonable and necessary to feel this way. The important thing is that underneath all that fear and those doubts, we "feel" as if we had obtained the result we wanted, our objective accomplished, and we are happy that we did it and that it turned out as we expected. If so, great, you have to act despite the fear of uncertainty and failure, but if that fear is so great, it does not make you feel a little uncomfortable, but it is so great that it causes you anxiety or you feel bad, do not do it, maybe this is not the time to do it. You always have to take your limitations to the maximum, to be able to jump into the pool, but if you see that it has no water and is buried, fill it fast.

What "Price" Are You Going to Pay for Not Taking It

You may not see them immediately, but trust me, they do. And he does not refer to price as the economic aspect, no, that is one of them. For now, we were in the same situation we found ourselves in, the one we wanted to change, remember? Because at that moment, it does not seem so bad (well, I stay as I am), but the day's pass, the month's pass and even the years. .and we continue to regret everything that we have ceased to be, to do or to have, because we believe that we remain the same, but what happens is that we continue to be more frustrated, with more discouragement, less security in ourselves for not having done what a day we believed that we could have done and we stopped believing in our capabilities. Then we realize the opportunities that we have missed and everything that we have not enjoyed and lived for not having made a decision. Think about the price you will pay for not doing it.

What You Have to Lose

That point is what we usually value when making a decision. We have to know what risks we are taking. It is closely associated with the previous point. Think: And if it was the wrong decision, what consequences can it have? What is the worst that can happen? Human beings have a natural tendency to always bet on the worst, so it will not be difficult for you. Usually, the worst that can happen is not so severe. I know it sounds a bit frivolous to say that you may think that you lose an investment, money, that it delays achieving your goal, that you get angry, you get mad at yourself, that you lose self-esteem because you think that you no longer know or make right decisions. Many things can happen, but still, you have to see it as a learning, as an experience, as something necessary even sometimes to take the next step, although we value that when we take a little distance. But sometimes daring to take action, decisions make us strong, teaches us that we are capable of many things, that if we have been able once, we will be able to do ten and that something important is that we surpass ourselves every day a little bit. So is it worth taking the risk? Only you can know.

If you still feel insecure, consider asking someone to help you because the support, the guidance, and the fact that they believe in you, even more than yourself, works miracles. The only thing you need is to trust you, trust that you will make the best decision according to the moment and circumstance and that whatever happens, you will live it intensely because this is how we build our life based on passion, effort, perseverance, and commitment.

On the off chance that you need assistance, you realize where to discover me, believe you, and everything will come. Brain science ideas, for example, the law of fascination and self, are additionally investigated in this segment of the book. From that point onward, there is an underlines on the significance of energy and positive intuition to move one's outlook away from a negative and foolish one. The book closes with an empowering message that asks the watcher to act naturally and go with their heart throughout everyday life.

In general, this book is an investigation of human brain science, particularly our frailties, tensions, and worries, just like a manual for fathoming it. It hits near and dear since it handles psychological wellness conditions that a considerable lot of us battle with and endeavors to comprehend it in the advanced world. Eventually, to defeat overthinking and its basic states of mind, one must be eager to change themselves and contact others, in particular. Personal development is something that requires guides and backing to achieve; else, an individual feels segregated, giving them no motivation to leave their brains and become a piece of society.

PART III

HOW TO DECLUTTER YOUR MIND

Introduction

Do you ever feel like if you were to look inside your brain, you would see a big, chaotic, messy pile of disorganized thoughts? When it feels like your brain is being hit left, right and center by a never-ending stream of thoughts and stimuli, it feels like your brain is in serious overdrive mode all the time. Hence, the stress that you feel. Your brain is sending you an SOS signal, waving a red flag frantically hoping you'll read the signs and free your mind. Imagine that your mind is like the closets and cabinets you have in your home. You spring clean those once in a while because you know that if you don't, the mess is going to build up and start overflowing. Well, your mind works the same way. We may not possibly see it physically, but we can feel when our minds are in serious need of some decluttering. Except instead of getting rid of physical items, what you're going to do this time is clear out the excess, non-essential baggage in the form of unhelpful and unnecessary thoughts. Clear them out and your mind will be free to focus on staying productive and motivated.

Say the word "clutter" and most of us immediately think about the physical clutter in our immediate environment. Unless someone points it out, you probably might not have thought about the fact that your mind is capable of hoarding habits too. You'll know when this is happening when you notice the following signs:

- You're ruminating too much
- You spend too much time focusing on negativity
- You're obsessing about what is beyond your control
- You have a hard time letting go of negativity
- You have a hard time letting go of your anger and resentment
- You have a hard time letting go of sadness and any unpleasant emotion that is weighing you down
- Your mental to-do checklist has a lot of unfulfilled goals and dreams on it
- You're easily distracted by your external circumstances
- New sensory stimuli and input makes you feel overwhelmed too easily

Unfortunately, no good can ever come out of a cluttered mind. You're using up too much mental energy and wasting a lot of precious time. A cluttered mind leads to disorganization, distraction, confusion, and makes it hard for you to make productive decisions, maintain a clear focus, sort out your priorities and in general, be productive. Dealing with a cluttered mind is not something to be taken lightly. Let it go on for long enough and you'll eventually lose touch with your present. You'll feel disconnected with yourself and your environment until finally, you begin losing touch with the relationships you have too. It can feel like you're walking around in an unfocused daze, unhappy and stressed, not knowing what you're doing with your life and why.

The only solution to this conundrum is to declutter your mind. The mental habits that you have right now could be stopping you from achieving your full potential. You must declutter to build the mental toughness, strength, and resilience you need to never let yourself be fazed by stress again. Once you let go of the mental baggage that is weighing you down, your mind will finally be free and so will you.

What Contributes to A Cluttered Mind?

The world we live in. That's the main reason why we've got so much clutter going on upstairs. From the moment you wake up, incoming information from social media sites, Google, news channels, content sent to you by friends and family, text messages, email notifications, everything comes at you all at once the minute you open your eyes and pick up your phone. This is a huge change from the lifestyle of several decades ago when people just woke up and enjoyed a few moments of silence before they got out of bed and began their routine. Today, the human brain is busier than it has ever been throughout history. We've turned it into a constant information-processing machine with the relentless notifications and stimuli that keep coming at us.

By the time we leave the house, arrive at work and sit down behind the computer to begin the day, we may find ourselves too mentally exhausted and overwhelmed to concentrate on our job. We're stuck and we don't know what to do about it. Even when we try to focus on our job, we're threatened by distractions all the time. A chatty colleague in the next cubicle. Phones ringing off the hook. Conversations flying back and forth. The frequent beeps of your phone, indicating a new notification has just come in. Emails that keep "dinging" in on your computer. Trying to fight it and hold onto your focus is exhausting in those conditions, even worse when you're mind is more cluttered than it should be. Outside work, a lot of factors try to pull our focus in different directions too. You're worried about paying the bills this month. You don't know if you'll be able to meet your deadline when you've got two kids to look after at home. You're considering applying for a new job but you don't know if it's a good idea. There will always be something to fret, worry, or think about. Where does it end?

When you don't give your poor, overworked brain enough time to focus on one thing at a time, you feel mentally stuck. Today, the expectation is you need to multi-task if you want to get several things done at once. But here's the thing: Our minds were NOT built to multitask. We're supposed to only focus on one thing at a time, but the pressure of juggling multiple responsibilities has turned multi-tasking into an acceptable thing these days. Truth be told, multitasking is ineffective in the long-run and rather than promote better concentration, it promotes anxiety instead. When you're chasing the clock instead of focusing on what you're supposed to be doing, you get nervous, flustered and anxious when you notice you're running out of time. In the haste to meet the self- imposed time deadline you've set for yourself, your focus dissipates. Mistakes get made, crucial information gets overlooked, and you're feeling emotional from the pressure of rushing to meet your time goal so you can move onto your next task. Multitasking is counterproductive, and you're contributing to your cluttered mind when you do that.

It's not only negative thoughts that you have to worry about cluttering your mind. Your environment has a large part to play too. Clutter happens when two types of disorganization are present: Situational and chronic. The first one happens when major life events or transitions take place, like getting married, having a baby, moving to a new home, having a loved one, starting a new job. In the hectic of change, chaos and clutter can ensue. The latter happens when we're our disorganized external environment doesn't show signs of improvement. As the clutter gets worse, so does our emotional state. Being in a messy environment can affect our mood. The amount of clutter that's sitting in your home right now could have such a deep emotional and psychological impact, and it has a way of affecting you mentally and physically without you even knowing it. Being surrounded by disorganization makes it hard for anyone to concentrate. When it becomes impossible to have clarity in your life, you begin to question and wonder what you're doing with your life.

We have become so conditioned to a life of materialism that we genuinely believe the decisions about the purchases we make are based on careful thought and sound logic, but in all honesty, that's far from the truth. There's a classic sales quote which sums up this scenario perfectly: "People don't buy products. What they're buying is better versions of themselves". The purchases that we make represent our hopes, what we believed to be true when we bought them, and what we hope it will help us achieve down the road. We make purchases in the hopes that we will finally become a happier version of ourselves, the way those ads promised. Unfortunately, when that doesn't happen, we become unhappy and the emotions we feel only aggravate the mental clutter that is already there.

Harboring negative emotions is a sign that your mind is cluttered. You know it is affecting you and causing you a great deal of stress, but you find it hard to let go anyway. It begins with one thought, one feeling and before you know it, you're sliding down a very slippery slope of unhappy emotions and you don't know how to slow down anymore. Negativity puts your mind in a bad place you never want it to be in, and it can quickly strip you of any possibility of happiness. It holds you back from living your best life. The trouble is, negativity is not entirely avoidable. There will be some moments in life that are less than pleasant. To say we're never going to feel the stressful effects of negativity would be a lie. We can't always avoid them, but what you can do is learn how to process these those, assess them, and deal with them in better, healthier ways so they don't linger and clutter your mind.

Release your negativity because your cluttered mind is begging for it. Release your negativity and watch how things begin to take a turn for the better.

Your mental habits are the reason you're not reaching your full potential right now. You've got a busy mind and it's stopping you from doing what you need to do. That's why you feel "stuck", stressed, anxious and overwhelmed so easily. A negative mind is one of the most debilitating hindrances you can have, taking up more than its fair share of space in your mind that there is no room for anything else. Like an unwelcomed house guest who has overstayed their welcome, the toxic behavior must be kicked out and the first thing you need to do to organize your mind once more is to develop self-awareness about your thoughts. How do you usually talk to yourself? What do you believe or think about yourself? What you're going to do now is be on the lookout for the warning signs, like victimizing thoughts or toxic self-talk tendencies.

Organizing your mind is going to require a lot of change. Decluttering was only the first step. The harder part would be trying to change the mindset you have, switching it from negative to positive. Overcoming negativity for good is probably going to be one of the hardest challenges you might have faced in a while. Breaking out of old habits that have been around for years is never an easy or straightforward process. It takes weeks of exercising positive thought replacement, which is where you replace your bad thoughts with good ones. It will take time before you begin to feel the visile shift from a heavy, cluttered, disorganized and chaotic mind to a mind that if happier, lighter, and feeling free.

CHAPTER 60:

Why Do You Have To Read " How To Declutter Your Mind?"

You must learn to declutter your mind, removing all of your unnecessary thoughts to make room for the most important thoughts. Practicing minimalism in your life can help you to focus on what matters the most to you and what it is that you truly value in life. You must, of course, learn how to focus. There are a few techniques to do this and make it easier for yourself to focus your attention properly. Finally, you can make focusing easier by prioritizing your tasks. Determine what you want to focus on, and put it in writing so that you have a visual representation of your most important tasks. By doing so, you will be able to start focusing and stop being so overwhelmed by all of your thoughts.

You must declutter your mind; it is essential to your mental health, productivity, and ability to concentrate. If you have a cluttered mind, you may struggle to sleep at night. You won't be able to focus on work. It will be difficult to enjoy your life when you are weighed down with countless thoughts and filled with worries. There are a few techniques that you can work on decluttering your mind.

One way to declutter your mind is to declutter your physical space. You may be mentally overwhelmed because of the physical environment that you are in. If your workspace is unorganized and filled with junk, your mind will reflect that; you won't be able to focus on your work. For this reason, it's necessary to go through your space and get rid of anything that you don't need. This will let you concentrate on what's significant. Your home, workspace, and car will all influence your mental health. Put yourself in an environment that helps you to focus instead of distracting you from focusing.

Another way to declutter your mind is learning how to not multi-task. It is instinctual to multi-task, as it seems like a way of getting more done.

However, you are splitting your focus among several tasks instead of giving your total focus to one. This will result in work that is of a lower quality, and you will feel overwhelmed as a result. Start making it a habit to completely finish one task before moving onto another. You will notice a dramatic change for doing so, as you will accomplish more and feel better about your work.

Be decisive. Instead of putting tasks off until later, decide what to do at the moment. If you have an important task that will take less than five minutes to accomplish, get it done first. Overthinking frequently occurs as a result of a potential decision; you don't want to make the wrong choice. If you struggle with making a decision, jot down all of the potential pros and cons of making each choice. Use that to guide you in the decision-making process. You will receive the opportunity to make more decisions each day. If you let those collect by constantly putting them off until later, you will inevitably reach a point where you feel extremely overwhelmed. Decide what to do with the tasks that matter and eliminate those that don't. For decisions you must make, get it over with instead of procrastinating.

You can also declutter the number of decisions that you must make. You may also prepare ahead of time so that you feel better when it happens. For instance, you may plan your outfits for the entire week on Sunday night. You may meal-prep so that you don't have to worry about food when it's mealtime. Make a schedule for routine tasks you must complete, such as doing laundry every Thursday morning or vacuuming every Wednesday night. Reducing your decisions and coming up with a schedule can help you to stop overthinking and have a schedule that you already know works for you. You may make a schedule for yourself so that you know exactly what to do.

Sunday	Monday	Tuesday	Wednesday	Thursday	Friday	Saturday
Pick clothes	Bathroom	Dishes	Vacuum	Laundry	Trash	Meal-prep

Minimalism

Minimalism, typically regarded as an intense trend, is quite helpful for those who wish to focus more clearly. Although some take this to the extreme and live out of a suitcase and choose not to have a home or car, minimalism is a practice of being more mindful of what you choose to keep in your life. It can be applied across your life to help you to only have what matters to you in your life. You may find yourself practicing minimalism in one area of your life, only to have that affect other areas of your life.

Minimalism isn't simply getting rid of things or decluttering your belongings. It is seeing what truly matters to you and sticking with that. Marie Kondo's technique is to only keep items that "spark joy." This means that you only keep the items in your life that truly make you happy. It can help you to go through each item that you own and decide what matters to you. You may have some things that you keep out of guilt or for "someday in the future." However, these items will only disappoint and frustrate you each time you see them. It is important to eliminate any items that bring you negativity or remind you of failure. Surround yourself with items that bring you joy and make you a better person. Your surroundings should be a reflection of you and what you love. You may choose to go through the items in your house, at work, in your car, and any other places that you may have. Do not bring any other items into your life that bring you down in any way; only own what you truly need to make you happy. This can help you to have a much clearer mind.

Additionally, you may practice digital minimalism. You may feel overawed with the amount of information that you have coming in. There will constantly be e-mails, texts, and other notifications. Social media can also be quite overwhelming. Delete any apps that in no use or that you don't find happiness from. Get rid of that educational app that you "should" be using but never do. Get rid of those storage-sucking apps that you don't like. Clean up your phone. Turn off notifications that you don't need. You may go through your e-mails and delete all the ones you don't need. Unsubscribe from e-mails. Create a labeling system. For social media, unfollow those who don't have a positive effect on your life. You may even choose to stop using social media or go on a social media detox. Limit yourself to a certain amount of time for social media. There is so much information that comes from social media, and much of it is unnecessary. Ensure that you are spending your time the way you want to be spending it.

Learning to Focus

Take some time to reflect on where you stand with your focus. Are you happy with your ability to focus? Do you wish that your focus could be improved? Figure out what your goal is. Do you find yourself getting distracted easily, unable to continue work after taking a break, or simply can't finish tasks with ease? Determine what it is that prevents you from focusing.

You may try to challenge yourself. Give yourself a task and a time limit and do your best. Become aware of how often you become distracted and how easily you can regain your focus. Write down anything that distracts you; you may not even realize how distracted you become! Afterward, ensure that you rid yourself of any distractions. Turn off your notifications. Make sure that it's quiet and peaceful. Whatever distracts you, make sure it's eliminated. You may plan a schedule for your breaks. Figure out what works best for you. Perhaps you like to work for fifty minutes and take a ten-minute break. Perhaps you prefer to work longer and have a longer break. It will depend on your personal preference. Regardless, it's important to remain completely focused while you are working and to also take breaks regularly. This will increase your productivity and allow you to be less stressed. Determine what sets you off. What is it that makes you lose your focus? Why do you overthink? Reflecting can help you to determine the causes of your actions.

There are a few ways that you may work on your focus. Meditating can help you to be calm and live in the moment. You will learn how to become a master of your mind. This can help if you feel overwhelmed by your thoughts. Although meditation can seem difficult (and even frustrating) at first, it provides you with a great boost to your mental health, ability to concentrate, and overall emotional well-being. Another way to help you to focus is by doing some physical activity. Taking a quick walk can help to get your blood flowing, ease up the tension, and wake you up. It can be quite refreshing to get moving, even if you simply take a moment to stretch. Remaining active can also help improve your physical health, which affects your mental health. This is why keeping your physical health in check is so important. Getting the proper amount of sleep, consuming the right foods, and drinking enough water every day is crucial. You will not only get better, but you will also be able to focus much better.

Overall, it's important to take care of your mental health so that you can focus better. By having some time to yourself and doing small things that you love, you can help yourself to feel better. When you are happy, you will feel much more motivated and find it much easier to focus.

Prioritizing

Perhaps you are able to focus, but you simply aren't focusing on what you should. It can be quite easy to get caught up in tasks that are unimportant or aren't urgent. You may think that you're being productive, but you're procrastinating. This is why prioritizing is so important. You must be able to recognize where you should be placing your focus. Otherwise, it's like you're on a treadmill. You're still getting something done, but you aren't going anywhere. You must learn to prioritize your tasks properly so that you focus first on what's important. After that, you may move on to other tasks.

One way to prioritize is by making to-do lists every day. It is helpful to make them the night before so that you can plan ahead of time. At the moment, you may not feel as motivated to accomplish every task. It helps to have one to three major tasks for the day. These are your top tasks. Even though they are the only things you accomplish for the day, you would be happy with yourself. These are your top priority, and it is what you should focus on accomplishing first. This will allow you to focus on what you must.

CHAPTER 61:

Health Benefits of Decluttering

Decluttering the physical space around you will help you create more space for more important things. In addition, such tidiness will also have an impact on your mind since everything will be organized and you will know where everything is. There are few things reminding you that they need to be arranged. Likewise, decluttering your mind also has its benefits.

Benefits

A Decrease in Stress and Anxiety

Clutter will stress you out. Feeling like your mind is messy may make you feel tired since there is a lot to do yet so little time. Similarly, mental clutter will also make you feel unconfident. You will rarely be confident about your abilities. Repeatedly, you will notice that you second-guess everything that you do. All this is happening because your mind cannot think straight. There is a lot that it is focusing on and therefore, finding practical solutions to the little things ahead of you may seem impossible.

By using the recommended strategies mentioned herein to clear clutter from your mind, you can be more equipped to lower your stress and anxiety levels. Your mind will feel more liberated. The new space that you have created will give your mind the energy it needs to think and make smart decisions. Therefore, you will feel more confident about yourself and the decisions that you make.

An Improvement in Your Productivity

Clutter can prevent your mind from achieving the focus it needs to handle the priorities that you have set for yourself. For instance, instead of waking up early and working on an important project, you might find yourself paying too much attention to the emotional burden that is weighing you down. Frankly, this thwarts your level of productivity. You are unlikely to use your time wisely, which affects your productivity.

Eliminating unwanted thoughts and emotions will help you focus more on what is important. You will find it simpler to set priorities and work towards them. You will wake up feeling motivated and goal-oriented. In the short run, you will notice an improvement in your efficiency. Over time, you will realize that you're more effective than ever before since there is more that you can do in less time.

Enhanced Emotional Intelligence

There are numerous situations in which we allow our emotions to affect how we perceive things in life. One minute you love someone and the next minute you think that they are the worst and you regret ever meeting them. In addition, these emotions cloud our judgment and we end up making conclusions that are not valid. Normally, this occurs when there is a lot on our minds that we have to handle. The result is that we fail to deal with these emotions in an effective manner.

Decluttering your mind requires that you get rid of negative thoughts that would lead to negative emotions. As a result, decluttering more often implies that you will master how to deal with negative feelings. You are less likely to allow negative feelings to weigh you down. This is because you understand that they are just emotions and letting them go is the best course of action you can take.

You can make a transformation in your life by choosing to declutter your mind. You will end up making better decisions that lead your life in the right direction. However, it is important to note that the decluttering process will only be successful if you know where the clutter is coming from. To start, you can evaluate yourself and find out why there is so much clutter in your mind. Is it because you overcommit yourself? Is it because you are overwhelmed with the challenges that you have to handle? Is it caused by your fear of making mistakes? Knowing the reasons for clutter ensures that you can control clutter in the long haul. In addition, the digital age that we're living in should not be an excuse to fill your mind with unwanted information. Feed your mind with quality information that drives you to achieve your goals. Curb your information intake and free yourself from clutter.

How Decluttering Your Mind Helps Stop Overthinking

Thinking about how to declutter your brain? Having a bustling personality can make you feel pushed, restless and overpowered. Fortunately, we've assembled a rundown of approaches to declutter your brain. The best spot to start to declutter your life is from within. Numerous individuals neglect the advantages of a sound mind can offer. The mind can move towards becoming hindered with psychological weight and indeed sway an individual's capacity to work. Necessary leadership can turn into a test and adapting to issues may feel unthinkable when you don't have a clear mental state; in this manner, it is imperative to figure out how to free your mind of excessive clutter. Since everybody is different there is nobody size-fits-all strategy to clear your mind of clutter; nonetheless, coming up next are some basic methods that can start you on your adventure to decluttering your life by first decluttering your mind!

Put pen to paper

At the point when you're attempting to keep mental tabs on everything that is going on, your contemplations are probably going to get confused. Keeping in touch with them down will assist you with prioritizing what's most significant, which will make you feel less focused.

You can check significant dates and updates on a schedule or in a scratch pad, and scribble down your musings on anything that is stressing you in an individual journal. It doesn't make a change whether you utilize an application or simply get a pen and paper.

Keep at it

Work a portion of the tips recorded above into your regular day to day existence to enable you to offload mental mess.

Ensure you get a touch of 'personal time' each day with the goal that you can slow down appropriately.

Much the same as tidying up your room keeps it from transforming into an all-out dump, reflecting, composing, ruminating and conversing with others consistently will help anticipate the development of messiness in your brain.

Be careful

We've all heard that reflection is a decent method to clear your brain and unwind. What you might not know of is that there are a huge number of approaches to be careful. This implies you can search for a way that suits you.

Some regular things to attempt are yoga, exercise and profound relaxation. Some not normal approaches to rehearse care are washing up, snuggling up or chilling by the seashore. Do whatever works for you.

Identify the issue

It's difficult to fix something if you don't know what's up. Know about admonition signs that your psyche is getting to be stuffed. Some normal things to watch out for are issue resting, poor fixation and not able to unwind.

When you've perceived that your psyche needs a spring clean, the following stage is to discover what's adding to the messiness. Invest significant time to think about how you're feeling. This will assist you in identifying what's worrying you, and why. After some time, you'll improve at detecting the notice indications of a jumbled personality and have the option to halt things from the beginning pleasant and early.

Converse with somebody

Conversing with a confided in companion or relative, regardless of whether on the web or eye to eye, can be an extraordinary method to clear your psyche, discharge a few feelings and get whatever's irritating you out into the open. It additionally gets a new take on an issue that is got you puzzled and is worrying you. If you're truly battling, recollect that you don't need to handle your issues without anyone else.

There are loads of different experts accessible to chat with about whatever's stressing you.

Keep in mind your past and develop from it

Help yourself to remember how you adapted or left a specific circumstance before. You can expect that on occasion you will slip once again into old examples. This is ordinary—those examples have been developing for a considerable length of time. Stress and blame specifically are difficult feelings. At the point when you get yourself in an old example, ask yourself, "How's my self-talk?" If you wind up drenched in tension, separate your stresses into two classifications: those you can control and those you can't.

Guide yourself to Stop!

Whenever stress rings a bell, or you verbalize it for all to hear, guide yourself to stop! Supplant negative considerations with positive ones. One case of a positive idea is a token of what you do have rather than what you need. This isn't just about cash yet, Also, your aptitudes, gifts, capacities, companions, family, and supporters.

In essence, making a concerted effort to clear your mind of clutter is a tremendous first step that you can take to handle on overthinking. As you begin to sort out through the fluff, you will be able to make better sense of your life and, most importantly, about the people around you. Bear in mind that if people are feeding that clutter, then it might be time to move away from them.

All in Good Time: Schedules

Lighten a portion of your mental stress by making a timetable. When you have all assignments sorted out and arranged out, with spare time included between, a significant measure of strain will be lifted. You will live more proficiently and suffer from less overpowering minutes. You can't get ready for everything, and calendars must be changed now and again. In any case, having a reliable schedule for the things you realize you should do, organized by significance, can have a considerable effect on your mental stress. Besides, this is an incredible method for promising you to have time put aside to rehearse your mind-decluttering systems.

Meditation for a Clear Mind

Meditation is a famous instrument to help declutter your life and your mind. You don't need to think as it was done in the good 'old days. Attempt this increasingly modernized system: start with music you appreciate. A few people profit by uplifting tunes or great songs, while others may prefer something edgier. The class is altogether up to you, and it doesn't need to be unwinding music. Next, locate a place where you can disengage yourself from others and diversions.

Use Words to Remove Tension

Written words are an integral asset to declutter your life. How you utilize them is up to you. A few people prefer to write in a diary. This can be private, and nobody else needs to see it. If you are anxious about others discovering your written musings, consider writing them down on a piece of paper and after that discarding it or demolishing it after you are finished.

Start to declutter your life presently, beginning with your mind. You will feel better, work all the more adequately, and suffer from fewer misfortunes. Besides, when terrible things occur, you will be better prepared to deal with them when your mind is clear of clutter!

CHAPTER 62:

Causes of Mental Disorder

Environmental Factors

Stress is a normal incidence in everyday life especially for those in stressful environments. Nevertheless, if the stress is prolonged and overlooked it can be a source of even more than anxiety. Long term toxic relationships or being in a place you have always disliked for a long time can lead to an increase of anxiety.

Those people that battle with stress are prone to struggle with anxiety, too. Stress may stem from work, a relationship such as marriage, and financial difficulties among others. They will contribute to the emergence of the anxiety attack which may progress to mental disorder if no correct measures are put in place. Stress is capable of weakening the part of the brain that is responsible for combating stress itself. Not only can it affect that but also the brain's neurotransmitters. It is believed to affect the production of these neurotransmitters and hormones responsible for maintaining the wellness of the brain. The mind will, therefore, become overwhelmed and lack the power to fight anxiety as it used to. As weak as it becomes it means that the individual will be shaken even by the mild instances of stress since the stress-fighting mechanism is so weak to protect against the attack.

Our life depends so much on the millions of experiences that help us understand life and eventually see the circumstances of life from different angles. Our upbringing has a lot to do with the way we see life. It determines our personality and also plays a key role in the development of anxiety. We react in much the same ways that we saw our parents while growing up. Having overprotective parents or limited social interaction while growing up may lead to social anxiety. If your parents are overly concerned about a situation, they may inadvertently instill fear in your mind. Fear can also grow as a result of abuse and bullying during childhood.

Indeed, the experiences lived during childhood play a pivotal role in shaping our personality and the way that we act in response to the world around us as adults. That is why childhood traumas tend to be the leading cause of anxiety in adults.

Trauma can be seen as a life-altering situation that remains engraved in the mind of an individual. Severe trauma includes situations that can be horrendous, violent or even life-threatening experiences. Post-traumatic stress disorder (PTSD) is an end result of traumatic experiences that go unresolved. Nonetheless, kids find it easier to cope with the trauma, and recover from it, as they join adulthood. Although, the scars may still be there, and the anxiety may still be persistent. With trauma, an environment that has a close resemblance to that scene of trauma or any linkages to the same has the potential of causing an anxiety attack. Their anxiety handling system is altered in that they no longer process extreme cases of stress or low-stress levels as they did process before the trauma.

Change is good but has its own challenges. Change has been put forward also as a contributor to mental disorder. It is true we feel unsafe in new places, though some people are able to adapt faster while some of us take longer to make adjustments to change. New environments put stress on us with emotions that may be unfamiliar. Anxiety develops from this uneasiness. Change however is not only environmental but also emotional, like the loss of a person you were so close to. Significant changes can grow out of proportion is they are not addressed in time. They may lead to full-blown mental disorders.

It is significant to put in mind that catching traumatic events early on can lead to a healthy recovery. For instance, if you have been in a car accident, you may have recollections, flashbacks and memories of this distressing event. Therefore, you may find yourself suffering from the long-term events of such a situation unless you are able to do something about it early on. If so, then the chances you have for a full recovery are far greater than if you simply let such feelings fester over time. Ultimately, your ability to deal with issues early on will enable you to find the right way of addressing them before they become a serious psychological condition.

Did you know that being afraid of being anxious will also make you even more anxious? In panic attacks, individuals are in constant fear of when the panic attacks will strike again. That constant fear keeps them very anxious. Why should we fear being anxious? The anxiety now becomes a vicious cycle. It is not only in people with panic attacks, but normal people may also experience this. Maybe there is an activity that you did some time ago and made you feel chills and when the activity is brought to you again the chills resurface. You fear getting chills again as a result of that first encounter. For instance, if have an unpleasant experience while riding a motorbike ride, it means that in future you may never try it again. The mere thought of this situation is enough to cause anxiety. Some are afraid that the severity of anxiety they felt at the time may strike again at the sight of the object that makes them. That constant fear is what fuels anxiety thereby creating a vicious cycle.

Healthy living and positive lifestyle habits will help cope with stress. Exercise and good living habits will keep every organ in your body healthy and in good working condition. When we practice poor lifestyle habits, we become prone to getting anxiety attacks. The brain and body will produce fewer neurotransmitters and hormones capable of coping with stress which makes you vulnerable and weak in warding off an anxiety attack.

Genetics

It is generally believed that mental disorders can be hereditary. In some cases, children may inherit overly nervous and anxious emotions from one or both parents. Research is still ongoing to identify if genetics are truly behind all this excessive expression of fear and anxiety. It has however been proven that anxiety disorder can be passed down to the next generation as a result of upbringing and conditioning more than genetics.

As such, upbringing also has a link to genetics; some children will obtain anxiety attacks or disorders after much interaction with their parents or being close to them. The brain capacity and how it works can be linked to heredity, therefore, the coping capacity to stress can be inherited making the mental disorder a gene-related condition. There are some pieces of proof that were put forward and they showed anxiety disorders run in families.

Studies have also shown that some individuals have a great tendency for getting mental disorder compared to other just because of their physical makeup. Therefore, it can be inferred that in some cases, there are underlying physiological conditions that cause anxiety such as lesions in specific parts of the brain. These lesions may inhibit an individual's ability to deal with stress and anxiety in a natural manner.

Medical Factors

However small the percentage of diseases that are capable of causing an anxiety attack, they do exist. These conditions will further the mental condition since they disrupt brain functionality hence affecting how the brain reacts to stress factors. Neurotransmitter and hormone production may be affected as well. The brain is a critical organ and an attack on it may bring about significant harm. Nonetheless, there may be no evidence of a physical condition. Still, the symptoms persist in such a way that the sufferer shows all signs without there being a clear physiological cause. Under such circumstances, the causes of anxiety may be related to a purely emotional cause or another undetected physical condition.

Biologically, the human body is wired to cope with any situation that arises. The thyroid gland is a body part that produces hormones that help deal with stress. These hormones are released to the bloodstream and are useful in controlling body metabolism and energy levels. Thyroid glands contribute to anxiety if there is more production of the hormone than necessary. Too much thyroid hormone in your body will bring about anxiety symptoms like sleeplessness, rapid breathing, fast-beating heart, the nervousness among others. It is important to have the gland checked if you feel more anxious especially while faced by a situation and the anxiety does not calm down even after the exit of the stimulus.

Medical conditions like hypertension, cancer, hormonal imbalances, among others, can also produce anxiety as one of its effects. In the case of cancer patients, anxiety can become severe due to the stress of the uncertainty surrounding the illness itself and its corresponding treatment. But not only cancer patients can be prone to anxiety. Any person who is going through a serious illness may have to deal with severe anxiety, as well. A serious condition can lead to constant fear and nervousness.

Allergies can also contribute to anxiety. For instance, if you suffer from food allergies, you may feel anxious about your diet. You may not feel totally safe when consuming food, especially if you haven't prepared it yourself, since there may be traces of the element which may cause a severe allergic reaction.

Alteration In Brain Chemistry

Imbalance in the normalcy of the brain's chemicals has been linked to anxiety attacks. The individuals suffering from mental disorders are likely to have some problems with their brain chemistry. Hormone and neurotransmitter production may be faulty. Neurotransmitters include serotonin, norepinephrine and gamma-aminobutyric acid (GABA). Nevertheless, research is yet to determine if chemical imbalances cause anxiety, or the other way around. Through therapy, be it through medication or some other drug-free treatment, most individuals are able to regain their proper flow and production of neurotransmitters.

Brain imaging devices have shown a discovery that those with anxiety disorders have more brain activity compared to those who don't. In general, anxiety keeps most of the brain active for a longer period of time. This explains the high levels of fatigue that arise as a result of anxiety. Mental disorders linked to brain functionality may occur due to anomalies in the blood flow in the brain or metabolism. These changes, however, have been proven to be temporary. Therefore, under treatment one can regain the power to cope with stress.

Mental disorder is treatable regardless of your circumstances or upbringing. All you must do is try and understand where it comes from and why. Once you establish the source of your disorder, the path to recovery becomes a lot clearer. The hardest part of a mental disorder is to pinpoint the exact cause of the feeling. It requires internalizing and spending some time with yourself to establish what makes you so nervous since mild anxiety here and there is part of life. What is not normal is anxiety that never subsides or dies a natural death after the exit of the stimulus.

CHAPTER 63:

Usual Remedy in Localized Deep Breathing

Breath is one of the calmest and most versatile items to use, so the breath will be the primary focus for the purpose of this exercise.

Start by taking just a few moments (about thirty seconds) to observe the breath, particularly the rising and falling sensation that is produced as the breath passes through the body and out. At first, just remember where you feel the sensation most intensely in your body. It may be in the belly, the diaphragm, the chest or even the shoulders. Wherever you feel it most clearly, just take a moment to notice the rising and falling physical sensation of the breath in this way. If the breath is very shallow and hard to detect, it may be helpful, and even reassuring, to place your hand lightly over the area just below the belly button on the abdomen. You can feel the ups and downs of the stomach very quickly as your hand goes back and forth. You should then return the hand to its initial position, resting in the lap, until the exercise begins.

Since the air and mind are so closely linked, the position of the air cannot make you happy. For some of you, that may sound very odd, but it's just a very normal phenomenon. People sometimes say that they are not "breathing correctly," that the movement in their chest can only be heard. And still, they add, they read books and went to yoga classes where they were told to take great deep breaths from their stomachs. This makes sense at first glance, we obviously equate the occasions when we've been really comfortable, maybe feeling tired on the couch, or lying in the toilet, with long, deep breaths, evidently coming from the stomach. Likewise, with quick, rapid breaths, obviously coming from the chest area, we associate periods of anxiety or concern. If you're sitting down and having similar feelings to the nervous form of breathing, then it's normal that you may think you're doing wrong. Yet you're really doing absolutely nothing wrong. Remember, there is only conscious and unconscious, undistracted and distracted— in the context of this exercise, there is no such thing as incorrect breathing or bad breathing. Specific relaxation techniques, of course, maybe part of yoga or any other practice, but that's not where we're going with this exercise.

If you've done it in your life to this point and you're reading this book right now, I'm assuming you've breathed perfectly well up to this point. In fact, I would say that unless you have done prior activities with relaxation or maybe yoga, you have not even been conscious of how you breathe much of the time. The breath is autonomous, our control is not required for it to work. Left to its own natural intelligence alone, the breath normally works very comfortably. So, instead of trying to exert your power (notice a pattern building up somewhere else here?), let the body do its own thing. It's going to be managing itself in its own time and way. Occasionally, in one spot, it may seem more apparent, and then change as you watch it. Sometimes, the entire time will rest very comfortably in the one position, whether it is the neck, chest or somewhere in between. Your only job here is to listen, observe and be conscious of what the body, of course, is doing.

So focus your mind on that physical movement, the rising and falling feeling, without any attempt to try to adjust the position of the wind. You can slowly begin to notice the rhythm of the breath as you are doing this. How does the air make the body feel? Is it fast, or slow? Take a few seconds to try and respond. Were the breaths deep or flaccid? You will also see whether the breath feels gritty or smooth, close or open, warm or cool. These may sound like crazy questions, but the same concept of adding a gentle curiosity to your meditation follows. It should take the cycle just about thirty seconds.

Getting a strong understanding of how the body feels these sensations, now concentrate on the air as it comes and goes every time. The simplest process to do that is to count the breaths as they pass (silently to yourself). As you know that the upward feeling counts 1, and as you know, the downward feeling counts 2. Keep counting to 10 in this way. Switch to 1 until you hit 10, and repeat the exercise. It sounds simpler than it really is. Before I get going, if you're anything like me, you'll find that each time you count to 3 or 4 before your attention wanders off into something more interesting. Alternatively, you may unexpectedly find yourself counting 62, 63, 64... And realize you've forgotten to stop at 10. Both are very common and are part of the meditation learning cycle.

You are no longer disturbed; by the moment you know that you were disturbed, the mind has drifted away. So all you must to do is gently put back the focus to the breath's physical feeling, and keep counting. If you can remember the number on which you were, then just pick it up from there and, if not, just start at 1. There are no incentives to make it to 10 (sorry to say), and so it doesn't matter whether you continue at 1 again or not. In reality, it can be very amusing in how hard it is to make it to 10 each time, and if you feel like laughing, it's okay to laugh. Meditation can look very serious for some reason, and it can be tempting to start approaching it as "hard work." But the more you can put in the sense of humor, a sense of play, the simpler and more pleasurable you will find in.

Continue to count in this way until you've set the timer that lets you know it's the session end. But just still don't get up from your chair. One very critical element still remains to do.

Finishing-Off

This part is frequently ignored, and yet it's one of the exercise's most important aspects. When you have done counting, just let your mind be totally clear. Do not attempt to regulate that in any way. That means not concentrating on the breath, not counting or something else. If you want to retain your mind, let it be occupied. If it needs to be silent with no thoughts at all, then let it be silent. This needs no effort, no sense of control or any kind of censorship at all, only letting the mind be totally open. Does that sound like a fantastic idea or a terrifying one, I wonder. Anyway, just let your mind loose for about ten or twenty seconds before the meditation comes to a close. Often when you do this, you can find that in reality, the thoughts are fewer than when you tried to concentrate on the air. "Why could this be?" You may well inquire. If you think back to the example of the horse that hasn't been broken in yet, it is always more relaxed and more secure when it has a bit of room, when it tends not to cause too much trouble. Yet, when he is a little too firmly bound up, then he appears to jump a bit. And if you can put some of this spacious consistency into the breath-focused portion of the practice, then you'll really start seeing a lot more benefit from the meditation.

Having let the mind wander free for that brief moment, gradually carry the focus back to the body's physical sensations. That is to put the mind into the physical senses. Note the tight contact between the body and the chair again beneath you, between the feet and the floor sole, and between the hands and legs. Take a moment to note some sounds, intense smells, or tastes, getting slowly grounded with each of the senses through touch and perception. This has the effect of getting you back entirely to the world in which you live. First, open your eyes gently and take a moment to readjust, refocus and be mindful of the room around you. Then, with the intention of taking the sense of knowledge and presence into the next part of the day, get up from the chair slowly. Be sure where you're going next and what you're going to do because that will help keep the sense of understanding going. Maybe it's going to make a cup of tea in the kitchen, or maybe it's going back to the office to sit at your computer. It doesn't matter what it is, what matters is being open enough in your own head that you can continue to feel with your full consciousness at every moment, one after the next.

CHAPTER 64:

Usual Remedy for Meditation

"While meditating, we are simply seeing what the mind has been doing all along." - Allan Lokos

Researchers attempting to study the psychological benefits of meditation attempted to analyze a wise Tibetan monk's brain waves through a scan. After the scan ended, the researcher said to the monk, "Sir, the machine has revealed that your brain indeed goes into an intense relaxation state, and that is a validation for meditation." The monk simply said, "That's not correct. My brain validates your machine."

Meditation can be practiced in several forms from deep breathing to chanting to tuning in to the sound of birds. Much as people like concise definitions, meditation can't be boxed into or narrowed into a particular type of discipline. It is what works wonderfully well for you. It can be as easy and straightforward or as intricate and complicated as the practitioner wants it to be.

The term meditation is borrowed from the Latin words meditari (to contemplate upon or think intensely about something) and medari (translates healing). Medha also translates into wisdom in Sanskrit. Meditation is seen as a spiritual practice that seeks to accomplish a sense of unison between one's body, mind, and spirit.

It is believed to instill a sense of calmness and tranquility within our sense of being or existing. Meditation clears our mental cobwebs from damaging and self-destructive thoughts, thus making the way for newer and more productive ideas. For some, it an essentially spiritual or religious practice, while for others it is a therapeutic activity.

Meditation instills a sense of inner calmness or peace to disciplined, committed and consistent practitioners. It paves the way for balanced, calm, objective and wise thinking.

A regular and consistent meditation practice leads to greater awareness about your feelings, thought patterns and actions. Being in a focused state of awareness leads to positive thoughts and better decision making. It offers greater clarity of thoughts and action, apart from putting a person in a serene yet energized mental state. Meditation has the capacity to completely transform your thoughts, actions, behavior, and life.

Our brain's prefrontal cortex region stores all the information linked to experiences and memories. This area is interconnected with our physical sensations and nervousness/anxiety regions via neural pathways. Meditation weakens the connection between your neural chains to reduce stress and anxiety, thus leading to a calmer state of mind.

A disciplined and regular meditation practice breaks neural links that cause us to experience fear or overwhelming physical sensations. We are in a better position to identify and be aware of these sensations without being intimidated by them. For instance, when our body aches, we automatically assume that something is wrong with our physical health or body. However, a well-cultivated meditation practice allows you to react to sensations more objectively. You will simply acknowledge the pain without being bogged down by it. Rather than stressing about or reacting to the physical sensations, you will simply accept it with complete detachment and without being overcome by its imagined implications.

There are plenty of meditation techniques based on various disciplines, spiritual goals, and religious philosophies. Pick the ones that best meet your objectives and make you feel a sense of oneness with yourself (or have principles you can closely identify with).

Metta or Loving Kindness Meditation

Metta translates to kindness and benevolence in the ancient Pali language. The practice is fundamentally practiced in Buddhist philosophy, where showing compassion for oneself is known to have several benefits. It has also been proven scientifically that a person's ability to display kindness to himself results in greater empathy towards others, more positive emotions, greater self-acceptance and a general feeling of living a rewarding and fulfilling life. There is an increased sense of purpose in living.

Loving-kindness is practiced by sitting down in a comfortable position with eyes closed. You start by gradually generating feelings of kindness and magnanimity in your heart. It starts by showing love, kindness, and compassion to yourself, and slowly progresses to showing love towards other life forms. The progression usually moves from oneself to a loved one to a neutral acquaintance to a difficult person to all four. Finally, you show loving kindness to the entire universe.

The feeling to be experienced while practicing this meditation is wishing for the happiness, joy, and well-being of everyone. The practice can be made more relevant by chanting matching words and affirmations that evoke feelings of boundless love and warm-hearted kindness for everyone. Visualize the pain of others, while reaching out to them and offer kindness. Imagine the state of being someone else and wish them peace, kindness, and happiness.

Mindful Meditation

Mindfulness of breathing is one of the most important components of Buddhist meditation rituals.

Much like mindfulness itself, it comprises intentionally or purposefully drawing your attention to nothing but the present moment. It is not just about being aware of the present but accepting it in a nonjudgmental manner by focusing closely on every sensation, feelings, emotion and thought as they surface and fade away.

Find a quiet, distraction-free space that gives you positive vibes. It can be anyplace from your backyard to a favorite corner in the home.

Settle the mind before you begin. Detach from everything happening in your life. Empty your mind, especially if you've had a stressful day or your mind is preoccupied with anxiety about things happening in the future. Feelings and thoughts will do a little jig in your mind before they settle down. Observe them dance and give them time to sit still.

Sit on the floor or chair in a relaxed position (use a cushion required). Keep your back straight, and hands loose. Mindful meditation practice starts by paying close attention to the movement of one's breath. Start by taking deep, long breaths.

Draw complete awareness of the breath, observing the process of inhalation and exhalation for every breath. Notice how every breath flows inside and outside the body. Experience the sensation of air moving into your lungs and being released through the mouth. Taking deep breaths helps bring a sense of calmness and relaxation to the body and mind. It helps bring a sort of serene stillness to thoughts.

Remind yourself you are in complete control of your thoughts, feelings, and emotions. When you notice feelings or thoughts that you don't want to flame or indulge, just release them without choosing to focus on them.

This is a valuable technique when it comes to decluttering the mind. Mindful meditation gives you the power to be in control of your thoughts and emotions, without being overwhelmed by them. You learn to gently notice your thoughts without judging them and develop the power to release them without being affected.

Whenever you find your mind wandering to distracting thoughts, acknowledge them fleetingly and return to the breathing. Start observing your inhalations and exhalations mindfully. Focusing on your breath lets you focus on neutrality. There is no good breathing and bad breathing. The breathing pattern remains the same irrespective of your thoughts and emotions.

Similarly, avoid passing judgments on your thoughts or even how you are meditating. Judging interferes with the purpose of mindfulness. It is all right to get distracted or have thoughts that occupy your mind. Simply acknowledge and let these thoughts or feelings pass. Mindful meditation is not a presentation and performance in which you are ranked or marked. Don't go hard on yourself and do what works for you, while retaining the basic principles of mindful meditation.

Guided Visualization

Guided visualization is one of the more modern meditation techniques that is used for decluttering the mind, reducing stress, fulfilling goals and spiritual healing. The inspiration behind it is still rooted in the Buddhism philosophy of "the mind is everything. What you think you become."

One of the key factors of this form of meditation over others is that guided meditation primarily focuses on the fulfillment of a goal. It can be anything from a job promotion to greater inner peace.

The meditation practitioner is guided into visualizing or imagining relaxing, calming and positive situations by an instructor. When you imagine positive and serene experiences in the mind's eye, the body reacts by releasing a bunch of chemicals that stimulate feelings of well-being and positivity.

The situations can range from anything from relaxing on your favorite beach to being in a bountiful park to lying down in your dream home.

You can use guided meditation audios in the absence of an instructor to help guide your focus to reach a relaxed, meditative state. Make generous use of the power of imagination and visualization to stimulate the brain into viewing objects, scenic wonders journeys and positive entities. You can combine guided visualization with affirmations to make imprint the message more powerfully on the subconscious mind.

Many meditation practitioners employ guided meditation practices for relaxation and refueling. It is identified to reduce blood pressure and stress-inducing hormones. The body and mind are calmed, and you feel a greater flow of relaxation and energy. There is a tendency to feel more rejuvenated and invigorated to take on fresh challenges.

Guided visualizations or imagery can also be used to fulfill personal/professional goals. For example, you visualize the entire presentation before you actually present in front of a huge audience to the minutest detail. Similarly, an athlete can imagine his performance on the field just before the upcoming race.

Others use guided visualization to gain a deeper understanding of their inner selves. It helps them tap into their intuition and strengthen their connection with the inner voice. They receive through the medium of these images answers to questions or feelings or thoughts that they were struggling to discover with the help of their conscious mind. For instance, if someone isn't sure about their career path he or she can use guided meditation to channelize their inner voice into guiding their decisions, thus eliminating a lot of mental clutter.

Qi Gong

Qi Gong is one of the most ancient forms of meditation originating from China. Essentially, it involves the use of breath and movements to spread energy throughout the body's various energy circles. Unlike most meditation forms, it combines breathing and martial arts movements to gain control of your thoughts, direct your body's energy and experience relief from stress.

Multiple Qigong exercises are categorized using more than 80 distinct breathing patterns. Some are martial arts based on energizing and strengthening, while others help meet mediation and spiritual objectives. Qi Gong can either be done in a static seated or standing position or through a flow of movements (most YouTube videos have movement-based Qi Gong practices). The meditation focused practices are done in a seated position without any movement.

It begins by being seated in a relaxed and comfortable position. Ensure the body is centered, relaxed and balanced. Ease your muscles, vital organs, tendons, and nerves. Regulate the breathing pattern by taking deep and intense breaths.

Lead the mind into a state of calmness.

Gather all your attention to the body's gravity center (lower dantian), a couple of inches below the belly button. This is understood to be the root of your vital energy, thus making it the energy center.

CHAPTER 65:

Usual Remedy for the Reformulation of All Negative Thoughts

Cognitive distortions represent the negative thoughts we have about the world around us, everything that happens in our life we interpret it as an unfavorable and irrational fact. There are several studies on these distortions and it turns out that there are about thirteen types instead, other people indicate that there are about fifty types. It is therefore impossible to confirm with certainty the precise number since some of these distortions add to each other.

Let's take an example, a person who asks too many questions about a problem and does not know whether to choose one path or another means that he is much undecided about his life and that he is creating "useless" problems even for a simple thing. Another case is that in which an adolescent is highly emotional and afraid of being among other people, he is certainly a person who hates talking to people and being in public and he prefers solitude because he feels agitated outside his "Natural and comfortable habitat" and the greatest terror is to do something for which people will laugh at.

The first major block that comes before him is the anxiety of being laughed at for how he behaves in these situations, this will then be the main cause of his biggest doubts that will knock on the door such as, for example, giving a speech for his employer.

They are people who believe that all their experiences without a doubt turn into error, they don't even give themselves a chance that this time maybe it's the right one to have excellent results. The most important struggle that they are/must face is that of emotion, emotional control because many of their behaviors derive from the non-control of their strong fear. To get out of this limbo there is no simple way, it is easy to fall into it but it is difficult to get out of it.

It is difficult but not impossible. We will have to get help from a therapist who takes care of emotional behavior and who makes us fight with the way we approach people. In time you will see the change and it will be easier to relate to them and as we said it would not be simple but, with a little practice and goodwill, you will overcome everything.

You will live with a completely different lifestyle together with the way of thinking no longer focused on fear, anxiety, feeling inadequate but, you will live in complete tranquility.

All this is possible through the sessions of emotional behavior therapy. It causes the mind and negative feelings to navigate different paths so that we can study them in-depth and eliminate what creates these feelings. Through this method, the irrational, unhappy, and instinctive flows are changed with the logical, venturous, and controlled ones and it is possible to understand what the true way of thinking is and what the false one is.

Taking the example of a person who asks many questions, just imagine that he is going to make a speech in front of his employer and only with the therapy sessions and the tools he has learned he will be able to get away and make a good speech without fear to be wrong. It must ensure that the mind is covered with positivity, tranquility, and relaxation and eliminate fear.

Once eliminated, you realize that it does not have a solid foundation on which it rests and can no longer "take over" your mind. To understand that they do not have a basis, one can ask: do people think they are incompetent, if so, why? Why do people laugh at them if they make even the slightest mistake? Is all this correct? No, it is not. When you understand this, you are more aware that you don't have to worry about anything.

But people will continue to ask questions, make you feel inadequate, and that you don't know how to do anything or that everything you do, you do it wrong. But this doesn't matter much to you, who asks you these questions is because maybe in his life he is not able to do anything and therefore makes fun of you but he is the first one who is afraid.

The next step is that of indifference, only this way they go on because you have the strength to go on and not to think and rethink the things they said to you. They also make themselves ridiculous when they see that you don't care what you're told and then, maybe, after this, they start thinking that it's not so important to make fun of people when they suffer from it.

Everyday people fight against cognitive distortions, our mind is a traveler sometimes thinking with concrete facts and other times without a real reason makes us feel panic. For this last reason, we have to challenge it, break down the panic we feel, and letting our minds be invaded by positive thoughts and tranquility. By learning this we will no longer be attacked by bad thoughts.

This also makes us understand how much we are influenced by emotions and at the same time we care about how we can get rid of them. Day after day, hour after hour, training and talking with our therapist will get better and better; you will start to take care of yourself, to love yourself, to support other people and to live life step by step, you don't need to get excited about the things that will happen if they have to happen you can't do anything, you have to live the way as it comes. To this, we must add a peaceful sleep, cope with the stress that we will experience, keep anxiety and depression under control together with a healthy lifestyle.

Emotional behavioral therapy is a cure that is based on the analysis of your negative emotions and the relaxation of the body/mind.

This is through a healthy and peaceful meditation, relaxation of body muscles, breathing techniques, etc. Everyday people suffering from emotional-behavioral stress must carry out this activity to better face the day and feel better with their body.

Furthermore, you will thereby cast aside negativity and live only in harmony with yourself and with others. This remedy is used by many people who suffer from this stress, and they are told to imagine these concerns as a hypothetical fact not as reality because our thoughts take into account our past feelings and make the same thoughts as the results we achieved and not as if they were reality.

Having to do this allows you to be aware of the world we live in and to react to things that happen; it makes you able to live life better without fear and it improves it if there is something to change.

Not always, however, but in a few cases, this therapy has helped people indeed made them suffer more and more, locking them up in themselves and bringing them other types of problems such as in the worst case, suicide.

This is "normal" for highly emotional people, who are unable to control their behavior, are unable to manage anxiety and fear. One thing that is certain and very important is that you should not be afraid of being laughed at when you say you have a mental illness, it is the same as a fracture in the arm, you must not be embarrassed.

People suffering from these psychological disorders tend to underestimate the problem and postpone it, but it could get worse and worse if they continue to delay it. With the therapy mentioned above, they become able to defeat this inner malaise on their own, with tools and "training".

They manage to get out of the edge of pain by getting out of fear, stress, anxiety, terror, and letting in tranquility, relaxation, and positivity. Continuing every day to train to enter these new sensations will flood more and more with positive thoughts so that it will become normal to feel this way.

If, for instance, during a discussion, you are proposing ideas, you do not dare to propose something because you are shy, you are afraid to speak, and then we start thinking "I cannot" or "I am not able" so you will never succeed to defeat this fear. But if when you are in this situation, you are calm, you analyze the case and you think that you are not there that you have to talk to people but you think it is all a hypothetical fact, you will see calm reaching and you will be able to speak, perhaps not by making a speech but commenting on something.

All you need is training, putting yourself in these situations does nothing but help you overcome the terror you feel. Only by doing this will you see that you will be able to bring out the best in you and people will appreciate you, and also you will show that you are good at something.

A very vital thing that many people often do is that they mix opinions and ideas with reality and this is very dangerous for yourself, for your victories and for your self-confidence. Divide things, differentiate them using other reasons.

The thoughts to be made when experiencing these situations, referring to the example of the person who is unable to express their opinion in a discussion, are:

- This is one of the many discussions that will happen, you will have to be ready for the others, take them as an example and go on
- The first one will surely be the most difficult, because you didn't have the opportunity to prepare yourself psychologically and you didn't know what was waiting for you but, in reality you eventually realized it wasn't so complicated
- If it is difficult for others, it doesn't mean it is difficult for you, everyone faces difficulties in his way
- Everything is tough at the beginning but, with excellent tools such as breathing for anxiety, for example, you can calm down and it will get better time after time
- If only for me it is difficult and for others not, I have to understand why it is not like that for them, understand how they take this situation, how they deal with it and take inspiration from it for the following times.

Thinking about these things while you are in a situation like the previous example, is useful for overcoming it as well as going to therapy. It isn't easy for sure but, if you do not try, as always said, you will never succeed. If I Try I Can! Only after trying, can you say firmly that you did it, after which you will feel much better and improve your lifestyle.

CHAPTER 66:

Get Rid Of Negative Situations of the Past

Have you ever found yourself frequently thinking about how you screwed up in a meeting that happened the previous day? Or, perhaps you handed in a report, and it wasn't your best work, so today you are continuously thinking about what will happen if your boss isn't pleased with your work?

When something negative happens in our lives, our minds often become stuck with thoughts on that particular event, even when it has passed, and it is already time to move on. This is very common and also a bad habit. There are many reasons why your mind keeps going back to something that happened yesterday, and this is causing thoughts to be present in your mind that are leading to cluttering.

I don't want you to simply forget about everything that happened in your past. We all have happy memories. That moment you met your partner. Your wedding. The moment you first held your baby. Your graduation day. These are happy memories that are locked in our minds, and we should really remember them for as long as we possibly can.

The memories and thoughts I refer to here are the negative ones that you are constantly thinking about. Those events that happened that are causing you to feel distressed, worried, and making you experience a lack of productivity the next day.

The second you get up in the morning, your mind should be fresh, and your thoughts should be cleared out. No need to concern yourself about yesterday. Think about the present – what is going on today, what will happen today, how you will tackle your assignments in the present, not in the past. Yesterday has already gone by, and no matter how much you think of an event that happened yesterday, and no matter how much you worry about it, there really is nothing you can do to change the past.

Why It Is Important That You Let Go of Yesterday

I already explained how constantly thinking about something that happened yesterday will simply clog your mind and further contribute to cluttering, but there are other reasons why you need to stop ruminating about things that occurred in the past that caused you distress. In particular, these activities have been linked to poor mental health and adverse effects regarding mental performance.

Harvard Medical School prepared a paper where they explained how one of their recent studies found a negative link between rumination and mental health. The study was conducted among a group of adults, as well as adolescents. There were 1,065 adolescents involved in the study and a total of 1,132 adults.

The study followed a self-report setting where all participants were asked about past stressful life experiences, along with a questionnaire about symptoms related to depression and anxiety. Participants were also asked about rumination.

The study made two important conclusions that I want you to take note of:

1. Exposure to stressful events in life increased the risk of rumination
2. Rumination was linked to symptoms of anxiety among adolescents. Among the adult patients, those who reported ruminating about past life events that were negative had a significantly higher prevalence of anxiety, as well as symptoms related to depression.

What this all means is that constantly thinking about those negative events that took place yesterday is going to make you anxious, and it is going to increase your likelihood of developing depression. Now, if you already experience depression and anxiety symptoms, this may only make things worse.

There is yet another study that I want us to take a quick look at. The study was conducted by the Psycho oncology Research Unit at the Aarhus University Hospital in Denmark. The study looked at how rumination affected both sleep and mood among a group of participants. There were 126 participants in the study, all of whom were current students at a university in the local area.

The study found that rumination, in other words constantly thinking about past negative life events, was linked to angry moods in the students. This provided proof of mood alterations caused by constantly thinking about the past. Additionally, the study also provided evidence of a link between rumination and depression.

Furthermore, another interesting thing was observed. Those students who reported frequently dwelling in the past had also reported poorer sleep quality compared to the students who were able to let go of the past and focus on the present.

Figure Out Why You're Holding On To Yesterday

To let go of past events and start focusing on what is happening in the present instead, which will ultimately help you unclutter your mind, you first need to discover why you are frequently thinking about that particular event. You also need to notice this. Sometimes, your thoughts can be so cluttered and unorganized, that it can become difficult to know when you are holding on to the past, and when you are simply confused about the present.

There are many diverse reasons why you may find yourself ruminating about what has happened in the past, whether it is something that happened at work or at home. In many cases, you may constantly be wandering back in your mind to that particular event because you are trying to work out a solution to a problem that the event posed. If you still haven't figured out a solution, then you are constantly thinking about that life event, and you are trying to figure out what to do.

Sometimes such an event might cause you to fear similar events in the future. Thus, you constantly ruminate about the event as you are scared that the same thing will happen when working on a different project.

There are times when it is a really bad habit that you developed. There's nothing wrong with you, it is just something that you do, and that you will need to stop doing if you wish to move forward, unclutter that mind of yours, and become more productive at the same time.

Looking at what specifically your mind is constantly going back to can also help to give you a sign as to why you are thinking about it so much, and why it is meddling with your ability to concentrate on tasks that you need to get done today.

How to Stop Thinking About Yesterday

When you are able to identify why your mind keeps traveling back to the past in order to ruminate about something that has happened, then you might be able to tackle this problem by focusing on the reason and the specific event that is causing you so much distress. Unfortunately, there are many cases where we keep thinking about something that happened in the past, yet we are not sure why. You try to focus on your work, a specific task that you need to get done, yet your mind is somewhere else, and your productivity is suffering in the meantime.

Luckily, there are a few ways that you can stop thinking about yesterday and start focusing on the present.

One of the effortless ways that you can alter your focus and get your mind back on track is to allow a couple of minutes to participate in an activity that will help to change your current emotion. When you think about a negative event in the past frequently, then it can make you feel negative thoughts, become anxious, or maybe even angry. You might feel frustrated with the fact that you cannot get your mind to focus on something specific.

By focusing on something that will "lift" your mood, you can get distracted from thinking about that event and start focusing on something else. Every person is different, so you'll have to determine what is useful for you. If you are able to get into a more positive mood, you might also start to look at that negative event in a more positive way, start to understand that it is part of your past, and that you can do certain things that will help to avoid it from happening again, or to correct the wrong that the event has caused. This way, you'll be able to stop thinking about it.

If your mind gets cleared and you start to feel positive after a run, then take a quick run. If you enjoy meditating, then have a quick session of meditation. You must find what works for you.

If you find that this particular strategy does not help you out, then get out a piece of paper and a pen. I'm going to walk you through a quick activity that might seem a little weird and even irrelevant, but you'll surely feel better afterward.

You see, the problem with thinking negative things about something that has happened in the past is that the brain will often continue to think so because it is difficult to just experience a shift in emotion without any positive reimbursement.

What I want you to do is contemplate about that thing that happened yesterday, or whenever it happened, and write down exactly the issue at hand. Perhaps you handed in a project that you are not very confident about and haven't heard anything back yet. Write down what is bothering you.

Now, write down what you think is going to happen. Since you are constantly thinking about a specific event, your brain is clearly also thinking about something negative that would happen as a consequence of your actions or the event.

Once you have written down the problems, you need to start writing down WHY this "consequence" will not happen. The thing is, when we think in a negative way, we quickly start imagining the worst-case scenario – and oftentimes, this is not what is going to happen. Thus, I need to write down a couple of reasons why you are overexaggerating.

For example, if you handed in a project that you are not too confident in, your reasons could be that you did adequate amounts of research to support the data you presented, and you presented the project in the requested format. You also followed the instructions that were provided to you.

Once you have written down a couple of reasons why the worst-case scenario will not happen, I want you to think of what happens if the worst-case scenario becomes a reality. Chances are, you will still be alright even if it happened. Write down the reasons why you don't really need to be so worried.

Maybe your boss will complain about the report you handed them, but after that, they will simply ask you for some alterations. You can easily add a timeslot for those alterations to your new schedule and pin it onto your to-do list – problem solved.

Remind yourself that, even if things turn out for the worst, solutions are possible, and everything will still be all right. You have no reason to remain worried about what has happened, as you cannot change the past. Remind yourself that you do, however, have the power to change what happens in the future, by thinking about what you are doing in the present.

CHAPTER 67:

Improve Your Decision Making

Your life decisions are what determine the quality of your life. They determine if you are living your life as per your potential or not. From a young age child are taught to have good decision-making skills.

My mom would send me to the store and if I did not find what she sent me for I would have to make a decision on whether to purchase the alternatives or not buy anything at all.

When I had a situation in school with other students or my teachers, my parents always made me see how my decisions affected my outcome. I would hear my mom say that it is good to have wisdom so that you make the right decisions because good decisions often lead to successful outcomes. And now that I am all grown up, and I have a family of my own I can attest to this.

You are your decisions just like your body is what you eat. Decision making was not always my stronghold. I have shared how I was in toxic relationships with my friends who led me to make poor life decisions. I struggled with picking the right choice and at some point in my life I was convinced that I would never pick the right side to anything.

It feels like I had some sort of bad luck because the path I often chose when making decisions frequently had me suffering. Something bad would come about and I would end up with losses. And I was so jealous of people around me who made decisions and good things happened to them almost instantly.

As a lad starting his career I did not understand when people said, "give me time to think about it and I will tell you my decision". In my head I would question what they needed to think about. It is just a yes or no answer. Pick a side and let us work with it already.

But as I grew older and suffered in the hands of my poor decisions I decided that it was time for a change. I needed to turn things around in my life and find this wisdom my mom always spoke that would help me make better decisions.

As I was changing my life to a minimalist one, I read in one of the many minimalist books that I engaged with that it does not matter what decision you choose, the decision-making process, and your position on your decision is what matters.

Having already come up with a decision, are you assertive? Do people around you trust that you have made that choice and you are not changing?

I came to realize that people never took me seriously when I did not make wise decisions. This often reflected in how I never stood my ground. Today my stand would be one thing and the next day or hour I would have changed and taken another path. I probably wasn't confident in my decision and my constant change of stance probably highlighted this to my mates.

On your journey to declutter your life you need to have impeccable decision-making skills. You also must be assertive. You have to make the decisions firmly and ensure that people respect your decision by standing by it.

There are plenty of decisions that you have to make in a day. Making your decisions assertively will make sure you do not have to go back to something you had already made a decision on.

This will lead you to greater heights because you will not spend much time constantly rethinking your decisions and going back on them. The time that would have been spent on this will be put into use elsewhere. Your assertiveness will also make people around you know that your word was final and you are not going back on it.

When I became a father I understood the importance of assertiveness in decision making. My easily wavered decisions made it impossible for me to discipline my children. I would tell them something but they never took it seriously because they knew I did not stand my ground.

Parents, you know what I am talking about. Children know from a young age when you are the softie. They never take you seriously because they know a decision you have made can be changed at any time with a little persuasion.

My sons would give me the cutest puppy eyes when I gave them a time out that made me pity them and reconsider the punishment. Those eyes had me hooked and I felt guilty for punishing them. But in doing so I was failing in my parenting.

If you do not make decisions, or you do not make them assertively, you will have a backlog of many things you are required to make decisions on. You will have a decision clutter box in your mind of the many things that require your decision but you do not know what you want.

When you constantly put off a decision you are tiring your mind with all the clutter. Because that is one extra thing that will linger in your mind and you will end up thinking about it, among other things, from time to time.

I have mentioned that I did not understand why people would want to be given time to contemplate a decision before making it. Because at that time I was a hasty decision-maker. Never did I take time to think about something and the repercussions it would have on my life. This is what helped me realized to change my mind all the time. Because a few steps into the path I had chosen I would realize that was not the way I wanted to go.

For you to make a smart decision you have to think about it. Think about the effects the path you take will have and other things that are related to the decision. This will guide you make an informed decision that will see to it that all sides are getting the best from it.

But do not put off the decision. There is a distinction between taking time to think about something and putting it off. When you take time to think about it you are actually going through the decision making process. And this will help you make your decisions assertively because you are confident that the path you have chosen is the best way to go.

But if you put off a decision you are not thinking about it. You are waiting until the time when it is absolutely necessary for you to make the decision. And then you will make a hasty and misguided choice that you will later have to change or look for remedies for the loss suffered.

When you constantly put off making a decision your brain becomes tired of the clutter. You are giving it so much to think about and you will end up with misinformed and misguided decisions.

Be Assertive Not Overly Aggressive

There is a difference between being assertive and being overly aggressive. The latter means you are being confident in an unwise decision. You are making choices to show your superiority and not to help the people involved.

Assertive decisions are those made out of confidence. Not overconfidence to show your supreme decision as the decision-maker. Take into consideration that the choices you make will have some impact on other people's lives. You want to be vigilant to make sure you are not causing them harm.

An assertive decision is made firmly but respectfully. You are acknowledging the interests of all people that are going to be affected, and laying out an all-rounded decision that will cater to all their interests.

Assertive people are:

- Self-assured
- Confident
- People with clarity of mind
- People that are persistent with their goals and work on achieving them

Decision Making Skills

You need to practice your decision-making skills. These skills should be impeccable for you to have enough confidence in the decisions you make. You will only achieve assertiveness when you are confident in your decision.

Imagine the embarrassment of making a decision and causing a lot of damage from it. People get fired from their job positions because they made the wrong choice in something they were entrusted with and caused the company a loss.

Even in your own life, you will be faced with so many situations that will require you to make decisions.

There are always three types of decisions, the wrong choice, the right choice, and your best choice. Always aim at discovering what your best choice is.

And this will need you to be innovative. Most times life may seem to have dealt you only two cards. The wrong choice and the right choice. But the great thing is you can be innovative and creative in your decision making. You can tailor both situations and take the good thing about each to make your best decision.

Your decisions will determine your growth. When I finally got my decision-making skills better I got promoted. I have shared that I was promoted after I walked away from my toxic relationships. At this time I had started making decisions that would better my life and bring me growth.

I had come up with payment plans for my debts and I made the decision to no longer live my life beyond my means. My good decisions made me eligible for the promotion. Nobody wants a poor decision maker handling many responsibilities in their company.

Your Emotions

Do not make a decision when you are emotional. This negates assertiveness. With assertiveness you should have confidence in your decisions. But when you make decisions based on an emotion you will not be confident that you are picking the right path. At that moment it will feel like the right thing to do, but when you are calm and not under the influence of your feelings you will see the mistake you have made.

I know how making decisions when you are extremely happy gets you entangled in situations you do not want. But when you were happy it felt like the best decision to make. This also happens when you are sad and angry.

When I am angry I am faced with the temptation of making a decision that will injure the people that have angered me. The temptation is powered by my anger and it feels good at that time to get back at the people that hurt me. Revenge seems like the best way to go. But each time I decided to go down this road I ended up hurting myself more.

Because I could not imagine that I was the person that caused other people to hurt like that. It was an uncomforting feeling that left me in a worse state emotionally that I was when I was angry. I end up questioning why I decided to take that path because I gained nothing from seeing others hurt.

Wait until you are sober and not under the influence of anything including your emotions when making the decision. You will be more in control of the situation and the repercussions of whatever choice you make will be clearer to you. This will make you confident in the decision you make and to be more assertive when doing so.

CHAPTER 68:

Identify Your Core Values

What Are Core Values?

Core values are thoughts, routines, rituals, and manners that you were taught to follow. They are ideals and customs observed by family, friends, and leaders taught to you as you grew up. They are beliefs also called personal values that you usually share in common with people in your social circle.

Core values help guide people in how they live their lives, interact with others, and guide our decision making. Core values are not strategies or operating practices. They are not part of competencies or cultural norms. They don't change with the market, administrative, or political changes, and are not used individually. Core values guide us in personal relationships, teaching others, conducting business, and in decision making. They clarify who we are and for what we stand. They can help explain why we conduct business the way we do and are a platform for our businesses.

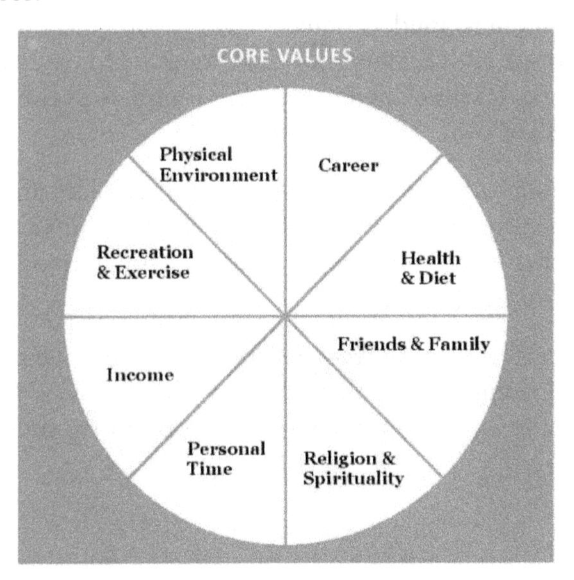

These often differ from culture to culture, family to family, and person to person. A guide will be provided for you, but you are encouraged to enhance and change the example to include your core values.

Many businesses and government departments craft core value statements and mission statements to guide employees. They are meant to instill a strong sense of purpose and standards for which they strive to achieve. To determine where your concerns might lie, let's try another exercise. You want to be in a calm, clear state of mind for this exercise. Phone off, family in bed, dog walked. Take a number of deep breaths and clear your mind.

Grab your pad and pen and create a pie chart with as many slices as you need. I have selected eight areas with values familiar to the average American. You may need to add or alterat some of the headings to fit your life.

You may need to sub-divide some of these slices because I am going to ask you to rate your contentment on a scale of 1 to 10 where 1 is totally discontented and 10 is totally contented. You might be happy with one aspect of your career but wish to improve upon other areas.

Now let's look at some sub-categories of these slices.

- Physical Environment at work. Think of the surroundings in which you work, is the atmosphere uplifting? Are your surroundings loud or quiet, chaotic or controlled, bland or bright, energy-draining or uplifting? Do you feel safe or exposed, is there anything living around you (fish tank, plants, terrariums)?

There isn’t much you can control about work, cubicles and beige paint being what they are. But is your own area clear of clutter and neat with a few items that (if allowed) help create a positive mood to make the most out of your workday?

Now, I realize if you are a K-3 teacher, there are some chaos, loud noises, and draining energy that comes with the job. However, is it controlled chaos, and happy singing at the top of their lungs?

How contented are you with your work environment on a scale of 1-10?

- Physical Environment at home. You may be controlled to what you can do if you are renting, have roommates, or family who's taste differs from yours. But think about your work and play spaces in the home, inside and out. Do you like them? If not, what would you like to amend? Think about your ideal living environment.

How contented are you with your home environment on a scale of 1-10?

- Career. Is your career moving along at the pace you expected or wanted? Are you happy with your projects, clients, co-workers, and company reputation?

How contented are you with your career on a scale of 1-10?

- Health & Diet. Are you in good health? Are you carrying extra weight or stress? How are your blood pressure and blood sugar? Do you smoke? Do you drink?

How contented are you with your health & diet on a scale of 1 to 10?

- Family & Friends. Is there anything you would like to change about your relationship with your family? Too much scheduled time and not enough quality time? Empty nest syndrome? Stressed about someone else's situation? Lack of input into family get-togethers? Holiday nightmares?

How contented are you with your family and friends' relationships on a scale of 1 to 10?

- Religion & Spirituality. Are you spending enough time feeding your inner spirit? This could be services, volunteer work, or more time in nature.

How contented are you with your religious and spiritual experience on a scale of 1 to 10?

- Personal Time. Do you get to take a nap once in a while? Spend time with friends, perhaps without spouse or family? There should be time during the week that isn't dictated by work or others.

How contented are you with your amount of personal time on a scale of 1 to 10?

- Income. Are you satisfied with the amount of income you have available? Are you able to save a set percentage of your pay? You may not be where you want to be financially, but are you on track with a plan to accomplish your financial goals?

How contented are you with your financial goals and savings on a scale of 1 to 10?

- Recreation & Exercise. For most people, there is not a significant amount of time available for recreation or exercise, so we skip it. This is not a slice to put on the back burner. The old saying: "The more you do, the more you want to do" is very true.

How satisfied are you with your time for exercise and recreation on a scale of 1 to 10?

Let's see what some answers might look like. For this example, I'm using a 37-year-old mother of two, divorced for eight months. She has a career and shares custody of the children with her ex-husband:

- Physical Environment - Work - not much we can change, but it is a quiet work environment, with working equipment and up-to-date software.
- Physical Environment - Home - I can't seem to keep up with everything. I'm doing fine to get the clothes washed and folded, and dinner cleaned up. The main rooms get vacuumed regularly, but the other rooms only once or twice a month. And I never have time for yard work, since the divorce. I can't even think about painting or replacing anything right now while the kids are still in grade school.
- Career - I'm lucky to get to work in my field of choice, and it is a good company for working mothers.
- Health & Diet - Doc says I'm 40 lbs. overweight. I think I eat well, just too much.

- Friends & Family - Friends - I have several circles of friends: parents from the children's school/events, some from college, and some from work. Although, again, since the divorce, my time has been limited and I don't see them very often. Maybe it used to be an eight.
- Friends & Family - Family - Quite a bit of our family is still in this area, and when I first started dating seriously and graduated from college, I had to set some boundaries for my own free time. Maybe that number should be higher; I'm actually seeing more of both sides since the divorce.
- Religion & Spirituality - for both. I keep the kids close to nature and their father and his family share their religion.
- Personal Time - The only time I have to myself is getting ready for bed, sleeping and getting up in the morning.
- Income - It's tighter now that we have two households, but we are okay.
- Recreation & Exercise - Again, time is tight with all of the children's events. I don't have much time for my exercise - those are some of the friends I've been missing. I haven't even thought about vacation, given all the changes.

As you have time, change the chart to suit your life and think about your answers. Fill out this exercise and set this chart aside in your work folder.

How Do You Worry?

As we have already noted, everybody worries. But you may have noticed that some people handle worries better than others. We've all seen the person who always seems to be concerned about the problems of everyone they meet. Then there some people who seem to never be worried, never show that emotion. There are different types of worrying.

- Time-sensitive catastrophic: If an event doesn't happen in the time specified, he or she begins to fret. The longer the delay, the more time and energy this person spends worrying. Their worries are overblown and center on the worst-case scenario. For this worrier, life can seem impossible to manage.

My grandmother was one of these. If your plane was supposed to land at 5:30 and you were expected home by 6:30 - if you were not in the house by 6:31 she started worrying. By 7:00 she wanted us to call the hospital to check for accident victims, it was exhausting for everyone.

- Victim: Everything is out of this person's control. They have no power and no one understands. They don't trust people, feel taken advantage of and cheated or abused.
- Avoidant: With low self-esteem, this worrier is a people-pleaser and worries about not being good enough. There are trust issues and this person seeks reassurance from others.
- Compulsive: This person worries about their work and productivity keeping tight schedules. They are overly devoted to work and set very high standards for themselves and others.
- Obsessive: This person is triggered by anything that goes wrong or not according to plan, for anyone in their direct vicinity. This type of worrier expends way too much time and energy on things he or she cannot control. It's a full-time occupation for this person. They put every situation under a microscope and repeat all outcomes in their head. You know this person, if something happens on the news, in the office, or even to a celebrity they worry over it all day long. It affects their work, and their co-workers.
- Controlled: Yes, something bad may have happened but there is nothing they can do about the situation. They carry on with their day even with the worry. Fortunately, we all also know some of these people and should follow their example. They don't let worries consume them to the point they can think of nothing else.
- Histrionic: The queen bee, people are attracted to this person's charisma and imagination. Constantly attracting drama to keep people interested in their calling card. They don't want to be out of the spotlight.
- Dependent: This worrier is worried about abandonment and shows devotion and loyalty to the point of being clingy and needy in relationships. He or she will do everything to keep connected to friend or lover.

- Narcissistic: This person believes that he or she deserves special attention. They crave admiration and worry about keeping up the appearance of perfection. Status and position are everything and they worry constantly about others finding chinks in their armor.

CHAPTER 69:

Clarify the Priorities in Your Life

You will likely have several essential things in your life that you want to do. So, make a list of all your goals and then select the three most important goals. These three goals are referred to as your tier-one goals: those that can alter your life. They aren't necessarily the goals that will help you in churning money effortlessly or earn you fame, but they are the goals that will provide your life with some meaning. These goals can be big or small and could be something like changing your profession, completing your education, or paying off your student debts. The only condition is that the goals mentioned in tier-one should be of some significance to you.

The next step is prioritizing once you have come up with a list of the goals you want to achieve. Some of the things on your list are more important than others. Prioritizing helps you sort them out in the order of their importance. A regular way to do this is to classify your goals into three categories, first tier, second tier, and third tier.

Goals in the first tier are attached to the highest priority. Imagine these goals as "must be achieved" for success in your head. Goals in the second tier are essential but are not crucial to the ultimate vision you have set. In other words, you should want to accomplish this goal as it will only help boost your sense of self-worth. Lastly, those in the third tier may not be as essential, and you could get away with not accomplishing them. This is the "sure, why not" Part of goals. Let me give you an example. A dancer decides to sit and write down prioritized goals to improve their success. This particular dancer is at a point in their career where they feel the need to broaden their repertoire of dance styles. A first-tier goal may be to attend a specific amount of dance conferences across the country in the span of six months. This goal is paramount to improving their success because it offers a chance to learn more styles of dance. A bonus is the networking opportunity to meet other dancers who specialize in their respective styles.

A second-tier goal would be to eat healthy foods or go to the gym four times a week along the way. Do you see how this type of goal serves as a supportive role to the first tier? Healthy food provides nutrition. Nutrition offers stability and longevity in an athlete's career (or anyone else's career for that matter). Going to the gym supports muscular endurance and joint strength, which are primary keys for the body to stay active at peak performance. Lastly, a third-tier goal would be to learn social media marketing to build a following. It's a great goal to obtain. However, it is not directly related to the art of dance or this person's burning desire to improve as an overall dancer.

To summarize, goals not only act as guides in life but also help you realize what is essential in your life. They help give you a laser-sharp focus that will make you impervious to distractions and setbacks that will inevitably come along the way.

Avoid unnecessarily

The following thing to do is to get rid of the things that are "up for what we talked," which are things that are causing you anxiety and stress and taking up too much of your time. You have to get rid of these things because they will suck the energy and life out of you. Items that are not on the non-negotiable list can be considered holistically and given priority when you need it to be.

CHAPTER 70:

Focus On Setting the Conscious Goal

How to Set Goals?

From the beginning, your focus should be to lessen and manage stress.

You should have five major goals in mind:

1. Work on Stressors
2. Employ Stress Management Techniques
3. Develop Stress Relieving Habits
4. Develop Emotional Resilience
5. Try to Find Avenues of Happiness

The goals have been sequenced as per priority, but you can work on all of them together if you can. Simply remember that while you work on the fifth, you can't ignore the first. No major change would come until you meet the first three goals.

Work on Stressors

Identify and Remove Major Stressors

Writing this book, it is impossible for me to guess the things that can stress you the most. But, whatever they are must be identified and removed. This is a top priority task. If you are thinking of moving ahead without accomplishing this task, you are not likely to succeed.

Here, it is important to understand that we are talking about changeable stressors. If deadlines stress you the most, then, does it mean that you would completely stop working and leave your job? That's an impractical solution.

If deadlines cause the highest amount of stress, then start taking less work. Don't let the pressure build whenever a task is assigned to start working from the very instant. Don't wait for the next day to start. It must be your target to finish the task much ahead of the task.

Whatever the stressor is, you will have to ensure that you identify it clearly and get rid of it. Stress management is not about working with stress. There is no way one can work with stress. You simply need to work around it.

Practice Meditation Regularly to Face Stressors Calmly

Meditation is one of the most trustworthy ways to counter all kinds of stressors. Meditation helps you in broadening your perspective. It gives you a positive outlook, and hence, you are able to remain neutral even in negative situations. This isn't a fight to get victory over stressors. We are simply trying to dodge or remain undefeated.

Your most sincere attempt should be to avoid stressors as far as possible, but life as we know it always works in an unpredictable manner. This means that there can be times when you come face to face with the stressors. In those circumstances, you must not lose your calm and composer, and this is what meditation helps you with.

- It gives you a broader perspective
- It lowers your reactivity to stress
- It makes you very calm and resilient

Adopt an Optimistic Attitude

Stressors affect us the most because, in the first instance of stress, our threat perception starts running lose. It is a negative attitude that we have towards life. One way to deal with this attitude is to develop a positive attitude.

Become optimistic in life. Optimism is an approach, and it takes time to make it a part of your life. You simply can't choose to become optimistic in the face of the problem. Therefore, this moment is the best time to start.

Optimism gives you great confidence that will be helpful in several ways to fight stress in life.

Exercise Daily

Physical exercise also makes you better equipped to deal with mental and emotional stress. Any kind of stress is ultimately going to affect your body. If you hear a piece of bad news, your heart will start pumping fast;

your blood pressure would increase, the release of stress hormones would take place. All these are physical reactions, and hence, you can't say that any stress is purely mental or emotional.

Now, suppose you are physically fit and have been exercising regularly. The impact of stress on your body would be less. This also means that your reaction may not be extreme. Remember that stress is simply your body's reaction. The greater the effect, the stronger the reaction. It is a simple third law of motion. If your body is absorbing the shock better, it won't react that aggressively.

Make exercise a part of your daily routine.

Employ Stress Relief Techniques

There are plentiful techniques that can be employed to lower stress in daily life. Here, we'll simply stick to basic explanation and use.

Deep Breathing

Deep breathing is a very powerful technique to lower stress. You can employ this technique anywhere. It would help you in battling current stress very effectively.

Meditation

Meditation is a great way to lower stress in day to day life. You must do meditation at least once a day and, if possible, twice.

Body Scan Meditation

This is also a procedure of meditation that can help you in relaxing your body and mind. This meditation technique uses breathing with your awareness to relax individual muscles in the body, and the effect is amazing. It can help in lowering physical and mental stress significantly.

Loving Kindness Meditation

This meditation is very helpful in developing a forgiving and grateful attitude towards others. If you practice this meditation every morning as you get up, you'll experience a tremendous change in your attitude. You'll find yourself feeling less angry and anxious.

Laughter Therapy

This is again a process of simply laughing out loud as often as possible and especially when you are feeling stressed and frustrated. The laugh would be forced and but the impact would be genuine.

Visualization

Visualization is a meditation technique that helps in calming the mind and lowering stress by imagining various scenarios. This meditation also has a very positive impact on lowering stress.

Mindfulness

Mindfulness is the practice of becoming aware of everything that you do. A mindful person starts making conscious decisions about every aspect of life and takes every action consciously. Living in the present is the quality of this style of living.

Develop Stress Relieving Habits

Ask for Hug from Loved Ones

Love has stress canceling power. When feeling stress, ask your loved ones to give you a tight hug. The personal touch has a very healing quality. If someone who loves you gives you a hug, your body releases a hormone called oxytocin; this has a very helpful effect on your mood.

Aromatherapy

It has been found that certain scents can trigger positive areas in the brain. Taking aromatherapy regularly can keep these areas triggered, and hence, you will be able to counter stress in a much better way.

Creative Hobbies

Creative hobbies give your mind a positive engagement. This also means that your mind gets less time to ponder over useless things, and hence, you will be able to manage with stress better.

Eat a Balanced Diet

Eating a balanced diet can also have a very positive impact on your stress levels. A positive diet keeps your stress level low, and hence, you are easily able to remain happy.

Everyday Take out Time for Leisure Activities

You should take out time for some kind of joyful physical activity daily. From gardening to swimming, you can choose whatever you feel like, but there should be something that keeps you pleasantly occupied for some time every day. It should be an activity you look up to. This will instill positivity, and there would be things you would eagerly wait for. Not only this, in case of any stress in the day, you would get an avenue to positively take out all the stress.

Do Positive Self-Talk

Positive affirmations are great as they keep you inspired, but positive self-talk is even better. Most of us can do self-criticism. It comes naturally. However, positive self-talk is something you need to cultivate. But it has a great impact on your personality. Once we start acknowledging our own abilities and qualities, our confidence increases several times over.

Express Gratitude More Often

Our failure to acknowledge the efforts of others is also a major cause of concern. We simply keep trying to find fault in others and don't really appreciate their efforts. This creates another problem, and that's discontent. We become difficult to please, and as much it is a problem for others, it is also a problem for us. Unknowingly we develop a grumpy attitude, which leads to stress.

We keep feeling resentment against others even for obsolete things, and it all keeps adding up. Expressing gratitude more often in public can get us out of this trap. It can also fill us with more positivity as we would feel happy about others and wouldn't have reason to be angry.

Maintain To-Do List Religiously

Maintaining a To-Do list is very important. If you want to reduce you stress, you must make a to-do list daily and keep updating it. This will inhibit you from missing out on important things and the resultant stress.

CHAPTER 71:

Boost Your Motivation

If you don't give yourself enough credit and believe in yourself enough, you will never find the extra push and the drive you need to break the cycle of procrastination. Sometimes a negative mindset can cause you to start believing that you're not good enough, or you're not deserving enough and thus, it would be better to just not do it at all rather than to try and risk failure.

Why Motivation Doesn't Work

Why doesn't motivation work to overcome procrastination? Simple. Motivation is based on human emotions and human emotions, like people themselves, are highly changeable and susceptible to a plethora of outside forces. Motivation is not a dependable bedrock to build your house upon. It is more like sand on the beach, changing with every tide.

This is the same reason that the gym is so crowded in January, but back to normal by mid-February. People are carried along by the emotional high of a New Year's resolution. They say to themselves, 'This is a new year, and I am going to be a new me.' They feel proud of their resolve, and their short-term accomplishments spur them on for a month or so. Then life happens. The "high" of the New Year disappears and with it, the positive emotions that were carrying them along. When the emotional tide wanes, so does their ability to stick it out in the gym.

Laziness –A motivation Rut

Just as motivation pushes us up, the lack of motivation (demotivation) leads us into a downward spiral. Or, to put it more plainly, causes us to descend into laziness.

If you are someone who believes the lie that you must feel motivated in order to act, you are especially susceptible to this danger. This is because laziness is simply a lack of motivation that has becomes habitual. In order to understand this phenomenon, let's go back to the diet example and the downward spiral of demotivation. The demotivation-spiral happens when an emotional crisis robs us of our feeling of motivation. Your boss yells at you, and interrupts the emotional high of your diet success. As motivation wanes, you eat to comfort myself. Then you feel guilty for cheating, and your motivation completely dries up.

You start to think, "if I fail at my diet once, that is not such a big deal". But what tends to happen to someone relying on motivation, is that after a few of these inevitable downward spirals, a person can become fatalistic in their thinking. "Why should I even try to diet?" they may think: "It never works. I always fail." Their self-esteem is damaged. The demotivation has become routine for this individual.

In other words, if we wait for motivation, we become inert, and lazy. This leads to boredom, depression, and low self-esteem. In such a state, motivation, with its ties to our emotions, never show up. This forms a cycle where we wait for motivation, no motivation appears, and so we do not act. Inaction makes us feel bad about our own laziness and these bad feelings make it even less likely that we can summon motivation. We are now in a habitual state of inaction and laziness.

Motivation Follows Action

Ask yourself these questions:

Do I feel motivated to write a 30-page paper when I am staring at a blank computer screen?

Do I feel like mastering a new language when I can barely stutter 'oui' and 'non'?

Do I feel like running 100 miles as a warm up my cold muscles in the first 100 meters?

More than likely, you answered 'no' to the above questions. For the paper example, after researching and banging out a few paragraphs, the writer feels more comfortable and confident. This is when he begins to feel motivation on a project.

The language learner feels motivation as she learns a few more words and successfully stings together a few phrases.

The runner feels motivation rise with each passing mile.

Another quirk of motivation is that is often follows. When action is taken, small victories spark motivation, such as the language learner speaking a few phrases or the runner putting a few miles behind him. As the writer gains knowledge on the subject she gains confidence and motivation follows. This is the right three-way process:

Action => positive emotional state => motivation

Motivation's Rightful Place

So, is motivation a bad thing? Of course not. Motivation may start you in the right direction or show up once you have already begun to act. It can even help you accomplish smaller, short-term goals. When we understand how motivation works, we can use it to our advantage.

In order to master motivation, we must remember that it is based in our emotions. We can take advantage of positive emotional states or outside motivational factors to get us started on small projects. The positive emotions that accompany setting a New Year's resolution might be enough to get us started at the gym if we look to other strategies to keep us on the right path. The thought of an upcoming visit from your in-laws may be enough motivation to get you to clean out the shoe closet.

We can also remember that motivation often shows up late to the party. We can remind ourselves that even if we don't feel motivated now, we may get a needed boost of motivation once we get going.

However, at best, motivation is an unreliable friend. At worst it is a lie that can trap us in a holding pattern of laziness and inaction. For the long-haul we need something more dependable, something that is present when motivation is nowhere to be found.

Once people start to realize what happens when motivation fails because it is not under their control, they turn their attention to something that they know they can control – willpower. Merriam Webster dictionary defines willpower as energetic determination. Willpower is what helps us get out of bed in the morning and go to work, even though that is not a place most of us want to be. Nobody wants to spend 8 hours of their day cooped up in a cubicle doing a job they're not passionate about, but they do it anyway because they know that they have to do.

Will Power- A Closer Look

Scientifically, there is a specific part of our brain called the prefrontal cortex that controls our decision making, our ability to plan for the future, and make choices that benefit us in the long run. This portion of the brain is found in front of the skull behind the eyes. Studies have mapped the brains of those with weak and strong will power and have seen actual differences in activity of this area of the brain. This little section of the brain could be thought of the 'willpower muscle.' And it activates in much the same way that your other muscles do: if you do not use it you lose it! The good news is that, like muscle, it can be rebuilt, retrained and strengthened.

Training the Willpower Muscle –Start Small and Build

If willpower is essentially a muscle, the only way to get more of it is to train over time. As we take small steps to go against our natural inclinations, we can slowly build up our willpower. If you struggle with writing academic papers, you could start small. Practice writing a page at a time, despite your feelings. Then two pages, then three.

If you are grappling with making healthy food choices, you could by sheer force of will, eat one healthy meal a day, and grind your way up to two, then three. You get the idea. As you train your willpower on these smaller tasks, it will become strong enough to tackle bigger and more complex challenges.

If you aren't sure where to being building up your willpower muscle, or if you realize that you have naturally weak willpower, a good place to start is by taking up a sport or going to the gym. The mind/body connection is a influential one and training them together can have a synergistic effect. If you begin at the gym by walking a half mile, then a mile, then jogging a little while, then running a 5 k, you will see a progress in your physical strength as well as the strength of your willpower.

Remember the adage, 'A journey of a thousand miles begins with a single step.' Think of tasks that you struggle with regularly and break them down into smaller chunks. Practice flexing your willpower muscles by achieving one small chunk of a task. As mentioned, motivation will often follow these small steps to action and help carry you along.

Willpower: Is it Really the Solution?

Willpower is beginning to look like the ultimate solution to our procrastination problem. It is something you can train up, unlike your emotions you can control it, and so there is no downside, right? Think about this scenario:

When your alarm clock goes off. Your brain says "Just 5 more minutes of sleep." Exerting your willpower, you counter with, "I MUST get up and go to work. I WILL do it." You begrudgingly get out of bed and go to eat breakfast, where you are faced with a choice. Pastry or oatmeal? Again, you will yourself to make a healthy choice. "I NEED to eat better. I promised myself I would. Don't even think about it that chocolate donut!" You eat the oatmeal, resenting each bite. In the first ten minutes of your day you have already engaged in two battles of the will. It's going to be a long day.

Just visualizing this scenario is enough to make one weary. There is a reason why. It takes a great deal of energy to go against our natural inclinations and exercise the power of our will. In the last decade or so, scientists started studying willpower and the idea that willpower is a limited resource. Scientist exploring this idea refer to it as willpower depletion. One such study conducted by psychologist Roy Baumeister in 1996 tested the concept of willpower through what was called the Chocolate-and-Radish Experience. In this experiment, Baumeister enticed a group of test subjects with the scent of freshly baked cookies. They were then led into a room with a plate of cookies and a bowl of radishes. Some test subjects were asked to eat the radishes while others were allowed to eat cookies. Afterwards, both groups were then given a complex geometry problem to solve. The group that ate radishes gave up on the math problem twice as fast as the group that ate cookies. The scientists behind this experiment concluded that the subjects who ate radishes depleted their reserves of willpower in resisting the cookie aroma. When they attempted the math problem, they simply had less willpower left that the group who got to eat cookies. The big problem with wanting to use willpower to win the procrastination battle is that if you're going to spend your life fighting against your procrastination tendencies, just like in a real battle, it is only a matter of time before you tire out and fatigue. Because of this, very few people manage to successfully overcome procrastination through sheer willpower alone.

CHAPTER 72:

Simplify Your Life

Most people think that they can accumulate a lot of possessions and still be able to live the dreams that they have. But the issue is that this is not true. Things only get in the manner of being able to live a life where you are able to do what you want and free to do as you please. You need to understand that if you're going to live a life of freedom, then you are going to have to take the time to look at how the things you own are holding you back. Once you come to this understanding, you are going to be able to get rid of the things that are doing just that. Then you are going to be able to spend more time on the things that matter to you.

The first thing that you need to do is look at everything that you own carefully and examine what is required and what is not. You need to see which items are used on a regular basis and which are rarely if ever used. Once you have achieved this, you are to throw out, donate, or sell all of the items that you do not use and keep all the things that you do use on a regular basis. This is the main and most crucial step toward you achieving the life you want.

Next up is to make sure that you do not bring any more clutter into your home. This means that you stay away from places where you usually buy things for the sake of it, like a shopping mall. Only buy an item if you see that you are going to use it on a regular basis. However, you will come to see that most of the things that you think were necessities were just impulse buys that would have resulted in more crap entering your home. That's all there is to lead a more simple minimalist life.

Get Focused

Before you ever embark on a new journey, it is important to get focused and clear on what you are doing. You want to know precisely why you are taking on a new adventure or path, and what this lifestyle will mean for you. Getting focused gives you the opportunity to completely understand what your motives and intentions are and why you should stay committed when things get difficult, which they always do at one point or another.

With minimalism, you should understand that the lifestyle is more than just living a life free of physical clutter. It is also about living a life free of mental, emotional, and non-physical clutter. You need to learn to stay focused on what you want and stop dwelling on things that do not serve you and have no purpose in your life. You can do that by getting focused and staying clear on what your goals are.

Initially, getting focused might be extremely simple. There are frequently two reasons why someone wants to become a minimalist: either they cannot stand looking around at clutter anymore, or they cannot hold all of the restrictions on their time. Because both of these involve stress and discomfort, people are driven to make a change in their life. However, it can be easy to stop making changes once you reach a place of comfort. Or, you may not want to begin because you realize that any difference will be less comfortable than what you are already doing. After all, we tend to stay in lifestyles that are most comfortable to us.

It is crucial that you learn that staying focused and determined takes effort on a constant basis. The focus is a balancing act that you must work towards regularly. The more you work towards it, the more success you are going to have with it. The following instructions are going to help you both with getting focused and clear on your path, and with learning to re-center your focus along the way. You will be directed through a couple of journaling exercises which will give you an excellent opportunity to get clear and provide yourself with something to refer to when it gets difficult. These activities are essential to your success, so it is a good idea actually to invest the time in completing them.

Decluttering Methods

Decluttering is essential to starting a minimalist lifestyle. It might seem a shame to get rid of perfectly good items, but there are several ways to justify decluttering so that you don't feel guilty. You need have no qualms about throwing away things that are worn, stained or no longer useful to anyone. Some of your quality stuff can be passed on to people you know. If a friend has often remarked that he loves a particular figurine, something that's not all that important to you, give it to him so he can enjoy it.

You can also donate items. Goodwill, Salvation Army, and other non-profits receive donations and accept almost anything. A Habitat for Humanity ReStore will be glad for your discarded household fixtures.

You can always hold a garage sale and make a little bit of money while getting rid of things you do not need; you can get to know the neighbors in the process. You can always donate or throw away anything left over. You will be surprised what people will take when it is free. Some non-profits will even pick up your unsold items at their thrift stores.

There are multiple methods available to help you declutter. I suggest you try out several and pick what works best for you. Decluttering does take time. Don't assume to get it all done in one day but do set goals to guide you through the process. I suggest you use a calendar to mark down each stage of your decluttering and assign them specific target completion dates.

It's easiest to tackle the process one room at a time. Be aware that cleaning out a closet will usually take most of a day or even two days; it's a big job! Decluttering the kitchen is also a one- to a two-day job.

Some experts say you should do a little decluttering at a time, giving one item away per day or filling one trash bag in a week. Others say it is all or none. They think you should go through every closet and drawer with clothing in it all at one time, so you don't forget what you have.

Remember, you make the rules. If you want to take it slow, take it slow. Just keep in mind that one item a day means it may take your life to complete the decluttering process! However, if you are enthusiastic about becoming a minimalist, get it all done in a week or two and start enjoying your clutter-free lifestyle?

The following are some popular methods of deciding what to discard, with techniques for staying organized during the process:

The 12-12-12 Method

Twelve is a nice round number. It doesn't take long to gather up 36 items and decide what to do with them. To work the 12-12-12 method, you collect things in your house, finding 12 things to put away, 12 things to donate and 12 things to throw away. You can frequently in a week. It's up to you.

The Four Boxes or Baskets Method

Acquire four large boxes that are nearly the same size or go out and purchase four of the same kind of laundry basket. One will be for trash you will throw out, one is for things to give away, one is for things you want to store, and the fourth is for things you want to keep. Take a room and start filling up the boxes or baskets. Once you fill them, get rid of the stuff in the trash box, box up the things you want to give away, then pack and stash what needs to be stored. Take everything out of the fourth plate and ask yourself, "Do I need this? Does it bring me joy?" If the response is yes, then put it in its proper place; otherwise, put it in one of the other boxes.

The Mapping and Rating Method

In this method, you make a map of all the rooms in your house. Mark where the doors and windows are located and draw in the closets. Draw where the furniture sets. Rate each place as to how cluttered it is, marking one for uncluttered, two for somewhat cluttered, three for very cluttered, and four for the last cluttered space. Start with the most cluttered room first and take that map with you.

Mark with an "X" the most cluttered area and start cleaning out there. You can use your 12-12-12 technique or the four-box method in conjunction with this plan.

Acquire Financial Freedom.

I know that many people argue that money is not everything or money is the root of all the evil... etc... But well, this is not true. According to several studies and research work on wealthy people from all around the world, it is now proven that if you are financially free, then you are happier than those people in your age/income group who are not economically free.

Of course, Money cannot buy happiness. But still, up to a certain level of joy, Financial Security is essential. Most of the people are scared of being broke or even bankrupt after their retirement or even before that because of the substantial debt.

In China, most people worry about their debt while sleeping at night rather than heart disease and diabetes. This is the scenario of people from everywhere around the world. But people that are Financially Free are not worried about these kinds of financial uncertainties, and that's why they are happier than others in the same age/income group.

Financially free doesn't mean that you should be a millionaire or multi-millionaire. It says that your monthly Passive Income from your various Investments such as Stocks, Bonds, Gold, Real Estate & Businesses or even salary is much more than your monthly expenses. Thus, also suppose if you stop working today, you can live for the rest of your life on the Income you generate from your Investments.

To obtain financial freedom, you must master your inner thoughts and spoken words. Your innermost thoughts are the start of everything that you create. What you focus on expands. Fear-based feelings will manifest themselves into reality if you allow them to grow in your mind. You should concentrate on the things that you want so that it expands and demonstrates in your life. Your words are also crucial as negative words such as "I can't afford it" or "I will never be rich" will send out the wrong message. The universe only responds to thoughts and words of abundance. Other things like creating a spending plan, setting financial goals, learning to invest or even simplifying your life all stem from this simple idea of mastering your inner thoughts.

CHAPTER 73:

Simplify Your Home

Decluttering your home is just one of the best things you can do to declutter your life i.e. you must declutter the home to have a decluttered life. The home mirror life as what we have in the house mirrors to a large extent whatever we have in the home. a little clutter here and there in our homes reflect the clutter we have in our life just as it has a way of affecting the management of our lives as regards living a cluttered life. Decluttering the home should, therefore, be an exercise carries out periodically to ensure that our homes are kept in perfect pristine conditions at all times.

The psychological effect of clutters in the house cannot be wished as having cluttered around the home gives the impression that our lives are not organized and that in itself causes a lot of stress. The psychological burden of knowing that your home is not in order gives you excuses for not inviting a person into your home out of fear that they may find out who you are by merely looking around. The choice that comes with decluttering your home spaces gives freedom and peace of mind, promoting the healthy well-being of the individual in the home.

Clutters do not just form in the home; in fact, it's oftentimes difficult to understand how clutters gather and it only becomes obvious when viewed from an outsider perspective. Little by little, gathering of times take spaces and only becomes evident when there appear to be little or no spaces left to arrange our things or to purchase other items.

Thus, it is essential that from time to time, we take a cursory look at our homes and eliminate things or items that constitute clutters around us. Most of these things appear valuable to us that they become very difficult to get rid of, thus encouraging cluttering.

Decluttering the home emphasizes the values of keeping the home clean always. It is very difficult to keep a home clean when we have clutters all around the house; kitchen, bedroom, sitting room store, etc. Having clutter around us makes it difficult to observe simple hygiene in our home as they take up spaces that should have been left bear so that we can effectively arrange and manage our surroundings toward as better healthy living condition. It is also important that we do this periodically so that clutter doesn't stock up to become a burden on the house.

With clutter lying around, things become very difficult to manage. They become energy zappers as they contribute to waste of inordinate amounts of time looking for things that we ought to find easily. For instance, having to spend a great amount of time to look for the remote control of your television or your mobile phone can be very frustrating. The indirect mental and psychological effect of clutter in some cases, tell of the physical condition of the home dwellers as several health challenges may arise such as depression and even obesity. In the worst-case scenario, the inability to manage our homes as a result of clutters in and around could lead to fires disasters and other hazards too.

You can be guaranteed that with a well-decluttered home, you will enjoy a relative peace.You are also guaranteed a stress-free existence because it relieves stress by providing a sense of control and accomplishment. More so, with a well-decluttered home, you realize that there are spaces that could be used for other things or that could be used to store extra things. This reveals the illusion that cluttered homes do give of a packed house. You do realize that after an effective decluttering, many unwanted items are gotten rid of, thus making the rooms and the home in general lighter and spacious. Except you are a minimalist, you do realize that you could still have some more valuable essentials in the house that the cluttered environment has prevented you from having.

One essential consideration put into focus is the reason why the home is cluttered. For an effective decluttering to take place, we must seek the underlying cause of the clutter in the house. Knowing and understanding the causes could help us understand how best to perfectly declutter and unclutter the home. Sometimes cluttering the home happens unexpectedly or unconsciously, this may be largely due to our carefree attitude to little things in the house as where we put stuff or how we gather trash. Most times we do not consider these things as liable to constitute clutters in the long run. However, as a little drop of water makes a mighty ocean, so does little carefree attitude or oversight causes our homes to clutter.

Therefore, to effectively achieve decluttering in our homes, we must understand why the clutters are present and measure out ways to avoid the reoccurrences. This will ensure our homes are well coordinated.

Part of the reasons could be having an overwhelmed life due to a hectic work schedule. W when we have an overwhelming life, we often do not care where things are or where we place them, we get too tired to arrange things as we have a lot of things on our hands that arranging our homes is not one of our immediate consideration.

Sometimes too, consciously or unconsciously, we hoard things, we get too concerned that we might need things in the future such that even if they are not needed immediately, we find ourselves buying and keeping them in the house. This is where budgeting and financial discipline plays a great part in helping to declutter. Sometimes too we find ourselves duplicating these things that we have bought out of oversight.

Whatever the causes of this clutter, we must understand that we need to know and find ways of addressing them so that after spending so much time decluttering, we do not end up cluttering our homes again, which indeed can be very exhausting.

Putting up a Cleaning schedule

To carry out decluttering, there is a need for us to put in place a cleaning schedule. Everything needs and takes time, and decluttering is not an exception. Decluttering requires patience and it is usually time-consuming, so you must dedicate time to clean up and declutter. That dedicated time should be one that is comfortable for the household. Probably a weekend affords the best opportunity as less stress in the week is experienced. However, such a time must be agreed upon by every member of the family. A time frame must be fixed too; an hour or two, with all hands-on deck, could be sufficient to declutter. Depending on the dimension of the house and the number of occupants, a time frame to achieve decluttering must be set as you have other pressing things to attend to.

Apart from the time frame set, the intervals must also be addressed. This exercise does not have to be once a year. It could be monthly, or quarterly, or it could even be as demands call for it. What is important is that a schedule is created at various intervals to address the need to declutter the home.

One Room At A Time

Doing one thing at a time promoted effectively, and you want to be effective in decluttering your space such that you don't come back to redo that which you thought you have done. It is therefore suitable that you take the task of decluttering a room at a time. For instance, you might want to address the sitting room first before others, or the bedroom. Whichever the choice, it is right that you stick with the plan of engaging a room at a time for maximum effectiveness.

Tips For Decluttering A Living Room

The living room is the most visible room and the most used room in the house. There is every possibility that this is even the most cluttered space in the house due to the number of times the room is used in comparison to the other rooms. The sitting room is also the most visible room for visitors. It is, therefore, significant that we keep this all-important room clean and decluttered always, for our comfort as well the comfort of the visitors. The sitting room has a lot to tell us about the state of our other rooms and a visitor can easily see how cluttered our lives are if the sitting room is cluttered. Thus, the sitting room requires extra care and effort in the act of decluttering and ensuring that everything is well arranged and in order.

Your sitting room should be a place of comfort and relaxation where you can enjoy private and bonding time with members of the family. It is the place where we spend most of our leisure time and this makes it, as stated above, an easy place to attract clutter knowingly or unknowingly, especially when you have little children at home. When the sitting room starts to become a mess, the home becomes very uncomfortable to live in. Thus, the sitting room ought to be given pride of place in the scheme of things as far as decluttering is concerned.

How best do you address the decluttering challenge of the sitting room?

Do not keep extra chairs in the sitting room. Always lookout for the temptation to have many chairs in the sitting room more than it is necessary. Replacing furniture material can be a delight as we try to make our living room look up to date with modern upholstery and furniture item. However, we must take due caution in making the sitting room look like a museum for old and new furniture. Keeping the room simple should become of the first concern we put into consideration when decluttering. Having many chairs and cushions laying on and around the floor makes the space look too loaded, and no matter how sophisticated we might want to appear, it ends up doing a great disservice to aesthetics of the room and also the health consideration too. A packed sitting room allows little room for adequate aeration. Simplicity is the keyword here.

Thus, to declutter your sitting room effectively, do ensure all unnecessary, or extra furniture items are taken away. You might be surprised that these chairs could harbor a lot of dirt and clutter which has made the room appear un-kept and stuffy, thereby hindering the good health of the occupants of the house. Ensure to do away, perhaps to the storage, that extra furniture that is not needed in the sitting room, or better still, give them to charity.

Get Rid Of All Paperwork

Paperwork should be packed out! There is the tendency and is indeed the reality the sitting room becomes the room for assignments and office work. This requires a lot of paperwork. Notably, a lot of files might have piled up calling for our actions over time. This makes space unavailable for relaxation and makes the room fuller than expected.

If you have a lot to do that requires a lot of paperwork, instead of stacking them up in the sitting room to make the room cluttered, you could create a room, such as an office within the house to cater for such assignment rather than have them in the sitting room.

CHAPTER 74:

Simplify Your Work

Decluttering your work might be more difficult if you work 2 or 3 jobs, or your job requires you to put in long hours.

But first, you need to sit and think about what you want for your career. Are you in your chosen career? How can you handle it in such a way that you will put in maximum productivity within the shortest possible time and get maximum pay for a job well done?

Concentrate on tasks that will promote you and give you a better chance at your place of work. Prioritize tasks that are important to your company. Also, reduce how much time you spend chatting and using social media while at work to save time. Most importantly, don't design your life around what you earn because your life is worth more than a paycheck. Instead, let your work fit into your lifestyle. There are many things that you may take for granted now that you don't need, and if you cut down on your expenses, you really don't need to beat yourself up for not doing overtime and actually having more time for yourself and your family.

You might not be able to make your work suit your lifestyle at the moment. You might still be living paycheck to paycheck and thinking, "I can't afford to leave my job. Your advice doesn't work for me."

Hold up. I'm not saying you should leave yet.

Your activities can get cluttered when you try to balance the multiple facets of your life. You not only have to contemplate about your body and health. You also have to plan towards work, commitments, daily actions and interactions, and digital life. Working towards these activities take up mental space. Before you know it, they clutter your actions and cause mental constraints.

Getting involved in too many activities leads to clutter. Cluttered actions are like cockroaches. At first, you see one or two roaches hiding behind your kitchen cupboard. If you don't fumigate the house, they breed quickly and multiply. Soon, you have a cockroach infestation to deal with.

Decluttering your actions fumigates your life and disinfects your mind.

Work hard at your job and get all the accolades you need. But remember that if you don't love your job, you'll have mental stress. Your best bet is to start working toward getting a job you like and one that will pay enough, and fit into your lifestyle. When you work hard and make good money from something you love, you'll fulfill your passion. You also see it as less of a chore, so it takes a lot of mental strain out of life. Caroline gave up her job in a fancy lawyer's office because it was making her stressed and grabbed a job that paid less but that she enjoyed more. It cut her stress levels in half, and she was able to adjust her life, so that she didn't need as much money to live on. In the process, it made her a happier and more fulfilled person.

CHAPTER 75:

Simplify Your Digital Life

How Digital Interaction Affects Us

The internet offers countless distractions for all the people using it since there are perhaps hundreds of millions of websites out there. The problem is that we are always browsing the internet using our PC, laptops or smartphones. The human brain is craving for information and the internet is the best source for it. We are checking our emails too frequently, we already have too many websites we visit on a frequent basis, we watch our favorite TV Series or movies on Netflix or Amazon Prime, or we simply listen to music from YouTube or Spotify most of the times.

This is how we "feed" our brain with information or keep it entertained, but the problem is that these activities are taking too long of our time. Do you really want to spend so much time on the internet? You probably don't want this, but you feel you can't help it. Your smartphone is full of all sorts of apps, most of them useless, but enough to cause you plenty of distractions. The apps you are mainly using are the email apps (e.g. Gmail, Yahoo), the social media apps (Facebook, Instagram), or instant messaging apps (WhatsApp, Viber, Facebook Messenger). Your banking apps are not that frequently used, but if you have other vices (like gambling or even gaming for instance), you definitely have some other apps installed (which you use to gamble money on sports bets, online casinos, or just waste an awful amount of time playing games on your phone).

The smartphone allows us to plenty of activities, if you are connected to the internet, from guiding you into unknown territories (maps, satnav kind of apps) to counting the number you step, the distance you covered (through running or walking), or the calories you burned. Perhaps, you use an app on your phone to pay for your bus ticket or for parking. These are just a few examples of very useful applications you can use on your phone, which you only have to use once per day, so you don't spend too much time on these apps.

However, these apps are not the problem, the ones for vices, or the ones with notifications are the ones that are giving us headaches. Obviously, there are social media apps, the email ones, or the instant messaging ones, on which you spend too much time online.

Minimizing Interruptions From Your Desktop Computer Or Smartphone

When your daily tasks or chores are affected by internet use, is time to do something in order to escape this madness. If you don't have a clue what to do in order to prevent these devices from constantly interrupting you, perhaps you need to check the following tips in order to be less distracted by these apps:

1. Turn off notifications

Nothing interrupts you like the notifications from your smartphone or desktop. How many times you immediately stopped what you were doing to check the notification? Probably too many times! Technology was designed to help us, but it seems like it has taken control of our lives. Notifications are making us interrupt our activities in order to check the information that just popped on your smartphone or desktop computer. Why do we let notifications dictate our lives? Do you think that it's really urgent to check the information from the notification? Somehow, we are set to be the first ones to find out about something, this is why we are always rushing to check them.

Easy to say, difficult to apply, right? Well, you will need to get over your obsession with notifications and go deep into the settings of your desktop computer or smartphone. Just turn off the notifications, as the information they provide will also be available later for you to check. Think of how many tasks you will complete if you just don't have anything popping on the screen of your smartphone or desktop PC.

2. Schedule your daily email checking time

You shouldn't check your emails every 15-30 minutes, this is not very productive for you, unless you are checking your work email, and this kind of information you need to access urgently. Personally, before starting your daily tasks at work or when you are having your morning coffee are the two best times to check your personal emails. You can check your emails in your breaks, but besides these periods, completely forget that you even have a personal email address. The ideal situation is to never use the work desktop computer to check your emails, you can just use your smartphone for it, but turn off the notifications on it, as you don't want to hear them too often and interrupt what you are doing.

3. Making content searchable

If you do want to search for content over the internet or in the memory of your desktop computer or smartphone, this definitely needs to be found easily. You don't want to spend too much time searching for documents or files, when you have something better to do. This is why you need to be very organized when storing files and documents on your smartphone or desktop computer, and we are browsing the internet to find something, make sure you use the proper keywords.

4. Do not disturb mode

How many times have you woken up when hearing a notification on your smartphone? Is this happening too often? If the answer is yes, then guess what? You don't have to be connected to the internet when you are asleep. Or, if you want to check the first thing in the morning the latest messages or news, you can just mute your smartphone/desktop computer. Don't worry! You will still hear the alarm, even though you muted your phone. If you don't want to be connected to anything during your sleep, just put the phone or laptop on flight mode. Set your own period when you don't want to be disturbed, it can be between 12 pm to 8 am.

5. Significantly decrease the use of your "vice" apps

If you are too passionate about gambling or gaming, then you need to spend a lot less time on them. Are you a hardcore gambler? Then, perhaps what you need is therapy and to stay away from online casinos or sports betting platforms. You don't want to lose a substantial amount of money on these platforms. I know, the online casinos are luring you with unbelievable bonuses, it sounds like they are just giving money away for free. But ponder where that money is coming from? It's money lost by other players, so how long it will be when your winning streak will turn into a losing one. How many gamblers can honestly say that they won more money than they spent on these platforms? Trust me, not too many of them, unless they are super professional!

Do you remember the hysteria provoked by a very addictive game called Pokemon Go? There are plenty of people who are hardcore gamers and play excessively video games on desktop computers, laptops, consoles, tablets, and even smartphones. These games are causing them addiction and they are so hooked into these games, so they really can't help it. Obviously, they need guidance and help to reduce the time spent on these video games, and possibly even eliminate the whole distraction. This is applicable to gambling addicts as well.

How Life Without The Full Use Of Your Smartphone Will Be Like?

To reduce your dependency on your smartphone, you will need to experience how life without it would be. Have you ever thought about going to some isolated mountain retreat, where you have the basic comfort of human life, but no wi-fi and mobile data are too weak to be used? If you haven't thought of this, perhaps you need to book a vacation. Why not have a full-scale experience of the outdoors, without having to use your smartphone, only if you really have to call for help?

Do you know where North is without using your mobile phone? There are some actions which you can use to guide yourself without having to use your mobile phone, but we're not going to explain them, as we don't encourage you to live like savages, to only use what nature provides, to hunt for food and drink water from the stream directly. However, what you need to put in mind is that when you go offline, the adventure begins and you can start living and engaging in amazing experiences in the real world. This should encourage you to take it easy with your smartphone, as you shouldn't be in a very close relationship with it.

Have you ever been on vacation when you started filming from the first second, or taking photos? It's like visiting a place but not actually being there, as you are behind the camera. You can't experience properly everything what that place has to offer if you are constantly filming and taking pictures. When you are doing this, you are too busy to experience what that place has to offer. You don't have the time to touch or feel your surroundings. The world looks a lot better through your own eyes, then through the lens of your camera, trust me!

We all like to brag about where we are, what we eat or drink, but let's not try to over exaggerate. The thing is, when you share the moment on social media, you are not quite living the moment, so the whole sharing part is blocking you to live the experience at full scale.

Use Technology Wisely To Simplify Your Life

Technology should be used to simplify our lives, not to complicate it. Make no mistake, technology abuse is complicating our lives, so why use technology so excessively? Here's how you can use one of the best features of technology, automatic scheduling. You can use it for:

1. Your emails
2. Your regular appointments
3. Updating apps
4. Compiling travel itineraries

CHAPTER 76:

Simplify Your Distractions

Attaining good mental focus and working with a decluttered mind requires ridding yourself of distractions. These production killers come in all shapes and forms, and each of us individually is the cause of our decreased productivity.

First, we need to get our minds right. Without that, the rest of the suggestions within this chapter are useless. Before beginning any task, you need to clear your mind of all other thoughts, so that you may focus only on the task at hand. We often get distracted thinking about things in our past and in our future, which are largely distracting to what is important: the present. While multitasking appears advantageous to most, the reality is, if you are not concentrating on the task at hand, you are not doing it to the best of your abilities.

Come up with a mental process before sitting down to do anything. This may include some physical things too. For example, you may decide that the tasks that involve quite a bit of concentration should happen after you have checked email, voicemail and anything else that might be intriguing your mind. Knowing that nobody needs anything pressing out of you helps ease the mind.

Depending on the task at hand, make sure you have gathered all of the tools, paperwork or information you may need before beginning. Having to stop and find what you are looking for disrupts your flow and jumbles the mind. For example, if you are scrubbing the bathroom, make sure you have your cleaners and sponges all set to go before even stepping foot in there. If you are filing taxes, gather all of your paperwork and a calculator. Try to anticipate what you need so that the task does not become so stressful.

Next, take just a minute to create a game plan for tackling the said project before you begin. Starting anything without a plan is a recipe for going off on a tangent, both in your mind and with the course of the project as well. Think about what you would like the outcome to be, and the basic steps for how to get there. For example, if you are making an ad to go in your local newspaper, think about a few things it should get across, what color scheme you need to work with and how big the ad will be. Once that's down, you have a great template to work with.

If writing is more your style, making a quick outline to gather your thoughts is the same thing. Any big project may seem daunting until you create the outline and tackle it one section at a time. For example, outline each of your chapters, briefly describing its direction.

This intense moment of concentration requires all of your attention, so once this small action is done, your brain is already primed to take action. Run with that focus and try not to stop until you are finished or out of ideas. Let the flow run its course. If you stop to take a break, look at email, or anything else, you will need to get yourself back on track.

Distractions come in all places, not just in a working environment. We often become distracted by things that keep us from developing relationships, enjoying our time and living our lives. It is sad to think that we spend most of our time at work, and when we are not, we find ways to distract ourselves from having a life.

Social media and television are two of the most detrimental distractions out there. A few minutes quickly becomes hours of downtime, in which your brain has largely been shut off. While it is a good thing to let your brain rest, the addictive nature of these things does hinder any progress in your work, personal and social life. The fact that meeting people and dating is now largely up to technology proves how far this has gone.

Not only are these vices distracting, but they are also the cause of information overload. Our minds are not meant to take in such an abundance of information all at once. It is more used to taking in the scenery of our environment and what is happening directly around us. Instead, we are getting news from all over the world, causing our minds to be in too many places at once. All of this noise is just adding detriment to your mental well-being.

How much time do you spend scrolling through social media or watching television? Do you truly get any joy or pleasure out of either of these activities or is it more to pass the time? The time you are passing is your life, which you are watching other people live on social media and TV. Why not put the phone down, get outside, and do something in real life? Instead of looking at social media, make a point to do something every day that you feel is praiseworthy of sharing on social media.

Limit your time doing these activities and replace them instead with other things that are more valuable, things that improve your life, your health, and your education. Read a book, get out for a walk, and get coffee with a friend. Switching up your downtime activities leads to a more satisfying, enriched life, and that is great for keeping your mind healthy and active. Plus, relieving stress with physical activities decreases mind clutter and improves all aspects of your life.

Everyone will have a diverse set of distractions. Take some time to pinpoint some of the big things that decrease your productivity, tax your brain and tire you out and do something about it. There is likely a workaround to any distraction that exists, so brainstorm some ways to make things run smoother, by taking time to address these problems now will save you time and energy in the future.

CHAPTER 77:

Decluttering Your Relationship

Decluttering is not without its challenges, and if you thought downsizing your belongings and clearing out your workspace was hard, wait until you try decluttering your relationships. The hardest part is probably going to be that period when you're adjusting to not having that relationship in your life anymore. Maybe you have to cut ties with someone you've known for years because you've come to realize that their behavior is toxic and adding to the existing stress you already feel. It's still not going to be an easy decision to make. You've become accustomed to having their presence around and there's bound to be an emptiness that is felt once they're no longer around. It's the same with every relationship you have. When you're so used to always being around or regularly seeing that person, their absence is not going to go unnoticed. From time to time, it will feel like you're all alone, even though you know your decision was for the best.

The Cost of Toxic Relationships

A toxic relationship can happen to anyone, even the most intelligent and confident people whom you thought would be too smart to make such poor decisions. There's no telling when and how you might find yourself in a destructive relationship because it often does not start out that way in the beginning. Toxic relationships will lie to you, hurt you, manipulate you, hence the need for social decluttering. Decluttering relationships and social minimalism are not something we give a lot of thought to since the focus usually tends to center around the physical aspect, the things that we can see in front of us. We're visual creatures, and it's only when we can see something that it starts to feel "real". Mental clutter, emotional, and even social clutter are an afterthought most of the time because we can't physically see the damage that it's doing and how it's taking a toll on us.

Buddha once said: "Holding on to a toxic relationship is like drinking poison, but expecting the other person to die". When you lose yourself in a toxic relationship, your judgment becomes clouded and it is harder to see what is best for yourself anymore. You forget who you are and what you want, and your happiness no longer becomes a priority. You start to get relaxed with it and make excuses for being in that toxic relationship because it feels healthier than having to deal with the pain of letting go of the person that you love (or think you love). Social decluttering is a very complicated issue. It is easy to get consumed in a toxic relationship because they tend to be overbearing. Toxic relationships cause you to lose touch with your own goals, passions, desires, ambitions, and purpose. They make you feel hopeless and powerless, unsure of what to do next. They add to your stress, feed your worries and trigger your anxiety. Their problems begin to affect you and start to become your problems. Eventually, you become resentful after a prolonged period of attending to everyone else's needs except your own. In short, the cost of toxic relationships is simply not worth it.

How to Declutter Your Relationships

Having negative people in your life or people that are causing you pain is pointless. That's not adding value to your life and it's certainly not sparking any joy to keep them around. Negative and toxic individuals will only drag you down and trap you at their level, so you become just as unhappy and as miserable as they are. You don't need to completely disconnect or disband yourself from them entirely (although in some cases that might be better) since it's not always a possibility. The toxic relationships could be within your family or the people you work with, making it hard to avoid them entirely. You don't need to ignore them entirely, but instead, aim to spend less time with them and direct energy and focus towards valuable people instead.

Keeping relationships around because you've known them forever or because you shared a few amazing memories together is not a respectable reason to hold onto them if they're weighing you down more than they uplift you. People change, grow, and develop in different ways. The relationships you have with people change because you change, and decluttering your social life might mean you have to learn to let go and be okay with that. It's going to hurt, but holding on because you feel bad is doing a disservice to yourself.

- Going Through All Your Social Media Accounts - This is probably going to be the easiest step in the entire decluttering process. Start with your social media channels and begin removing the people that you don't know all that well or no longer interact with, and the people that you feel no longer add value to your life. It's going to feel difficult at first, but as you begin deleting these "friends" one at a time, it starts to get easier. Don't worry about what they are going to think. If those "friends" barely put any effort (or none at all) into interacting with you on these social media platforms before, there's a good chance they're not going to care whether you're still connected in the digital space. Streamline your push notifications by turning off the unnecessary ones and only keeping the important notifications turned on. You don't need notifications from your social media apps, they're only distracting you. Notifications should only come from text messages, and your important reminders.
- Go Cold Turkey - This one is going to be a little challenging, but it's the most effective way to identify who are the people in your life that truly value your presence and your friendship. Go cold turkey and stop interacting with everyone. Now, "everyone" in this context is going to be subjective to how many people you know, so there's no fixed number and no specific quota of people you need to top communicating with. Avoid texting first, avoid reaching out first, and avoid being the first one to call or make plans. Go cold turkey, stop, sit back, watch and wait. Observe who are the ones who notice your "sudden silence" and reach out to ask if everything is okay. Who are the ones who value your friendship enough to maintain the interaction? Sharing or liking your latest Facebook status or Instagram post doesn't count. The friendships of value are the ones who check up on you randomly for no reason just to see how you're doing. Or the people who are there for you during your lowest and darkest moments in life. The ones you can really count on in the case of an emergency. You might be surprised to find that the ones who value you enough are a lot fewer than you thought. But at least this way, you're decluttering the unnecessary relationships and keeping the ones that add true value and happiness to your life.

- Knowing Your Relationship Values - What type of relationships do you value? What type of relationship do you think adds value to your life? That is worth investing in? More importantly, what makes you happy? Those are the key indicators you need to think about when downsizing your social connections. Always go back to one, simple benchmark. Is this relationship making you happy? Or is it adding stress?

Decluttering Your Romantic Relationships

Relationships that last a lifetime. Something we all want, yet not everyone is lucky enough to hold on to. Divorce rates in today's world are higher than they have ever been that holding on to a relationship that is strong enough to last you through the years has become a rare commodity. That's because love alone is not enough to sustain these relationships. There needs to be security, a deep connection on an emotional, mental, physical and spiritual level. Love takes work and if you're going to make your job easier, you need to remove the clutter from your relationship. Remove the physical clutter, remove the mental, emotional, spiritual clutter. Any element that is negative and causing a strain on your relationship needs to be evaluated.

When we feel detached from our partners, we don't realize that the clutter (physical and non-physical) could be the reason for it. We don't know how to take in these emotions of disconnect, and as a coping mechanism, some people resort to shutting down or shutting themselves off from their partners instead of working together to find a solution. Negative patterns and coping mechanisms remain prevalent, and until we start actively decluttering our relationships, it's going to get worse, not better. Declutter the negative patterns in your relationship by doing the following:

- Recognizing the Patterns - This is an obvious first step and it requires an exercise in self-awareness. Observe the way you handle things in your relationship right now. Do you shut down when you feel overcome with emotions? Or lash out? Do you work together or work against each other? Conduct a careful assessment of the way you and your partner currently handle stressful or emotional situations and take note of the negative, unhealthy patterns of behavior that need to be changed. Or decluttered, in this case.

- Unite Against Negativity - Nothing brings people together quite like uniting against a common enemy. In this case, the common enemy in your relationship is anything that is stopping you from functioning well together as a couple. Like negative behavior, for example. Think of decluttering as a bonding project you can work on with your partner. If it's been a while since you worked together as a team, this could be one exercise that starts bringing you closer again. Instead of working against each other, make this your goal to work on together.
- Let Resentment Go - A major source of clutter in any relationship is anger, resentment, jealousy, and ego. Everyone has some level of ego and pride within them. How clearly these traits get displayed depends on how well we can control them. Ego left unchecked can often cause tremendous havoc, especially in the relationships which are closest to them. Ego is a negative emotion which causes feelings of resentment, anger, fear, and jealousy. Bringing up past arguments, misdemeanors, and mistakes only clutter the relationship that you have with nothing but toxicity. So, let it go and let the past stay in the past.
- Learning to Ask - One unhealthy habit that causes a lot of clutter in relationships is when couples don't ask for what they want or need. If you're not having your needs met, you need to ask. Your partner is not a psychic and assuming they can pick up on the little hints, or read your mind or anticipate your thoughts is how fights, unhappiness, and arguments tend to happen. Assumptions are where a lot of communication breakdown tends to happen, and you'll be doing your relationship a favor if you learn to communicate and ask for what you want.
- Expectations Are Clutter - Expectations are yet another form of clutter that add no value to your relationship. Imagine if you started every conversation with your partner with the expectation that this is going to go badly. You'll probably sabotage the conversation and make it worse without even realizing it because the clutter is subconsciously weighing on your mind. Why? Because you expected it to go badly. Learning to communicate with an open mind takes practice, but first, declutter expectations and let them go.

CHAPTER 78:

Enhance Your Self-Esteem in Negative Moments

How you feel about your well-being is very important to your happiness in life. Having a high opinion of yourself, what you do, who you are and love for yourself is one important thing that people have little of in today's society.

Some benefits of developing high self-esteem include:

- Life becomes lighter and more straightforward.
- More inner stability.
- Less self-sabotage.
- You'll be a more admirable person to your close ones and colleagues at work.
- You'll be happier.

Those are the benefit of developing good self-esteem, but how do you develop this habit? Here's how;

1. Say "Stop" to your inner critic

An ultimate place to start if you want to raise your self-esteem is by learning how to change the voice of your inner critic.

Everyone has an inner critic that often spurs us to do things to gain the acceptance of others in your life and on social media. This need to gain acceptance drags your self-esteem down as you will begin to judge yourself by how others judge you.

The inner critic is that voice inside our head that says destructive thought in your mind. For example, it says words like;

- You are not worthy of this position; it's above your technical skills.
- You aren't worthy of that girl; she will leave you for another.

- You are a bad mother.

You don't have to accept the things the inner critic says to you; there are ways to minimize them and replace this thought with more positive thinking. You can change what you think about yourself.

One way to get over the inner-critic is by stopping whatever the inner critic pipes up in your mind is to create a stop-word or stop-phrase for it.

Once the inner critic brings a thought to you, shout "Stop" in your mind. Or come up with your word or phrase that can stop the train of thought. Then refocus your thoughts to more positive things.

2. Use healthier motivational habits

To reduce the inner critic's intensity, motivate yourself to raise your self-esteem, and take positive actions.

3. Take a break for self-appreciation

Self-appreciation is very fun and straightforward, and you will notice a remarkable difference if you spend just two minutes on it every day of the month. Here's how to go about it;

Gently take a deep breath and ask yourself this question "What are the three things I like about my life?" A few examples of answers you can give yourself are;

- What I write impacts a lot of people.
- I'm a good boss at work.
- I'm very caring and thoughtful when it comes to dogs.

These talks won't just help you build your self-esteem; they will also turn negative moods into a good one.

4. Do the right thing.

You will raise and strengthen your self-esteem when you do what you think is right. It might be a small thing like going to the gym in the morning or helping your child with their studies.

To make it more effective, stay consistent in the right thing you've made up your mind to do. Make sure you take action every day.

5. Handle mistakes and failures positively

It is normal to stumble and fall if you go outside of your comfort zone. It is necessary if you want to do things that matter in life. Everyone that has got to great heights in life did; you hardly hear them talk about it. So, remember that, and when you falter try to do these things;

- Be your best friend: Rather than beating yourself up and being angry about it, ask yourself this question. Who will support me in this situation? How will the person help me? Then start imagining the person advising you.
- Find the upside: one other way to be more constructive is to focus more on opportunities and optimism. Ask yourself these questions: What can I learn from this? What benefit can I derive from this situation? These will help you to change your viewpoint on failure.

How to Develop the New You

A lot of us build up images in our heads about who we are "I'm boring, I'm suffering from depression," etc. we reiterate this point to ourselves, and we convince our mind that it's the way we are. No matter the length of time that you've been repeating these ideas to yourself, in reality, it is always changing.

I do not have any uncertainty in my mind that anyone can change anything they want about their life, whether it is a mental image, physical appearance, or a bad habit. If you're ready to develop the new you read through the fifteen steps below:

- Have an understanding that there is nothing that can't be changed about your life, even though it looks permanent. Think of the changes that occurred in your life from a younger age until now.
- Learn that a belief is not the truth but a thought you keep thinking.
- Have an understanding that our beliefs are not the truth; they are just things we chose to accept. For example, my favorite color is blue, I'm shy, etc.
- Have a realization that unless you have the desire to change, the change will never happen.

- Decide to change. If your decision to change is not deep rooted, real change will elude you.
- Gradual steps towards the change you desire. You can run towards the peak of a hill in a flash. You have to take little steps that matter every day. Once you form a habit of taking those steps in the first few weeks, the pattern sticks, and it becomes easy from there on. If what you want to change from is an addiction, gradually reduce the craving. For example, if you're going to quit smoking, gradually reduce the number of cigarettes you smoke every day.
- Experiment with things that you find easy to change at first; this is to create the belief in yourself that if you can stop doing this, you can stop doing that.
- Choose something that you always had in mind to change, so that you will appreciate the effectiveness of your new mindset.
- To make the change real, tell some people close to you, and give yourself some accountability.
- Ignore the inner critic that keeps telling you that "you can't do this."
- When you mess up, don't give up. It's normal to mess up at times.
- When you slip back into old patterns, make a considerable effort to get out of it. Don't scold yourself; notice it, laugh at it, and try again.
- Don't let the opinions and suggestions of others weigh you down; it will take some time for them to notice the change.
- See yourself as the new you. Feel free to tell others that things are no longer the same. Say to them, "I now exercise everyday", "I will not smoke again" etc.
- See the old part of you as a part of you that is gone. If your mind or a friend suggests that you should go back to your old self, say, "That was the old me." If the thoughts keep coming back, distract yourself from them. If a friend keeps suggesting that you revert to your old lifestyle, cut every means of communication with them.

CHAPTER 79:

Simple Daily Practices to Overcome Procrastination

When talking about procrastination, everyone might relate to it because there's none who could deny it. Whenever you miss your deadlines, the level of anxiety rises above your head and you are forced to complete the project as soon as possible. But deep down, you know it is impossible to complete because there is so much to do. Yet, you try! Procrastination will make your life miserable, so try not to make it a habit.

Some people want to stop procrastinating, but they are unable to because they don't know how to do it. Or sometimes, they might be missing the motivation they need. And it can be frustrating, I know. You must understand the fact that procrastinating factors differ from individual to another:

A writer will procrastinate on the project he/she was assigned. And then, he/she must work day and night to complete the project.

A student will delay school work and then, complete it at the last moment.

An athlete will delay medications because he is so concerned about the current game.

If you evaluate each example above, you will understand that through procrastination every individual mention in the example will be affected. For instance, the athlete will have to deal with a lot of severe issues if he doesn't treat the injury right away. Likewise, there will be a lot of emotional drawbacks as well.

I'll share some of the practical daily practices that you can follow to overcome procrastination. These practices will help you beat procrastination even if you are feeling lazy or unmotivated. Before you begin reading the practices below, you must bear in mind that you can select any of the following practices. This means you are not forced to practice all the habits below. Let's get started!

1. Find solutions to potential emergencies

Procrastination is not a simple bad habit; rather it is dangerous. It will have a vast impact on your health. Sometimes, you might even lose the great bonds that you shared with your family members. They might even come to a point where they assume that you no longer care. There will be situations in life where you have to deal with unexpected priorities such as death, sickness, and much more. Such situations can't wait because you will have to address them immediately. In this case, you would have to drop all the scheduled tasks. Some other times, great family events might turn into dreadful situations, and you can't avoid them and get back to your work. Emergencies don't come with a warning, so you have to put up with the obstacles it creates. How can you avoid emergencies? Are you going to stop everything and address the issue? Or if you have already delayed the work and then, something urgent comes up, how are you planning to handle it? What might happen when you ignore the emergencies?

To handle emergencies, you have to have a clear picture of the type of emergencies that you are dealing with. You can think about the aftereffects of avoiding the emergency. Or think about the people who are related to the emergency, how will they feel if you ignore it? What are the activities that you can take to solve this emergency issue so that you can get back to work? Or can you put off the emergency issue because it is not life-threatening?

Before you dig in further, let me tell you. If you are working so hard that you don't even have time for your family, it means you are losing a lot of good things in life. You are not actually living your life — this where the concept of smart working comes into the picture. You can easily get busy and forget about the people around you. Or you can easily put off emergencies that you believed as not important, and those emergencies might actually turn out severe. Of course, you might be so busy that you don't even have time for important things, but it is all about your priorities.

No project, appointment, or meeting is worth ignoring for the emergencies that might affect the life of a loved one. I'd suggest stopping other things when something urgent comes up because procrastination is not only about work but also about life. If you address emergencies right away, you wouldn't have to deal with the worst cases down the line.

Most of the time, we think procrastination is all about work and how we delay work. But here I pointed out something that you should consider.

Basically, if you organize work-related activities and complete before the deadline, or if you have completed half the work already, unexpected priorities might not create a huge impact on your work life. What matters is being organized and knowing how to prioritize your life matters.

2. Do daily review

Another best way to avoid procrastination is through daily reviews. If you allocate ten minutes from your day, you can do the review. When you are doing the review, you will be able to find the priorities of your day. Then, you can analyze the tasks that will have a huge impact on your short-term goals. To make this review session simpler, consider carrying out a Q&A format. What are the scheduled meetings that you need to attend? Are there any emails that you must reply to today? Are there any official papers that need to be edited today? Are there any appointments that will take more time than you allocated? What are the tasks that require more attention?

Likewise, you must do a Q&A to find out the layout of the day. But you don't have to stick the questions that I have mentioned. Instead, you can prepare your own Q&A and follow it. If you do this daily review, you will be able to understand the layout for the day. When you have a layout, you will be able to stay on track. You will have proper knowledge of the tasks that need more time or a quick response. Hence, you will not procrastinate because you are aware that it will impact your goals negatively.

If you want to study one of the best concepts that beat procrastination, it will be the Pareto Principle. This is all about 80/20 rule. Try to learn more about this concept before you actually apply it to your day to day activities.

3. MIT's or the Most Important Tasks

It's tough to beat procrastination if you begin your day with a to-do-list that bursts out with tasks. You must have a simplified to-do-list if you want to get things done on time and correctly. How can you simplify your to-do-list? It is pretty simple if you focus on MIT's - most important tasks. You have to settle for the tasks that will have a considerable impact on your long term goals. This is recommended by many experts who focus on productivity.

My tips are to select the top three important tasks that need to be handled by the end of the day. It is better to pick two important tasks that have tight deadlines and another that will impact your long-term career goal. If you keep an eye on MIT's concept, you will be able to curb procrastination. Once you complete the two most important activities of your day, you will be interested in doing the other activities by the end of the day. And that motivation is very much needed if you want to succeed in beating procrastination.

4. The Eisenhower Matrix

Of course, who doesn't like productivity? You'd be glad if things happen the way you planned. But sometimes, things don't work as you planned. If your life is also like mine, filled with constant emergencies and changes, you must have the ability to make quick decisions. If you want to make a decision, you need the support from the Eisenhower Matrix. The founder of this concept was in the army. It was the reason why he invented this concept. It's not possible to work according to the plan when you are in an army. There will be sudden changes and importance. In such an instance, the Eisenhower Matrix concept was the guideline. If Eisenhower utilized this in the army, why can't you utilize this in your life to avoid procrastination? When you are dealing with this concept, you shouldn't forget the four quadrants related to it. By focusing on the four quadrants, you will be able to approach your day to day tasks accordingly. Let me mention the four quadrants in detail:

Quadrant 1: Urgent plus important

These are the tasks that required to be completed first because they are way important than any other tasks and they directly deal with your career goals. Plus, you must complete the tasks right away because they are urgent. If you complete these tasks, you will be able to avoid negative consequences. Once you get your Q1 tasks completed, you will be able to focus on other tasks. For example, if you have to submit a project at the end of the day, your complete attention should be given to that project because it is both urgent and important.

Quadrant 2: Important yet not urgent

The tasks under Q2 are important, but they are not urgent. Even though they might have a huge impact, they are not urgent. Compare Q2 to Q1, and then, you will understand the difference clearly. Typically, Q2 tasks will include the ones that have a huge impact on your long-term career or life goals. Yes, you need to allocate more time and attention to these tasks. But you seldom do it because your mind knows that the tasks in Q2 can wait. Meanwhile, you'll be focused on the tasks in other quadrants. Don't make this mistake because your long-term goals are the reasons why your short-term goals exist. For example, your health is one of the important factors, so if you don't spend enough time on it, you will regret it. Yet, when you get busy, you are unlikely to spend time on Q2 tasks. Especially, you are not obliged to answer anyone about Q2 tasks.

Quadrant 3: Urgent yet not important

The tasks under Q3 are urgent, but you don't necessarily have to spend your time in it. You can either automate or delegate the tasks to someone who can handle it. These tasks are not so important, so it is okay to delegate them. These tasks often come from a third party and the tasks under Q3 will not have a direct influence on your career goals. But when you are handling Q3 tasks, you must note down the tasks that you delegate. For instance, if you are doing a time-sensitive project and if the phone rings and if you attend you might get distracted. Or sometimes, it might not even be an important call. For such activities, you can assign someone. Even if it's an urgent call, you can still assign it to a person who can handle it. Through this, you will be able to manage your day!

CHAPTER 80:
Be Prepared Without Obsessing

People worry expecting something bad to happen. Worrying becomes a security blanket that deals with the threat that the future is likely to bring. It is not a bad thing, but when your anxiety grows as a result of your desire to control future events; you suffer the consequences of your obsession.

Worrying in this light becomes very tiring because no matter how strong you are, you cannot control the future. You can plan for various eventualities but at the end of it all, your obsession is only going to exhaust you.

Planning vs. Worrying

Are you planning or worrying? It is always good to be pre-plan and prepare because it helps you deal with the outcome. The negative aspect of planning is that when you become too obsessed with controlling the outcome, you get completely overwhelmed by your anxiety.

It is essential to plan because it helps keep things in order, but you should be aware of the boundaries so that you are sure that your planning is still proper:

Planning	Worrying
Listing ways to positively achieve a certain outcome	Listing how ill-prepared you are for a future catastrophe
Making step-by-step action plan	Trying to conquer all aspects of the problem simultaneously in an effort to overcome it

Recognizing things that you can and cannot control	You are obsessing over things you cannot control, in an effort to control them
Making plans towards fulfillment of things that you can control	You are getting frustrated and fearful about your lack of control
You are prepared to seek assistance from others, if necessary	You are certain about not needing assistance of others and if ever, you will delay asking for help till the last minute

Looking at the table above, are you a worrier or a planner? Are you crossing the boundary of obsessing over threats and danger, to the point that you are completely overwhelmed by your emotions?

Writing Your Worry Diary

A good way to assess if you are obsessing is to keep track of your behavior with the help of a 'worry diary'. You see, worrying is not a bad thing. But when you are already obsessing, it can be very disruptive and destructive.

A 'worry diary' is like any journal. It is meant to keep a record for the purpose of tracking. How do you maintain this 'worry diary'?

Step 1: Write down all your worries. Do not be afraid to elaborate on what you feel. Write down all your worries in an effort to declare them and by doing so, you will release it and unburden yourself.

Step 2: Clarify your thoughts. After declaring your worries, it is time for you to understand the depth of your problems so you can decide which ones are worth your attention. Are your fears and worries legitimate? What is the likelihood of these things to happen?

Step 3: Challenge your thoughts. After examining every single worry you have, face them head-on by challenging them. The problem with most people is that they let their worries destabilize them and thereby get stuck in a helpless position. Scrutinize your worries to understand them thoroughly and then you can effectively deal with them. Consider all possible solutions to your dilemma and imagine the worst case scenario.

Step 4: Reframe your worries. At this point, you should have a better understanding of your worries and anxieties, so that you can reframe them in a more realistic sense. Your worries do not have to be the endpoint of all things. There is a way that you can strategically conquer all your worries so that they do not win over you. By reframing your worries, you give them a different outlook, so that they no longer feel so threatening. Being prepared is always an asset. It allows you to be more systematic, but at the same time, it encourages obsession. When your planning takes this course, it is no longer healthy. It is therefore important, that you keep yourself in complete touch with your emotions so that you can handle them efficiently and effectively.

Live Today

The threat of dreadful things to come will always bring enough reasons to worry. Future events that threaten to come in a catastrophic manner have the influence to overcome us. The very definition of worrying is, "causing anxiety about actual or potential problems". It may not have happened yet, but it has already disarmed you. Is that healthy?

Some people will argue that worrying allows them a level of preparedness.

The threat of future events may be real but you should not allow yourself to be entrapped by them. The consequence of this is neglect of the present.

Threats of The Past

The future can be ominous but the past brings poison. Some people are not necessarily consumed by the unknown future, but they are unable to let go of their past. They become a prisoner of their past pains, defeats and suffering and consequently become reluctant to move forward.

Are you afraid of the shadows that hide in your past? Sometimes the past represents mistakes and errors that brought destruction—it causes someone to be wary and defensive. No one wants to repeat their mistakes. Worrying about the past is extremely natural because past events have a way of adversely affecting the present and future events. But dwelling on the past is not going to change the outcome of things that have already happened, so you should focus on the present because at the moment that is what matters most.

Living in The Present

Sure there are potential dangers looming over you. Of course, anything can go wrong. Being oblivious to the danger of the future is irresponsible and carelessness but abandoning the present is more ridiculous.

1. You have to make a decision to live in the present because:
1. It will teach you to be more forgiving. It teaches you to be more grateful for what you are today, and gently releases any regrets or grudges that you may carry from the past. Everyone has scars to show, but not everyone is able to forget the intensity of pain it represents.
2. It will give you a sense of fulfillment. Life can throw anything in your direction. If you do not acknowledge it, you will lose the opportunity. There is so much life has to offer but sadly, people miss most of it because they don't pay attention. If you keep your eyes open to the present—you will experience life in its best form, and always feel complete, happy, cheerful and fulfilled.
3. It will take care of your worries. Living in the present is like leading an active life that is focused on accomplishing day-to-day functions. You want to get things done and you focus on them, so your worries automatically get blocked, even without you knowing it.
4. It will make you feel free. People are too overcome with expectations, past happenings and worries. These things can weigh quite heavily and can greatly affect your life by confining you within the boundaries. By letting go of your past and living the present moment, you will feel much

better. It will feel as though your world just got bigger and less demanding.

5. It will help avoid disappointments. When you are living within strict boundaries and are compelled to meet expectations, there is a greater chance for you to encounter failures and disappointments. Living in the present makes you feel free, and also happier. It is easier to get upset when a list of unreasonable expectations is not met, so let go of all these boundaries.
6. It will open you up to better relationships. Positivity attracts people and living the moment opens your life to the world in a 'can-do' attitude that many people appreciate and admire.
7. It will make you happier. Worries can work as obstacles. When you allow them to inundate you, they can overcome you to the point of destroying you. Worry and anxiety are like poison and when you choose to live in utmost toxicity, your life will feel heavy. But when you choose to live in the moment, you will feel lighter and happier.
8. It will make you more accepting of the unpredictable. The future is extremely unpredictable. There is no way for you to know what's going to happen, and so you have the tendency to obsess on trying to control everything. When you learn to start living in the present, you become more accommodating and tolerant of events, irrespective of how unpredictable they are. Unpredictability is frightening but your fears don't have to overwhelm you.

The past and the future are significant, but your present needs your full focus and attention. The problem with most people is that they are either trapped in their past or are threatened by their future. Both distant but the obsession of the past and future takes the focus away from the present which matters the most.

How To Live Today

By now you should have a better understanding of how important your present is, so you will need to focus. The past is instrumental in formatting your present and the future is the fulfillment of your past and present days. The future is the final destination and it is natural to obsess on the details, but you have to learn to prioritize.

You choose to live today because your present is more important than the growing threat of your worries and anxieties.

Step 1: Let Go. You have to let go of your past, however painful it is; and you have to surrender your future. You have to accept that there are certain things that you cannot change or control, so just let go. There is no point in obsessing about these things.

Step 2: Savor It. Life is too beautiful for you to simply take it for granted. Your present life is there for you to savor and enjoy so allow yourself to experience it. Why are you so worried about the past and the future, when what matters is the present moment? Experience every moment as it unfolds and this will allow you to perceive life in a different way.

Step 3: Practice Mindfulness. Mindfulness represents your ability to clear your mind of the negativity so that you can focus. To live the moment, you have to be fully aware of the happenings around you. This will allow you to elicit appropriate responses for various situations, and it is achieved through meditation. Mindfulness is easily transformative.

Step 4: Take Action. Do what needs to be done today. When you talk about living the moment, it involves taking action. It means that you have to take the step that actually keeps the ball rolling. You shouldn't waste any more time dwelling on the past that you cannot change; and you should not worry about controlling the future because no matter how much you worry, the only thing that matters is the present.

You live today because the truth is that time is gold and if you dwell too much in the past and future, you will not be able to catch the present. If you are not careful, time will slip away and you will just lose it. To be able to exist at the moment, you have to be in complete control and find perfect focus.

CHAPTER 81:

Why Good Sleep Is Important

Good sleep is extremely important to your wellbeing. Actually, it is as essential as healthy eating and exercise. Sadly, a lot can interfere with normal patterns of sleep. 10 reasons why good sleep is critical.

1. Bad sleep has to do with higher body weight

Bad sleep has to do with an increase in weight. Those with a short period of sleep tend to weigh much more than people who get adequate sleep. Probably, short sleep is one of the most significant risk factors for obesity.

In one comprehensive research, children and adults with a limited sleep period were 89% and 55% more likely to experience obesity.

It is assumed that the influence of sleep on weight gain is mediated by several factors, including hormones and motivation.

If you are stressed to lose weight, it's absolutely essential to get quality sleep.

SUMMARY: short sleep time in children and adults is associated with a higher risk of obesity and weight gain.

2. Healthy sleepers prefer to consume fewer calories

Research has shown that people who are in need of sleep have a larger appetite and consume more calories. Deprivation of sleep causes normal appetite hormone variations and is believed to cause impaired appetite control. This involves higher ghrelin levels, the appetite-stimulating hormone and lower leptin levels, the appetite-suppressing hormone.

SUMMARY: Poor sleep affects appetite-regulating hormones. People who sleep well, eat fewer calories than those who do not.

3. Sleeping well will boost focus and productivity

For various aspects of brain function, sleep is essential. Cognition, attention, efficiency and results are included. Both of these are adversely affected by a lack of sleep. Scientific research gives a good example. Modern interns with extended working hours over 24 hours made 36% more severe medical errors than interns with a schedule allowing for more sleep. Another study found that short sleep can have a negative effect on certain aspects of brain function to a similar extent than alcohol poisoning.

However, good sleep has been demonstrated to improve problem-solving skills and enhance children's and adult memory performance.

SUMMARY: great sleep can increase problem-solving and improve memory. Poor sleep was shown to affect the function of the brain.

4. Athletic performance can be maximized by good sleep

Sleep has been demonstrated to enhance athletic performance. Long sleep was shown to significantly improve tempo, accuracy, reaction times and mental well-being in a study of basketball players. Less sleep was also related to poor exercise results and functional limitations in older people. A research performed in over 2,800 women has shown that poor sleep is related to slower walking, weaker grip and more challenging independence.

SUMMARY: longer sleep has enhanced many aspects of physical and athletic success.

5. Bad sleepers are more vulnerable to heart attack and stroke

Quality and length of sleep may have a big influence on many health risk factors. Those are the factors that are likely to contribute to chronic illnesses, including heart disease. A review of 15 studies showed that people who do not sleep enough are far more likely than people who sleep 7-8 hours per night to have cardiac disease or stroke.

SUMMARY: the risk of heart disease and stroke is increased when you sleep less than 7–8 hours per night.

6. Sleep affects the metabolism of glucose and risk of Type 2 diabetes

Sleep restriction increases blood sugar and decreases the response to insulin. In a healthy young men report, the sleep reduction of six nights in a row to four hours a night-induced prediabetes symptoms after a week of decreased sleep. Bad sleeping habits are as well linked to the general population's adverse effects on blood sugar. Those who sleep under six hours per night have consistently shown an increased risk of type 2 diabetes.

SUMMARY: sleep deprivation in healthy adults can lead to prediabetes within six days. Most studies show a strong link between short-term sleep and type 2 diabetes.

7. Poor sleep is associated with depression

Mental health problems, such as depression, are closely associated with poor sleep and sleeping disturbances. 90% of people with depression are estimated to worry about the consistency of their sleep.

Bad sleep is also related to an increased risk of suicide death. People with sleep problems such as insomnia or obstructive sleep apnea may have substantially higher depression levels than those without.

SUMMARY: poor sleep habits are closely linked to depression, particularly for people with sleeping disorders.

8. Sleep enhances your immune function

Even a slight loss of sleep has proven to weaken your immune function. A wide two-week study tracked the production of the common cold after people received cold virus nasal drops. They found that those sleeping less than seven hours were almost three times more likely than those sleeping for eight hours or more to catch a cold.

When you get cold sometimes, it might be very helpful to ensure you have at least eight hours of sleep a night. It can also help to eat more garlic.

SUMMARY: a minimum of eight hours of sleep will enhance the immune function and help to prevent the common cold.

9. Bad sleep is associated with increased inflammation

Sleep can have an important impact on your body's inflammation. Sleep loss is actually known to cause excessive inflammatory markers and cell damage. In disorders called inflammatory bowel disease, poor sleep has been closely linked to long-term inflammation of the digestive tract.

One research has shown that sleepless people with Crohn's disease are twice as likely to relapse as well as sleeping patients. Researchers also suggest sleep tests to help predict results for people with long-term inflammatory problems.

SUMMARY: sleep affects the inflammatory responses of your body. Poor sleep is associated with inflammatory bowel diseases and may increase the risk of recurring diseases.

10. Sleep affects social connections and feelings

Sleep deprivation limits the ability to socially communicate. This was supported by multiple experiments using facial emotional recognition tests.

One study showed that the capacity of people who did not sleep was reduced to identify signs of rage and joy.

Researchers think bad sleep has an effect on your ability to identify important social indicators and process emotional information.

SUMMARY: deprivation of sleep will reduce your social skills and ability to interpret the emotional expressions of people.

In the end, good sleep is one of the pillars of health along with nutrition and exercise.

Without taking care of your sleep, you literally cannot reach optimal health.

CHAPTER 82:

Journaling

Talk to any life coach or motivational speaker worth their words and the single most important tip they'll give you for decluttering the mind and creating more mental space for constructive thoughts is journaling.

Journaling is known to foster positive thinking by conditioning our subconscious mind in the right direction. It is a powerful tool for reflecting upon and out thoughts and keeping a close watch on our goals and priorities. The power practice of writing allows your subconscious mind to internalize those ideas, and thus lead to actions or behavior that adds to the positivity.

Everything that makes its way into the subconscious is believed to be real. When you keep writing something repeatedly and consistently, you lead the mind into believing it as true.

Everyone has a different style of journaling, and it's highly personal. There is no right or wrong way to do it. While some folks work with a gratitude journal, others stick to a goal journal, and still, some others go with penning down their reflections.

If you get into the habit of penning down your thoughts regularly, sometimes even you'll be stunned by what you write in a flow that is directed by the stream of consciousness. The thing is that thoughts stored in our subconscious mind are not immediately accessible to the conscious mind. When you write in a flow guided by the stream of thought of consciousness, you are basically accessing everything that is stored in the subconscious, thus unveiling powerful thoughts, feelings, and messages. This reinforces the positivity and creates more of it in the subconscious mind. It leads to a cycle of positivity that declutters the mind effectively. Here are some powerful tips to help you declutter the mind through journaling.

Maintain a Daily Journal

Maintain a daily journal, where you list 10 to 15 things that happened throughout the day which you are truly thankful. Try including new blessings each day. It can be the smallest possible of things like being able to enjoy the natural beauty on your way to work because of the eyesight/vision you've been blessed with or say bumping into a person you hadn't seen for long.

Go beyond the regular stuff to show thankfulness towards the tiniest things. Your subconscious mind will soon realize how fortunate you are, and guide your actions into creating more of these blessings. Practice this just before going to bed to create a deeper imprint on the subconscious (which is active while we are asleep). Show appreciation towards everyday things that you would otherwise take for granted. It is powerful for eliminating plenty of mental clutter and replacing it with positive thoughts and emotions.

Personalize It

Your journal equals to your mind and your individual, unique reflections. Pick a journal design or journaling style that is personal and connects with you at a deeper level. Some people find great fulfillment in accompanying their journal entries with drawings and sketches, while others find writing poems highly cathartic.

Use a bunch of relevant motivational quotes, visuals, stickers and comic strips. Add short stories that inspire you or images of people who've had a largely positive impact in your life. Include lyrics of songs that have a special relevance in your life or have inspired you at some stage. Include anything you can closely relate to. I know people who add ticket stubs to remind them of their travels.

Embellish it with evocative/inspiring images and affirmations. Make it an enjoyable and fun compilation of personal writings that you can instantly identify with. Personal reflections don't always have to be intense, serious or boring. You can keep it humorous and lighthearted too.

The act of writing should reinforce your goals or objectives in life, and drive your thoughts and/or actions. Though there are several applications and tools to write journals virtually, I would strongly recommend handwriting them to make it even personal and build a powerful link with the subconscious.

The process of writing with your hand keeps the writings more spontaneous, natural and unfiltered. It also gives you greater clues or insights about your emotional state of mind. Some people decipher their emotional state simply by looking at their handwriting.

Think of your journal as a tangible representation of your thoughts, feelings, and goals.

Devote a Fixed Place for Journaling

This isn't sitting inside an ivory castle for hours while you pen your thoughts. It is about devoting a single space to express your deepest thoughts. Having a regular place for journaling allows you to reflect on your thoughts more effectively.

There is a greater scope for introspection in a place you can closely connect with or are comfortable in. Your journaling space can be anywhere from a meditation room to your garden or backyard. The space should induce a feeling a feeling of introspection or reflection. Ensure the place you pick is free from distractions and positively energized.

Include objects such as candles, flowers, statues, and images to create a more pleasant, inspiring and positive sensory experience while writing.

Ditch the Edit

You don't have to craft your journal entries using impeccable prose. It doesn't need to be perfectly worded or have profound insights. Focus on personalizing and internalizing the experience rather than fixing the grammar or sentence structure. These aren't professional reports but personal pieces writings that convey your deepest thoughts and emotions.

If you emphasize too much on the writing style, your flow of thoughts will be disrupted. Rather than mechanical writing, think about it as a process of introspection or reflection. Focus on bringing your thoughts to the fore. Let your thoughts be guided by intuition rather than an obsession to write accurately.

Maintain a Dream Journal

Dreams are the pathway to our subconscious and intuition. They have the potential to lead you to solutions your conscious mind may not be able to. Recurring images and occurrences may point to a pattern that is developed in the subconscious. It helps unlock the deepest feelings and thoughts stored in your subconscious. Above all, writing a dream journal helps you develop a greater understanding of your thoughts and a higher connection with yourself.

Always keep your dream journal near the bed so you can quickly wake up and note your dreams before they fade away from consciousness. Keep reading these entries periodically. Is there a specific pattern, thought process or theme? Can you identify a predominant emotion such as a feeling of inadequacy, insecurity, fear, etc.?

This helps you uncover and analyze your innermost thought processes and how they eventually impact the thoughts that go into your conscious mind to create mental chaos. Our dreams are closely linked to events and circumstances that have happened or are currently happening in our lives. Tuning in to them is a great way to clear the chaos and focus on what matters.

Identifying and interpreting dreams can give you unbelievable insights about thoughts stored in alternate realms of consciousness. This helps in bringing about a shift in our thought processes and patterns. Each time you note a dream as soon as you see it, give it a relevant and easy to relate to title or theme. It will make it easier for you to identify a pattern or recurring themes.

Though you may be able to master the process of interpreting them right away, you'll get into the habit of recording it, which can always be referenced later. Interpreting your dreams and intuition helps you define your underlying thoughts and transform them.

CHAPTER 83:

How to Deal With Negative Emotions

Have you ever wondered why people possess different reactions all the time? Sometimes the results become so weird that other people become affected. In some circumstances, this can be even worse, while other times, it is just displaying positively. If you are not aware, then this is an emotion. It comes in different shapes and from different angles. That's, you can experience emotions the whole of your life span.

Therefore, emotion may refer to a mental state which relates to your nervous system and always caused by chemical changes. That's, the sentiment is an amalgamation of various feelings, thoughts, and even behavioral responses. Also, emotions can include some degrees of displeasure and pleasure. Therefore, you cannot define an emotion since it comprises many issues. In most cases, it talks much about our moods, disposition, and even personality. Again, emotions involve motivation and temperament.

Many scientists, day in day out are working hard in their various fields so that they can come up with the best definitions of emotions. However, this effort renders futile. Even though research is increasing daily, no better description of emotion will define the word emotion. Our feelings range from happiness to sadness, frustrations to fear, depression to disappointment, and all these have either negative or positive impacts on our life. Emotions are always complicated. They are biological and psychological, and because of that, our brains respond to them by releasing chemicals and hormones, which later send us into an arousal state of mind. Surprisingly, all emotions result from this process. That's both negative and positive emotions.

There are five primary emotions which control our daily life. We have anger, surprise, sadness, fear, and even disgust. Remember, there are other emotions, which include frustrations, confidence, cruelty, apathy, boredom, and much more. Positive emotions can increase our daily productions within the various companies we are. Again, they improve our well-being and instill in us that nourishment we dearly need in our life. These positive emotions include happiness, joy, kindness, and confidence, among others. Always we strive to have the best but believe me; no one will ever experience total pleasure forever in their lives. As a result, they, at one point, will get frustrated, annoyed, and at some point, engulfed in a quagmire of anger. More so, during this time, everything might come to a standstill. That's if it was a specific task being done, then its completion might be in jeopardy.

We are going to look at various ways in which we can use to tackle our negative emotions. You need to understand that negative emotions are part and parcel of our lives, and the best way to feel less affected is to control them. Below are several ways that you can eventually use to deal with your negative emotions to enhance self-esteem.

Develop self-awareness

Self -awareness is all about knowing yourself. You can't deal with your negative emotions when you are still in the darkness about yourself. Knowing your personality and many things that involve you is the first step to understanding yourself. In the end, you will realize that tackling every type of dark side of your emotions becomes even much more comfortable. Knowing yourself includes several attributes that reflect on your healthy being. That's your desires in life. Other traits include strengths, weaknesses, and even your personal beliefs.

Every time try as much as possible to keep in touch with every detail of your emotions, especially the negative ones. By doing this, you are creating that self-awareness of yourself, thus, increasing your personal touch within you. In most cases, even in various companies and other organizations, the most critical trait of a leader will always be self-awareness. That's the ability to be able to monitor the emotions and reactions of everyone. Mastering the skills of self-awareness sometimes proves to be problematically sophisticated or difficult tasks. However, when you can find good ways to help you learn, then this will be of great help.

Remember, being an effective leader needs much of self-awareness than anything else. However, it is good to note that many times, most of you will avoid self-reflection, which can emulate you in knowing your self-awareness. You will end up getting feedback about your personality from various people. Nevertheless, in most cases, this will be full of honesty and flattery since no one would wish to tell you the real truth about yourself if, at all, it embraces some harmful elements. Along the process, you will not get a good perspective view of your self-awareness from outside people. Therefore, because of these, you might harbor a low level of self-awareness without your knowledge.

As a result of these, it is better to take a look at the various ways you can use to increase your self-awareness. In the end, you will recognize that you are developing your self-awareness, and these will help you tremendously in dealing with your negative emotions.

You must know your skills, weakness, and strength

Skills are attributes that are inborn in some people. However, in some, they get them after a thorough practice in various fields of study. Training and learning also increase the level of skills someone is having. When you know your skills, strengths and look for ways to cope up with your weakness, automatically, you will realize an improvement in your self-awareness. Everyone has strengths, skills, and shortcomings that define our being. That's, telling much about ourselves. Always, these will help you to accomplish your goals in life. You can have short term or long-term goals depending on your abilities and the several resources that are at your disposal. In this case, your goal here is developing your self-awareness, which will help you deal with negative emotions that are threatening your life. On the other hand, we have weaknesses that will always pull you back each time you put an effort to fulfill some of your dreams. In this scenario, your strength will be available to help you out since all these are for your development.

Develop self-management

We all have negative emotions that are pulling us down. The more we try to solve each one of them, the more we get some problematic situations. Our weaknesses play a significant role in making sure we are not succeeding. Negative emotions derail our potentials in every field of industry. By developing self-management skills, you will be in a position to safeguard yourself in a way that you won't be affected by the negative emotions. However, in most cases, this will never be a comfortable ride. That's, some people lose self-control and end up causing much damage to others, while others may have a social withdrawal. All of this impacts, society or the people around you, in a negative way. Therefore, with self-management skills, you will be in a position to keep negative emotions such as anger, frustrations, sadness, dishonesty, low self-esteem, low self-confidence, and much more at bay. Fortunately, there are plentiful ways that you can use to develop self-management skills, which will enable you to deal with negative emotions in a better way. Below are illustrations on how to build your self-management skills.

Physical activity

Physical activities involve keeping your body in good shape. In the end, this will improve your body image. Improved body image leads to high self-confidence, self-esteem, and self-love. A person having improved body image will be able to experience positive emotions always. Even though negative emotions are not easy to avoid, their effect might be less or minimal. Physical activities make you have a reliable and robust nervous system. That's, you will be in a position to perform more tasks and hence, copping up with your challenges. In developing self-management skills, physical activities play a significant role here. Activities such as jogging, different types of sports, and even fitness always help in strengthening the body and keeping your muscles in good shape. All these will also enable you to be productive under your capacity.

CHAPTER 84:

Take Advantage of Positive Moods

You can choose to take advantage of positive moods to rewire your brain. Being purposeful in utilizing the opportunities the positive feelings provide can be the key to the rewiring process.

The utilization may lead to an increase in the functionality of areas in the brain related to moods. The same may result in a positive feedback loop.

Whenever you have a positive disposition, think back on what is causing the same. The act will solidify the connection psychologically between the cause and result. The same will present an opportunity for you to repeat the effect. In regard to neuroplasticity, you may end up forming new connections at the neural level, therefore, rewiring your brain to take advantage of positive moods. The effect is known as competitive rewiring. It encourages the negation of adverse neural networks. One should note that the opposite may result in a strengthening of negative neural connections.

The process may come in handy when you face anxiety or overthinking as you can shift your mindset to a predetermined method to attaining positive moods. The ending may be a reduction in your levels of worry and overthinking. The mental action may get rid of negative emotions. Changing how you react to circumstances can result in a positive shift.

Take advantage of positive moods to rewire your brain by supporting them with a healthy lifestyle. Some foods like walnuts boost neuroplasticity capabilities.

In adults, neuroplasticity requires an enriching environment to occur. The ingredients of such a setting include challenge, focused attention, and novelty. Practice focusing your attention on positive moods can encourage the right kind of neuroplasticity.

You can take advantage of your positive moods to rewire your brain by utilizing them to consider issues that you previously felt were evidence of failure. The mental environment that positive feelings create can aid in providing a fresh perspective on such matters. The actions you take during such moments can give you the confidence to try tackling obstacles. The new viewpoint can help you develop a higher capacity to endure adverse events. The same can translate to the development of perseverance.

There are recommendations for individuals to determine when they are most likely to experience positive moods. In the context of daily life, such emotions can connect to a specific time of day. Utilizing such periods to tackle your most challenging tasks can aid you in rewiring your brain effectively.

Use the atmosphere of positive moods to embrace a relaxed mindset for brain rewiring. The result may be a reduction in negative feelings like anxiety and depression.

Positive moods can help you recognize people or situations that encourage the occurrence of positive emotions in you. They can act as an identification signal that you can use to choose who you should align yourself with to achieve a rewiring of your brain. They can be an indicator of what your mind is considered as a haven.

Positive moods may also be a signal that you are succeeding in your journey to becoming more positive regarding your mindset.

Positive moods can help you take more risks without the drawbacks of negative emotions like fear and anxiety. The subsequent success from taking chances may heighten your levels of self-esteem and confidence leading to a positive feedback loop that rewires your brain.

Positive emotions can include excitement, happiness, hope, inspiration, and joy. Such sentiments can be the impetus that keeps you persevering on your path to a positive rewiring of your brain.

Positive emotions support leading a healthy life, which is an essential element in the quest to rewire your brain. The role of positive emotions in rewiring can be significant.

The words you use when under the influence of positive emotions can further support the rewiring of your thought process. They achieve this by simultaneously creating a positive picture in your mind regarding how you feel. The two-pronged effect causes a higher impact on the neural pathways as compared to a sensation that only has a singular pronged impact.

You can take advantage of positive emotions like gratitude to prevent the trap of exaggerating the negativity of a circumstance you are facing. The result will be the ability to remain objective, therefore, limiting the creation of negative neural networks via the process of neuroplasticity. Gratitude can help you develop your active listening skills as you are more open to hearing other people's points of view.

The positive mood of serenity can give you a calm outlook allowing you to observe situations objectively without influence from negative emotions. It promotes a feeling of self-acceptance, which can result in a higher level of self-esteem and confidence, which can bolster your positive neural networks.

Positive moods can bolster your level of interest, making you curious, which you can use to find out the root of your thought framework. Your understanding of such a foundation can help you create a pathway for improving the state of your neural networks. Interest allows you to improve your level of attention, which is a component of effective active listening, a skill that you can use to gain objectivity. The result will be an improvement in your thinking process, leading to rewiring your brain positively.

The positive mood of hope can help you be more optimistic about future possibilities, therefore, increasing your appetite for risk. The same can be a requirement for testing new pathways to your goals. Success after such risk can lead to the creation of new positive neural pathways.

Pride in the right context is a positive mood. Pride can help you to embrace yourself by negating negative emotions that cause low self-esteem and confidence. It can provide a pathway for you to enjoy your achievements without falling into the trap of perfectionism. The sentiment can provide you with the drive to further sharpen your skills resulting in the formation of a positive feedback loop. It can give you confidence in your capabilities.

You can take advantage of the positive mood of amusement in the context of rewiring your brain to handle difficult conversations. The emotion can diminish the sting of criticism while allowing you to delink from negative emotions. The effect can be the achievement of an objective mindset. The sentiment can be critical in forming supportive relationships as it may make you more approachable. Positive social relationships can assist you in rewiring your brain.

Use the positive mood of inspiration to tackle tasks that you find foreboding. The emotion can diminish the fear factor that can accompany such responsibilities. Your level of engagement is bound to be higher, which may lead to a better quality of work. The result may make you feel uplifted and ready to face the challenges that come your way. The effect can be the positive rewiring of your brain when it faces similar challenges. You may, in the future, not rate comparable obligations as tough assignments, therefore, decreasing the probabilities of forming adverse neural pathways. Inspiration enables you to learn from others, therefore, presenting you with a platform to work on your active listening skills. The result may be an improvement in the quality of your social interactions. Such a social environment may support the formation of positive brain rewiring.

Take advantage of the positive mood from the emotion of awe to work on your active listening skills. The feeling can make you more open to other people's viewpoints, which may improve the quality of your social interactions. The effect can be the support of the formation of positive neural networks via the phenomenon of neuroplasticity. Awe can make you more appreciative of life events, which may diminish the effect of negative thoughts and emotions.

Take advantage of the positive mood from the emotion of elevation to further extend the admirable qualities you witness in others. The possibility of scalability of the impact can change your social environment leading to the formation of positive neural networks. It can cause you to be kinder, therefore, developing empathy, which may change your viewpoint from negative to positive. You may be more generous, allowing you to be less critical to others and even towards yourself. The effect may be a diminishing of negative self-talk, which promotes adverse brain rewiring. Elevation can provide you with a new vision that can shift your focus from negativity.

The positive mood of altruism can help in feeding the positive feedback loop through the habit of generosity that it encourages. It can also help you develop empathy, which may be a pre-requisite of facing circumstances from a broader perspective.

Taking advantage of the positive mood that associates itself with satisfaction can help you rewire your brain. It can give you a feeling of pleasure that you may find rewarding, which may contribute to a positive feedback loop. It may result in contentment that will protect you from pleasing others at the expense of your core values. The same breeds self-belief and confidence, which can further strengthen positive neural networks, therefore, rewiring your brain.

Taking advantage of a positive mood arising from the emotion of relief can reduce your level of skepticism, which may work to shift your viewpoint of the world from negative to positive. Basing your decision on such sentiments can make you more hopeful regarding the future. It can strengthen your capability to trust others.

The positive mood that links to affection can help you identify the root cause of what you can use to build up preferable neural networks. The sense of pleasure that accompanies the emotion can help you grow your capability to persevere.

The positive mood that associates with cheerfulness can help you achieve a hopeful viewpoint towards the challenges you face, diminishing the possibility of creating negative neural networks. The emotion can help you tackle obstacles, which may increase your self-belief and confidence. The result can be a rewiring of your brain to believe in possibilities. It can make you upbeat, which can have a positive effect on those surrounding you, making them eager to support your goals. Such social interactions can be your pathway to distress, therefore, diminishing the possibilities of creating negative neural networks, and in essence, rewiring your brain. The emotion can cause you to be noticeably happier, which may make you appear more approachable, therefore, leading to stronger social relationships. You can be more open to taking risks, which, if they result in positive outcomes, can cause you to create positive neural networks.

Taking advantage of positive moods coming from pleasant surprises can cause you to be more believing in others. You may, therefore, change your viewpoint of the world from negative to positive. The result can be a rewiring of your brain positively. It may cause you to be hopeful of positive possibilities and make you more open to taking risks. It can help you move away from your imaginations and take circumstances as they come without making negative predictions. The result may be experiencing fewer situations involving self-sabotage.

Take advantage of positive moods that arise from confidence. Self-belief can be the catalyst that changes your viewpoint of the world from negative to positive. The result can be an increase in your capability to take risks as the confidence in your abilities soar. Such actions may feed into a positive feedback loop system resulting in a rewiring of your brain. Confidence will increase your levels of self-esteem, diminishing your need for approval from others.

CHAPTER 85:

Habits You May Take Up To Declutter Your Mind

Healthy Diet, Constant Drinking And Creative Cooking

This is one habit that can keep you in a dreamy land forever. Imagine returning from work with a very sour mood. But you stepped in the door and the mesmerizing aroma of a delicately made food hits your nose. Trust me, it will fill your senses and for a second you would be forced to erase the days' snag from your brain. You would look around and whatever plans you have made earlier would hang. You were probably planning to go straight for a shower, but now, your mouth must be showered with this yummy food. If you take up a habit like this, you would be dunked into it so much that the flavors would be the only drifting in your head, no mental clutter. Not just you, your friends, families and housemates would certainly share in your spirit.

Controlling Your Words And Manners With Everyone

This is another habit you must be aware of. How do people feel after you spoke with them?

Like a piece of old thrash or an amenable fellow who could do better with suggestions from their superior? How people, especially those who are inferior to you feel after you have just addressed matters a whole heap. It is a good habit that can indirectly contribute to your rest of mind among others. It can determine how settled your mind is to attend to other issues and persons.

If just to maintain your composure so you can handle all the problems you have, it is ideal to let things slide and not get mad at your inferior every time. Allow them a benefit of doubt, don't take it out on them. People naturally learn faster when corrected with respect to their personality, and trust me, you can do it.

Drafting A Workable Schedule

I cannot overstress how essential this is for everybody. 'I wish I can be so organized!' you would hear a man condemn himself in anger. But hardly would he ever take any steps towards that. Ideally, being organized begins with having a plan. A proper schedule of how you think things would be workable. You can do this by making a reflective study of things you need to do in the coming day or week. Fix them in a timetable and remember to add the duration of each of them. You should consider bunging up the timetable so well that the time is well spent. Once in a while, allocate a free time for emergencies.

Allocating A Personal Time In Your Schedule

Among other periods that must be reflected in your timetable, personal time is a must. It is a superb idea to make out time for yourself. This period will be spent resting and reflecting. You could curl up in your bed or bathtub, lay your nerves and shut your eyes. Then begin to think of your interactions with everyone.

What was the look on your secretary's face? What was the reaction of your boss? Could you have done something better for your child that night? Should you have chosen mortgage over auto loan? This is the reflection hour. Think about it and criticize yourself as much as possible, but make sure you remember that some things cannot be changed and there is no point in making a fuss about them. It's in the past already, the only correction that would be made to them is to forestall anything like them again.

Cleanliness

Cleanliness and orderliness are brothers. There isn't much sense in organizing a pack of litters all around your lounge. Make it a duty to clean yourself and your environment.

Your dress, teeth, hair, shoes and so on must be kept clean and neat.

The most important is your heart. Don't fill it with thoughts of harm and hatred for others, neither should you occupy it with assorted thoughts of complicated problems. By the same token, give your environment your best shot in cleaning. Your files, your drawers, the trays, the cups, they all deserve to sparkle at all time.

Timed Decisions

Making timed decisions is crucial to the success of your organization. It is not necessary to respond to every mails immediately. You should not fight back every bout immediately. Sometimes, it is okay to stop and wait for the right moments. That is time consciousness too. But you must commit to your memory that some decisions are best taken immediately. They would only lead to further complications when delayed.

For example, making a decision between staying with your kids or travelling for the business trip at night. If you remain undecided for too long, you might realize you have spent too much period thinking about it, and you could have spent that time on something productive. Now, your kids would be in bed by the time you return and you didn't obtain a ticket at the airport which means moving out is highly unlikely. Why not simply decide earlier and take the right steps?

Sharing Others' Problems

I will continually encourage you to listen to other people. Have listening ears. Even when you get tired of listening easily, find a way to make the speakers take the bull by the horn and get the basics. You will succeed more as a leader if you learn to listen to the problems, ideas and suggestions of your followers. Of course, it is not obligatory to follow them all, but you will surely know better if you listen to them. Study them, your friends and families want you to listen to them too. In the same vein, people are always willing to you, so there is no genuine reason you cannot share your worries with others occasionally.

Avoid Intimidation

The golden rule of human states that you should do unto others only that which you want to be done to you. Familiar with that rule? You have to use it now than ever.

You should remember that humans are only created to hold different positions, they are equal.

Of course, you remain superior at work if the tables don't turn, but on average, you could treat everyone equally without a smear or disrespect on your image.

Patience

Patience is a good habit that can win you medals and nothing less. Keeping it cool with everyone is a way to keep your nerves. If you ensure you are always patient in every case, you will hardly get worked up over anything, and you would be meticulous enough to make the proper choices only. This is a prominent reason you should consider being patient at all times.

Hold Sway On Your Emotion

Fair warning, you will come under pressure a lot of times. If you work in a customer relations or production unit, you already have some good idea what I am talking about. Your clients can be demanding to a point that you could almost lose your cool with them. You would get irritated by some requests and you might have to keep apologizing without getting angry in turn. It is hard to maintain your cool in cases like this, and that is why you must learn that habit.

In the wide world, we can metaphorically state that we all work in the customer relations, and everyone who walks up to us is our customer. They want to share ideas, information, and their disagreements with us. We must be able to keep our anger at bay when dealing with them all. No matter how offending their comments had been. This is a good habit that can save you from suffering a mental turmoil and at the same time, maintain a fair relation with the speakers.

Have A Positive Mindset

Having a positive mindset is a cure to many problems in the world today. Regrettably, positive mindset is exactly what more than half of the world does not have. It is what leads anyone to frustration, anger, impatience and finally suicide. Not having a positive mindset means you may never see the positive side of any action, decision or opinion anyone raises around you.

You are solely concerned about the negative sides. But where does that get anyone?

You will see progressive solutions rather than problems if you begin to use a positive mindset.

Setting A Goal And Working To It

This habit definitely belongs somewhere at the top. It is one of the first few habits anyone should cultivate. Setting a goal is crucial to having a direction in life. It determines your schedule, your friends, your enemies, your diets and so many other habits. You can only become successful if you have a goal you are working towards and you take on other habits as your progress in pursuit of your goals. Set your goals and grind hard to achieve them. You can do that without becoming mental cluttered, especially if you get engrossed in some other activities.

Relaxing

If you are someone who feels it is okay to overburden yourself because that is how you are inspired to do better, I hate to tell you that you are wide of the mark. You do not have to overburden yourself in order to achieve, you must learn to spare some time for leisure. If you are at some critical point that requires you to overwork however, be sure you compensate yourself afterwards. You should try to break out of the suits and ties for a pajamas or bikini. Find a way to relax your tense muscles and stressed brain. You need whatever could distract you from work for a while. You might go sun bathing, sightseeing, or stop at the cinema. We all have our diverse ways of relaxing, you should realize and don't wear yourself out before you use it.

Increasing Your Core Values

On an encompassing note, another habit you can look forward to creating is maintenance. You are certainly not an entirely bad soul. There are a couple of habits that are good about you originally, you will also pick many here and begin to inculcate them in bits, but inculcation is not enough, you must find a certain way to preach them and instill them into your life stream permanently.

There we are! The ways to create good habits and the most important habits that you can create; all at your fingertips. Let's go to the other side!

CHAPTER 86:

How to Make Your Emotions Work for You

How Do You Figure out the Kind of Emotion you are Feeling?

Typically we want to fully understand them, as we have had some strong responses. Did you ever get lost about what kind of feeling you felt? Our thoughts may also be vague to allow for an explanatory attempt. Another explanation for this is that a lot of ways use the same logic method. For example, feeling frightened to speak to your class may feel different from the fear of a lion. One explanation for confused emotions is that different forms of feelings that all sound the same. First, you're worried about frustration and fear, and quickly make your heart pound.

Also, when you experience an extreme reaction, you generally need to discover what the inclination is. You may see, for instance, that your heart is dashing, there's a lion before you, and you want to flee. You may then concur that "dread" is your best decision when you consider different sentiments that you may be encountering. As it were, you may have the idea that "I'm most likely frightened at present, since I figure this lion may hurt me." Even with such an alarming circumstance, you may not understand that you're frightened until you run off and have a second to consider it. Researchers have discovered that a few people think it's harder to make sense of what feeling they feel than others. Individuals who experience issues understanding their feelings likewise have more trouble in causing themselves to feel better.

Look out about your thoughts, and try to determine what you think. It will help you fix your challenges and make you look better if you feel bad. It's just a snapshot of what you've learned from the times you've been depressed, frightened, or confused about how you're going to deal with those situations should they happen again.

Steps To Allow Your Big Emotions Work For You

1. Recognize what you are feeling

That is harder than it would seem. We are generally bad to know how we feel. We may know that frustration tightens our chest or sinks into the stomach pit. Still, we haven't spent time contemplating the sensations associated with more complex emotions such as deceit or pleasure.

We have this exacerbated by an abysmally limited cognitive lexicon. Although a Berkeley study identified 27 distinct types of emotions, most of us recognize the 'primary colors' as sad, angry, fearful, happy (all Inside-Out characters, not coincidentally). However, for more nuanced hues, we don't readily have terms like frustrated or humorous. This means that we have no way of recording any feeling of emotion, but we know it when it happens. Changing it is the first move. Take a second, take an emotional wheel like this one (there are many online) and pick the expression that is closest to the emotion you 're feeling right now. Take a while and check the body. What do you look like? What triggered it? Where do you feel? Know the 'sensation mark' so you'll remember the same experience next time.

2. Expand the feeling to its maximum.

Now breathe in that feeling, and make it as big as you can. Rely on that love, yell, spring, or curl up with the grief, and then fall. Feel the anticipation in your bowl, or the nervous butterflies in your stomach. Feel the tightness of your jaw, or tingle your fingers. Turn your focus on getting the experience as broad as you can and being as welcoming as possible. Some feelings may pop up, like your left knee might throb, twitching your eye. Let's see what's in here for yourself.

3. Just let yourself know exactly how you feel.

Feelings are logical by nature, just a signal attempting to tell us something. If we assign judgment as thoughts of 'good or negative,' we promote remorse, 'what should be,' and if we dismiss the idea, we start the process of festering—moving into the role of curious spectator, instead of judging. Look at it as an unfolding picture. Interested, so watch out. There may be a warning for you, and maybe there isn't. This is good. Permit yourself if you feel like it.

4. Let it go

When you complete this process, your thoughts and emotions will shift and evolve. They will naturally pass past or taper away. Note the shifts, relax, and expel emotions. If it lingers, that's no judgment, okay too-mind. Thanks for the ride, and let go.

That's just uncomplicated! Give this one a chance. Be real. On average, according to a poll of over 11,000 people, people registered to have feelings 90 percent of the time. The more you use this loop, the better you can comprehend and regulate yours.

CHAPTER 87:

Create a To-Do List

A to-do list is one of the most basic, yet easily overlooked, task management tool at anyone's disposal. Essentially, a to-do list contains information about what you should be doing, how it should be done, and when it must be done.

The principle behind a to-do list is quite simple. It has also been around for so long. However, no matter how simple it is, the problem with a to-do list is that people tend to forget about them eventually.

Some find it too simple that they think it is not effective in serving its purpose. Others recognize the importance and merits of a to-do list, but they lack the discipline in maintaining one in the long run.

To better illustrate to you why you should create and keep a to-do list, the next section covers the effects of having no to-do list in your day-to-day life.

How Your Life would be Without a To-Do List

Life, by nature, is chaotic in itself. This is further complicated by the demands and complexities of the modern way of living.

With the mountain of tasks that you must accomplish day by day, things can quickly become overwhelming. When this builds up, the amount of stress in your life will increase exponentially. Many experts recommend the usage of a to-do list to manage one's activities and responsibilities better. However, some people find it hard to pick this up as a habit.

Studies show that without a to-do list, an individual's level of productivity significantly drops down. You may also experience the following scenarios when you do not create a to-do list of your own:

- Jumping from one task to another, thus decreasing your efficiency in finishing up your tasks;

- Missing out on important deadlines because you forgot that you have to do it in the first place;
- Being vulnerable to potential distractions around you;
- struggling to achieve a balance between your home life, your work life, and social life, among others;
- Having no sense of direction at all especially when it comes to what you should be doing next; and
- Lacking the feeling of accomplishment by the end of the day.

To resolve these problems, you should try incorporating the creation of a to-do list in your daily habits.

How a To-Do List Helps with Overthinking

One of the most important negative effects of overthinking is analysis paralysis. This means that you become stuck in your mind, mulling over the same issue over and over again, without anything to show for it. This then leaves you with little to no time and energy to act and carry out your other tasks. A to-do list can help you overcome this by keeping you focused and on track with what truly matters. Aside from boosting your productivity, it may also be beneficial to you in psychological terms. According to researchers, a to-do list can:

- Give you the drive to get things done;
- Prevents you from being distracted by your irrelevant thoughts and other unnecessary elements from your environment;
- Prevents you from doing unnecessary repetitive behaviors;
- Break down complicated tasks that may bring about feelings of anxiety and worry about failing to accomplish the said task;
- Improve your pacing, and therefore decrease your stress level;
- Relieve you of the pressure to finish everything all at once; and
- Relieve you of the worry that you have forgotten to do something important.

Ultimately, a to-do list can also make you feel happy and satisfied. A listing with all the items crossed out serves as proof that your day has been quite productive. You will be able to resist any feelings of doubt, especially those of self-worth and self-confidence. As such, your mind will have no reason to devolve into an endless spiral of anxiety, negative thoughts, and worry.

Maintain a To-Do List & Stick To It

Many people who do not—and cannot—maintain a to-do list view it as a burden. They think of it as a list of chores to do and deadlines to meet. Over time, this perception prevents them from making a habit out of creating and managing a to-do list.

Some people are natural at keeping things organized and on track. However, for those who are not born or conditioned to do so, here are some effective tips that will allow you to maintain and stick to your to-do list:

- Associate your to-do list with positive thoughts and feelings.

This is the first thing you must do to incorporate a to-do list in your daily life successfully. Remind yourself of the practical benefits of keeping one. Try to recall how good it feels whenever you get to cross something off your list. By doing this, your brain will be conditioned to put things in your to-do list to get it done and crossed off.

- Write the list for the benefit of the future.

You might not immediately realize the advantages of maintaining a good to-do list, but your future self would appreciate your efforts. Regardless of how great you are at remembering things, life may throw you a curveball at any moment.

This may leave you scrambling for direction and information. A to-do list that contains all the important details that you must keep in mind would be a lifeline during those challenging times.

- Categorize the items on your list, depending on their importance and your personal preference.

Many people skip the process of categorizing the items on the to-do list. This is a vital step to make because it improves your chances of getting things done. Through this, you would be able to prioritize your tasks better. One way to categorize your list is by arranging it according to what must be done, and what would be pleasant to do if you have the extra time. By doing so, you would not miss out on the all-too-important deadlines in your life. It will also remind you of the things you can do with your time, thus saving you from having to rack your head for something to do.

- Accept the fact that to-do lists are changeable.

Remember, a to-do list is only a tool. Its contents are not rules or demands that you must follow at all costs. Sometimes, you have to change the items in your list to suit your current needs.

Starting anew is perfectly alright. It shows that you flexible enough to roll with the punches. By learning how to adjust yourself and your to-do list, you will be able to better deal with the stress and anxiety triggers that may come your way.

- Treat your to-do list as a symbol of your accomplishments.

Conquering your to-do list requires a lot of time and effort. Therefore, it is normal to feel proud about finishing a task in your to-do list.

Similar to assigning positive feelings to your list, thinking of it as a record of your wins for the day will help you stick to this habit.

This will also do wonders for your mental health. Anxiety, negative thoughts, and worries will have little to no place left in your head when it is filled with your accomplishments for the day.

Now that you understand the importance of having a to-do list and how maintaining one can drastically simplify any feelings of anxiety, negative thoughts and worries you could have, it is also important to know how to create an effective to-do list.

Creating an Effective To-Do List

Some rush through the process of creating a list, thus giving them a one-word outline that vaguely describes what they must do. As a result, they cannot follow through the listed tasks, which then leaves them an impression that to-do lists simply do not work for them.

To help you write an effective to-do list that WORKS for you, follow these quick and easy steps:

1. List down three tasks, at most.

A shorter list containing the most important tasks that you must accomplish would allow you to get a sense of success by the end of the day.

2. Make each task actionable.

You can do this by using an active voice rather than simply indicating the outcome that you want to accomplish. For example, instead of listing down "detergent" in your list, you should write "go to the grocery store and buy detergent." The first part of the suggested statement may sound obvious and unnecessary to you. However, keep in mind that the more complicated the task, the more helpful it is to include these small details into your to-do list.

3. Assign the priority level for each task.

You may ask yourself which of these tasks would make you feel most accomplished. Your answer would then have to go to the top of your to-do list (priority list) with a "high", "mid", or "low" label depending on their relative urgency to you.

4. Write down the rest of your tasks in a separate sheet or file (overflow list).

They should go to another queue, so that you can focus on what matters. Since the top three tasks are considered as significant items, they may also take up a lot of your time. Therefore, keeping the other tasks in a different list would keep you from feeling overwhelmed. Ideally, you should store this where it is accessible but away from your sight. This would enable you to refer to the overflow list when you have run out of things to do in your priority list.

5. Make your priority to-do list visible.

You may transfer them in small post-it notes that you can stick in a location that you frequently see or go to, such as the refrigerator door. If you prefer to add more details, you may opt to use index cards instead.

6. View each task one at a time.

Many people feel overwhelmed whenever they see a list of things that they must do. To avoid this from turning into a negative feeling, you can impose viewing limits upon yourself. There are certain task management apps, such as the "Todoist" and "Omnifocus" that allow their users this viewing option. However, if you are using post-it notes or index cards, then you can just simply stack them over one another so that you can only see the topmost item in your list.

7. Record the status of your task

Recording the status of your task whether accomplished or not makes you accountable. It makes you have a deeper reflection on the position of your commitment to fulfilling your tasks for the day. It makes you ponder on how you can improve on your level of accomplishment and also gives you a sense of refocus to ensuring that unaccomplished tasks are executed and taken off your to-do list as quickly as possible.

Conclusions

What happens in the outside world is beyond your control, so you have to focus on what is going on inside of you. Your mind is the most important part of yourself to keep healthy. Whenever our mental wellness is suffering, it gets in the way of our ability to function in our daily lives. It will influence your ability to make good decisions, how you respond to hardship, and your overall quality of life.

You can get a handle on your thoughts. If you work for it, you will get to the point where you can say what thoughts have a place in your mind instead of being led by them. Keep at mind what is at the core- stress and burnout. Do the practices we talked about daily. Transform your life into a more tranquil experience. Focus your energy on what needs to be done right now, not your end goal, what occurred before, or what could happen in the future. If a thought comes in that falls into the unnecessary category, let it go. You make thoughts go away not by repressing them, but by allowing them to come and go without paying any mind to them.

If you feel yourself becoming overwhelmed again, that's not horrible. It just means you need to renew your commitment to decluttering your brain. All you need to do is make today better. Sometimes it is as simple as reminding yourself to breathe and that there will be a solution to whatever is troubling you today.

Changing your brain isn't an overnight process, but it isn't as difficult as you might think at first. At its core, it just means you need to reach for what you want, not just in thought but in action.

Don't engulf yourself with this task. Whatever instances you are in, think about making the best decision in it. When you wake up and make yourself breakfast instead of sleeping in, you are choosing to help your health. When you are at an occasion, and you say hello to someone as opposed to hanging at the side, you are already becoming a more social individual than you were in the past.

Any person can say, "I wish I had more company/money/etc," but no one gets anything they do not grasp for. This includes bad things. The people you devote your time with and the things you invest in are going to be the toughest forces in your life. If you want decent things, you have to become a person who interests them- this means you need to devote in and give to yourself. Think about how a canvas that was completed in five minutes looks compared to if the artist took five hours to finish it, paying attention to the details- where it is lacking in color, where it needs a change in hue, etc.

In short, changing your brain begins when you first wake up in the morning and ends when you go to sleep at night. You have that much time to change how you think, feel, and act, which is going to ultimately decide how your life goes.

PART IV

OVERCOME ANXIETY AND DEPRESSION

Introduction

Living with constant negative thoughts and intense fears can cause someone to wish for a way to ease their pain or to develop unhealthy habits that could make their symptoms worse. Anxiety is linked to a number of other mental illnesses, most notably depression. This combination can cause people to make poor decisions when they are motivated to find relief from their symptoms. Some people will turn to drugs or alcohol, which can make symptoms worse. Other people might find more dangerous ways to cope with their mental problems and cause physical harm to them.

Being anxious can also result in people refusing to leave their homes or to only go to certain places that they feel are safe. This can lead to a drastic decrease in social interaction, which can be counterproductive to their healing. Humans are innately social beings and when people take interaction out of their lives, they might experience even more negative thoughts or have trouble noticing when the thoughts are irrational. It can also impede their ability to relate to others socially when they are out of practice and thus further cement their fears that communicating with other people is too painful an activity.

Not all anxiety is likely to pose harm to the person experiencing it. Typically, it takes high levels of constant anxiety to lead a person to attempt to find relief in substances or to avoid things they believe cause their anxiety completely. Once a person starts going down that road, though, it is essential to get help as soon as possible. Exposing themselves to the things they are afraid of is often the best way for a person to understand that it is not an actual threat. If people do not have a strong support network to get them through their anxious periods, they may have a more challenging time getting back to a healthy place.

One of the hallmark symptoms of anxiety is having a sense of impending doom that you can't shake. This can make it difficult to concentrate during the day or find a moment of peace because there might always be that voice in the back of your head telling you something is wrong. Sometimes this voice can become so loud or be so relentless that it induces a panic attack. The sense of fear and stress that come with panic attacks make it even more challenging to move on with the day after the symptoms have passed.

A sense of doom is not the only negative effect anxiety can have on a person's life, Anxiety and depression are often closely linked and it is not uncommon for someone with one of these afflictions to also experience the other, whether it's from time to time or chronic. The combination of these two issues can often cause increased stress in people, which can then lead to headaches or high blood pressure. Sometimes having physical symptoms to match their emotional ones can be too much for a person to handle and might lead to avoidance coping.

The constant discomfort can also contribute to someone becoming irritable or extremely quiet. When you have a headache, it can be challenging to concentrate and having people at school or work demanding that tasks be completed can sometimes cause someone to react angrily because they cannot handle any more stress or focus on projects. On the other hand, some people might become extremely quiet because they prefer to fight their battle on their own or they might think no one would take their struggle seriously if they did voice their pain.

Anxiety can also cause people to believe they are having breathing problems when symptoms start to increase in intensity. When they are distressed their breathing might become shallow and rapid, which makes it hard for the body to get enough oxygen to calm down. Constant anxiety and stress can also contribute to stomach issues in some people. Stress is known to cause stomach ulcers and it is not uncommon for people with anxiety to develop ulcers or other gastrointestinal problems. Some people experience cramping, pain in the abdomen, or develop food sensitivities. Keeping a healthy balanced diet can help anxious people to avoid or alleviate some of these symptoms.

Depression not only affects you, but it affects various aspects of your life, including the people you love. Discovering how your life is affected by depression can be one more step towards making improvements and pushing you towards the help you need.

Many times we lose sight of those around us, the love we have for them, and how they can help us, when suffering from depression. Reminding yourself that you do have people who care and they are willing to be "bothered" by your troubles will help you in your treatment.

It is true that you will get into a vortex of limited sight when you are in the midst of depression. This cycle of negative thinking can battle between knowing people will help you if you ask, and being unable to ask. Perhaps you have had thoughts that they will not be able to help you or understand. Maybe, you have tried to reach out, but their own lives are busy and you feel like you are just in the way.

There might be times when you've thought, "I'm always the one to get a hold of that person(s), so obviously they do not care about me." Often the other side to the story is that your friends and family do care, but you are right—they are living their life and sometimes that life is just as bad as yours.

Your friends might be battling their own depression. It's even happened to me. I was suffering much from depression brought on by the loss of a loved one, although that loved one was not completely gone; his mind was due to dementia. My friend's said they would be there, but never called. It turned out that one of my friends was suffering from post-partum depression and another had been battling depression for 10 years. All three of us didn't tell each other of our struggles and battles with depression, but tried to keep it all inside. Instead of leaning on those that could help and understand we all chose to keep it to ourselves and try to deal with it in their own way.

The relationships could have been entirely and forever fractured. Some relationships were among the friends because of the depression and mistakes made. For example, one friend lost their spouse in a divorce due to the severe depression plaguing the friend.

Relationships and Depression

Each side has a view. This view may be in line with the other side or it might be completely different, and somewhere in the middle is the truth to everything. Depression and relationships are like this.

You are in a dark place. Everything you see around you is negative. The things that should be bringing you happiness are not. You also feel that no one can understand. You tend to withdraw. There is also a high likelihood that you are complaining about many things to yourself and to your friends. Every little thing can be a complaint, when depression has its hooks in you.

The deep relationships you have are often able to deal with this side of you. Their love demands that they help you work through the troubles, at least for a time. Those who love you want the best, they want to be able to fix your problems, but their own issues can start to stack up. They can begin to affect the person, so they are unable to help you.

If it gets to be too much or you are also unwilling to seek help, then the loved one may pull away. They may try to seek happiness elsewhere or simply give up on you. The strength that exists in your relationship will determine whether a person is willing to stay by your side, through the tough times and provide the care you need.

CHAPTER 88:

Stress

Stepping outside of your comfort zone and developing habits that push you towards your goals naturally come with some side effects, one of which is stress. We try to avoid stress like the plague, and that's partly because we understand more about the impact chronic stress can have on our bodies and mental health. However, not all stress is bad, and part of acquiring mental toughness is learning whether stress is helpful or harmful.

Nevertheless, before we get into that, we're going to look at what the stress response is and why it can be so detrimental.

What Exactly Is Stress?

The definition of stress when the term was coined stated it is "the non-specific response of the body to any demand for change" (Selye, 1936). These responses can be mental, emotional, or physical. Think about all the things your mind and body do when you feel stressed. Your stomach may twist up in knots, anxiety begins to build, and all you can think about is whatever it is that's causing the stress. All of these (and more symptoms) culminate to form what we know to be the stress response. Once it gets going, it's tough to return to a pre-stress state unless the stressor itself is gone.

There are three different types of stress, according to the American Psychological Association (n.d.).

Acute Stress

The most common and frequently experienced type of stress, acute stress is brief and results from your response to a situation. For example, if one of your kids is sick and there's no one to stay home to take care of them, you may feel tense and irritable, which are both symptoms of acute stress. This type dissipates fairly quickly and goes away entirely once the stressor has been removed or dealt with.

Episodic Acute Stress

This is the next level of acute stress, and you may have guessed by the name that it describes acute stress that happens regularly. If you frequently have tight deadlines at work, you may experience episodes of acute stress whenever the deadline is approaching that then go away once you've handed in your project.

However, this type of stress can also arise, no thanks to your mindset. If you worry a lot or are a perfectionist, you'll experience episodic acute stress. This may lead to stomach problems and emotional distress.

Chronic Stress

Chronic stress is the new buzzword in health and psychology circles. The more our world evolves, the more stressors we're exposed to and, consequently, the more often we feel stressed. Any situation that is not fleeting or easily solved can lead to chronic stress because the stressor never goes away. This can lead to chronic illnesses — aside from stress — that significantly impact your quality of life and shorten your lifespan.

Tips And Exercises To Reduce Stress

Knowing the difference between good and bad stress means nothing if you continue to let the latter beat you up. Instead of allowing yourself to fall apart whenever you feel stress kicking in, turn to these tips so you can acknowledge it without letting it take over.

Accept A Lack Of Control

One of the biggest reasons why we feel stressed is that we can't accept that we aren't always in control. We have an obsession with micromanaging because we think that's how we will achieve success. In reality, this tendency to seek control in every situation makes us too rigid and unable to see alternative pathways. Then, when we run into a roadblock, stress comes crashing down because our perceived control has been thwarted.

The key to reducing stress is acknowledging that you're not in control. Instead of letting that idea scare you, let it relieve you of the burden. You're not in control, and that's okay. You don't have to manage your entire life, and that's a good thing. Once you step back and let life happen on its own, you can adjust better when situations change and make decisions from a place of logic instead of fear.

Reduce The Stressors You CAN Control

While most aspects of life are not under your control, there are some things that you can influence. Wherever possible, reduce stress by eliminating your burden and learning to manage tasks better. If you have too much to do at work and it's causing you anxiety, learn to delegate. If you can make a change that will reduce the amount of stress in your life without sacrificing your goals, do it.

Breathe

One of our physical reactions to stress is tension in the jaw, neck, and torso, which can affect our breathing patterns. When our body is deprived of oxygen, it's not happy and will tend to let us know. Combat both the physical and mental reactions by remembering to breathe and employing a breathing exercise when you feel overwhelmed. Counting backwards from 100 while slowly breathing in for one number and out for the next refocuses your brain and takes you out of the situation momentarily.

Leave Room For Fun

Life is supposed to be enjoyable. Whenever possible, leave behind the worry, stress, anxiety, pressure, and seriousness and just let yourself have fun. Go to the park and watch the kids play, take a dip in a lake, or put on your absolute favorite movie and pay attention to every scene and line of dialogue. When we let stress take away our ability to enjoy ourselves, we're at the beginning of the end. No matter how bad things get, you're still alive, so let yourself appreciate that.

Avoid Vices

You may be tempted to turn to cigarettes, alcohol, or drugs when under a lot of stress—don't. No matter how good you may feel after indulging, the underlying issue doesn't go away. Once you sober up, you'll be right back where you started only with a massive headache and a higher risk for lung cancer.

These vices also go against the tenets of mental toughness. Learning to be strong enough to face your issues head on will serve you well in all aspects of life, whether they're stressful or not.

Take Care Of Yourself

You can't take care of your life if you don't take care of yourself first. A healthy body and mind are better equipped to deal with life's challenges, so don't neglect either. If you sit on the couch eating junk food all the time, neither your body nor your mind will be in optimal states. The moment a stressor occurs, you'll be knocked down.

Physical and mental strength (and toughness) start on the inside, so once you've built a firm foundation of wellness within yourself, stress from external sources will have a harder time breaking through and getting to you.

Eat well — whatever that means for you. Diet fads come and go, but truly healthy foods never change, so stick to the basics. Get up and move during the day, especially if you feel stress coming on. Do whatever you need to do to prioritize your health because you have nothing without it.

Know When To Ask For Help

You don't have to go through your entire life doing everything by yourself. That's not always a sign of strength. If you try to push through without asking for help, even when you really need it, that's foolhardier than anything. Know when you need help and don't be afraid to get it. Sometimes the stress in our lives is avoidable if we would only take the proper steps to manage our time and resources better. This goes along with reducing the stressors that you can control. Don't let your need to be a tower of strength is the cause of your collapse.

Beyond these tips, just remember that stress is psychological. It all starts in your mind. If you can learn to control your mind through mental toughness, you'll be able to address the root of the problem. Turn your mindset around so that you can turn your life around.

CHAPTER 89:

Negative Thoughts

Defining Negative Thoughts

Negative thoughts, at least for the purpose of this book, are more specific than just any thoughts that are negative. We are specifically discussing thoughts that are automatic and pervasive, that color your behaviors and understandings of those around you. Your negative thoughts are going to be the automatic thoughts that are responsible for guiding your behavior, so you do not have to think about it.

We all have automatic thoughts—these are thoughts that save your conscious mind from spending precious real estate, figuring out what to do next. For example, you likely do not consciously think about your reaction to slow down and stop when a light turns red when you are driving—you just do it. You have developed the automatic thought that tells you to stop driving when the light turns red. This is a good thing because if you had to think about each and every aspect of driving consciously, it would be incredibly tedious, exhausting, and challenging. You would suffer if your child bickering in the back seat with a sibling caused you to lose the focus necessary to drive effectively. However, sometimes, those automatic thoughts that you have learned become negative—they begin to encourage you to behave negatively because the thoughts themselves have been corrupted.

Challenge Your Negative Thoughts

Depressive thoughts are sad, self-critical, self-destructive and rather bullying. People that are struggling with this disorder are found standing on the edge of a cliff where the easiest way is just to let go than to work their way back to where they were. This mind frame of a total pessimism leads to an extremely negative lifestyle. And while it seems like putting a more positive spin on life, is an impossible task, depressed people can indeed generate alternatives.

The only way you can start to challenge your negative thoughts, is to change your point of view, and take a look at things from other angles. Even if it seems really hard to do so, know that it is the only way:

Wake up on the right foot. Instead of going back to bed, so you can skip your day, put on a smile, even if that means faking one in the beginning. Making this a habit will help you overcome depression completely. Every morning upon waking, rather than being overpowered by the feeling of loneliness and emptiness, tell yourself encouraging words such as, "This is going to be a good day", "I'd be better today than yesterday", "Today, I choose to be happy", and other positive thoughts. Doing this every day will help you start your day right, and there is a high tendency for you to stay positive for the rest of the day as well.

Stop being harsh on yourself. Being so self-critical, depressed people think the worst of themselves. Anytime a thought of worthlessness crosses your mind, ask yourself if you are able to share that taught with someone else. If your thought is so dark, that you cannot say it out loud, say the exact opposite to yourself. For example, if you think 'I am the ugliest person alive', say 'You are the most beautiful person in the world' to yourself. Eventually, this will boost your self-confidence and will contribute to overcoming depression.

Allow yourself to be imperfect. It is known that most of the people suffering from depression are perfectionists, trying to meet relatively high standards. Anytime you fail in your attempt to be perfect, challenge the negative thoughts that are occupying your mind by saying that you are only human, and no one is perfect, and try to set the bar lower.

Stop jumping to conclusions. How many times has it happened to you to think the worst of yourself when in the company of other people, considering how they must think the same things about you too? Have you ever gathered proof for this? For instance, you are at a party, and a person avoids you and shows no interest in chatting. What is the first thing that crosses your mind? Is it something like 'he/she must find me pathetic'? Do you have evidence for this? If not, then challenge that negativity immediately. Who knows, maybe that person has some serious issues and does not feel like talking. Why can't this be the reason?

Compare before you decide. Never, think about the cons of things, without comparing them with the pros. Let's say you want to buy a car, but cannot decide because you only think about their negative sides. It is time to weigh each advantage and disadvantage. For example on 'It is more expensive', respond with 'It is comfortable' and so on, until you make your decision based on actual facts and not on your pessimism.

Write them down. Every time a negative thought takes over, write it down. Do this every day and at the end of the week, only when you're in a good mood, review what you have written. Ask yourself if such negativity was indeed necessary? Or maybe you were exaggerating?

Stop labeling yourself. All of the attempts to overcome depression will fail if you continue to give yourself names such as loser, or idiot, whenever you think you've done something wrong. Everyone makes mistakes.

Ignore the temptations. At many points in your attempts to overcome depression, there will be temptations that would want to pull you back from your progress. For example, when something reminds you of that bad event in your life, you begin to over-think again until your emotions take over you once more. This is dangerous because it is possible that you will be back to zero. You will find it challenging to maintain a positive outlook in life this time. Thus, when temptations arise, divert your attention to other more positive things.

Jumping To Conclusions

This negative thought pattern involves you assuming you know exactly what will happen next before it actually does happen, even if you have no real reason that can justify your thoughts. You think that you will fail that test you have coming up because you are bad at the subject, so you do not bother studying for it at all. Of course, jumping to that conclusion basically guarantees it, then, because you fail after refusing to study.

Worst Case Scenario

When you think in terms of the worst-case scenario, you assume that the worst will happen, even though there are several others much more mundane, less disastrous explanations. For example, when your child misses your phone call, you assume that he was kidnapped and sold into human trafficking instead of thinking that his battery died, or he forgot to turn the volume on his ringer after school.

Personalization

When you engage in personalization, you are essentially assuming that you are the root cause of any suffering around you. If you see someone at the grocery store that looks annoyed, it has to be your fault. If you see someone having a bad day at work, you think that it must be caused by your own behavior or failure to provide the care necessary.

Control Fallacy

When this is your negative thought pattern, you essentially give yourself far more credit for influencing those around you than is actually true, to more of an extent than is present in personalization. In this case, it is your fault that there was an accident outside your house last night because you forgot to change your light bulb, so it kept flickering outside, and the flickering was enough of a distraction that it caused an accident.

Fairness Fallacy

When your thinking falls for the trap of the fairness fallacy, you are essentially hyper-focusing on fairness. You assume that everything must be fair, whether it is the case or not. When life is not fair, you believe that you have been wronged and therefore focus on that fact, emphasizing that you should have gotten the same as the other person. This, of course, keeps you hyper-focused on the past and ideals rather than looking at skills, personality differences, or the million different ways someone else might be more deserving of something than you.

Blame

When you engage in blaming, you are either blaming outward or inward in a way to explain away the behavior.

You may assign blame to yourself, focusing on the fact that it is your fault that something happened for reasons that you may not be able to explain rationally totally, or you may blame someone else to avoid blaming yourself, such as saying that the reason you failed your latest exam was because you were sick and therefore not in the right mindset.

Should Haves

When you focus too much on the should-haves, you lose track of the fact that many things in life do not play out the way they should. Parents should not lose children, but it happens. Focusing on what should have happened is essentially just keeping your mind focused in the past rather than allowing yourself to recognize the fact that things do not always work exactly as planned.

Emotional Reasoning

Think about the words involved in the naming of this negative thought pattern—emotional and reasoning. They are exact opposites. When you are using emotional reasoning, you are using your emotions to rationalize your thoughts. Essentially, you decide that your feelings are accurate and allow them to color your thinking. For example, if you feel like you failed, then you will call yourself a failure.

Change Fallacy

When you use this sort of negative thinking, you assume that other people will or have to change for you in some way, and when they do not change for you at all, you lose focus on what the reality of the situation is. You get so caught up in the fact that other people need to do things for you that you stop trying to do something for yourself.

Labelling

Labeling refers to assuming that something can be reduced down to one or two words. For example, you label yourself a failure because you have messed up, or you label a job you have as pointless because you feel like you cannot get anywhere. You fail to see everything else that whatever you are labeling actually is and reduce it down to the simplest terms possible.

Refusal To Admit Wrongdoing

When you engage in this sort of thinking, you essentially reject the possibility of being anything other than right. Even when someone provides you with concrete evidence about whatever has happened or contradicting your claims, you assume that it is wrong altogether.

There is absolutely no possible way you can be wrong, even if you have to twist things in order to have even a possibility of being right.

Heaven's Reward

When you fall for Heaven's reward thinking, you assume that you're good deeds must be rewarded. If you do something good for someone else, you feel like you have to get some sort of benefit for it, despite the fact that in the real world, no one cares to reward you for doing what is expected of you or what makes you a good person.

CHAPTER 90:

The Key To Relaxation

Sleep And Relaxation Techniques

We discuss aspects of sleep because sleep is an essential factor that influences and causes fear. This can be a vicious cycle: you feel anxious and cannot sleep. Your lack of sleep, in turn, causes more fear and emotional fragility. Is there a way out? No doubt. There are many.

Most of the methods in this book are related to sleep in a number of ways. Meditation improves sleep quality and physical activity, such as sports and yoga. Better breathing even improves sleep quality. Creating a sleeping routine helps. But what if all this isn't enough? How can you get the rest you need to keep your mind and body healthy and anxious? First, make sure your bedroom is free of distractions. Your bedroom must be a room in the house that is only provided for rest and relaxation. There is no television, radio, smartphone, or computer staring at you when you rest. If your alarm clock is a function of your cell phone, make sure that this is the only reason why this is used in the bedroom. Second, establish a solid bedtime routine. Wait for an hour to get ready for sleep and go to work, family anxiety, and other torturous thoughts during this time. Take a warm bath to increase relaxation. You can light a scented candle, which also has a calming effect. Don't do any tasks at this time. Relax!!

Third, when you go to bed, bring a good book or maybe a crossword. Don't record an interesting novel that you can't write, but take something more entertaining. Inspirational books and self-help books are great to read when you sleep, as long as they don't wake you up and worry about what problems you might have. Finally, make sure the room is dark, quiet, and peaceful. If, for some reason, you wake up and can't sleep, write a diary, or even breathing exercises. If your concern keeps you awake, write it down - and let it go. The better you sleep, the better you will feel the next day and overcome your fears.

Alternative Methods For Relaxation

We will look at some traditional methods that you can use to deal with anxiety and stress. These methods are less scientifically researched than the techniques presented so far, but they are recognized by many people who face the daily struggle of anxiety. Discuss significant diet changes or take herbal medicines with your doctor before continuing, because he can tell you about health or drug interactions. These methods, combined with previously introduced techniques such as meditation, yoga, and eating, can significantly improve your mood and create a more peaceful, happier life!

Aromatherapy refers to the ancient practice of using flavor essential oils to ease one's mood. Aromatherapy, now considered a form of "alternative medicine," can be useful in treating anxiety by creating a relaxed atmosphere of relaxation. In fact, aromatherapy is always used. When you think about it: we use perfume, soap and scented shampoo, air freshener, and scented candles without thinking about it. Why do we use this product? Because, of course, it smells good! Despite the lack of research to support aromatherapy as alternative medicine, it is difficult to deny the pleasant fragrant qualities for mood and senses.

You can buy essential oils at grocery stores, herbal stores, or online. However, you don't have to limit yourself to essential oils. If you find a scented candle that you like, use it! Aromatherapy is perfect for calming baths, meditation sessions, and visualization sessions - as long as you are well-ventilated, so you don't become overweight! The following list contains essential oils that are known to provide a soothing fragrance, and are recommended for people with anxiety:

1. Ylang Ylang: Ylang Ylang is an essential oil extracted from the flower of a plant known as the "perfume tree." Besides anxiety, ylang-ylang is used as an essential oil to relieve hypertension and normalize skin problems. This is an amazing ingredient from Chanel No. 5, so it can't be said that "Eastern Mystics" appreciate the aroma of the Ylang Ylang flower. Calming effect, this oil helps relieve stress.
2. Bergamot: Bergamot is an essential oil from the orange bergamot tree, which grows in southern Italy and France. Bergamot is often used in Earl Gray tea - another example of "aromatherapy" used in the main product. The smell of

bergamot is orange and raised; this works well to calm the nerves.

3. Sandalwood comes from sandalwood, which is harvested and cut and has been used as an essential oil for centuries. It has been primarily used in the religious rituals of many Chinese, Indian, and Japanese religions. It has a musky, woody fragrant and has a calming effect on the nervous system.
4. Chamomile. Not just for tea! Chamomile has a sweet and slightly fruity fragrant and comes from Roman chamomile flowers. If that happens in the garden, you should consider growing chamomile because the aroma really brightens the garden! Chamomile flowers can be broken down into tea or distilled into essential oils. Chamomile oil and chamomile tea are ideal for calming the mind and reducing stress.
5. Lavender: Lavender has been linked to stress relievers for years. Lavender comes from the mint family and grows throughout the world. It can be used in food, tea, and essential oils. Lavender is a popular stress scent that is usually found in scented candles, bath salts, and perfumes.
6. Geranium rises: This is another flower commonly used in the perfume industry, healthy roses, is mostly planted for refining in South Africa because of its oil. Geranium Rose is ideal for creating a calming atmosphere and is also used to balance hormones.

These are just a few of the essential oils that are commonly used to treat anxiety and stress. If you enjoy aromatherapy, I highly recommend that you investigate further and find more scents that will give you a better feeling of calm and relaxation.

Hypnotherapy

Hypnotherapy includes Hypnosis, as the name suggests. Dr. John Kappas, the founder of the Hypnosis Motivation Institute, defines the role of a hypnotherapist as follows: "Causing the hypnotic state of the client to increase motivation or changing behavior: contact the client to determine the nature of the problem. This prepares the client to enter the hypnotic state by explaining how Hypnosis is functioning and what the client will experience. Tests to determine the level of physical and emotional suggestion. Individual hypnosis methods and techniques based on interpreting test results and analyzing client problems can train clients in self-hypnosis.

Hypnotherapy can be used to overcome the basic fears and fears it causes. At the same time, like cognitive behavioral therapy, this is also used as an initial approach to exposure therapy. Nowadays, hypnotherapy is often used to manage and support anxiety, depression, phobias, insomnia, and addictions. To find a certified hypnotherapist, check online, ask a friend for help, and discuss this method specifically with your therapist. He can recommend a hypnotherapist, or at least discuss the pros and cons of this form of therapy.

Acupuncture

Acupuncture originated in ancient China and involved the use of small needles inserted into acupuncture points on the skin. The reason for acupuncture is to release the energy stored in the body's meridians, thereby creating balance. Current scientific studies show that traditional acupuncture effectively reduces pain and nausea. Although studies do not categorically show the effectiveness of acupuncture in relieving anxiety, many patients have found this practice useful.

If you are interested in testing the benefits of acupuncture, make sure you do adequate research on your doctor. Acupuncturists in the United States must participate in an accredited program and have a license for 3-4 years. Fake acupuncturists, like fake chiropractors, can do more harm than good. Make sure your acupuncturist is qualified, has a satisfied patient history, and is licensed.

Meditation And Its Benefits

Meditation is a conventional practice; however, it has been adopted to help manage anxiety for the past decades. According to research, Meditation is an effective treatment in addressing stress and managing the symptoms of anxiety, phobias, depression, and panic disorder. Meditation is an antidote to breathing dysfunction, and it is an effective way to position your mind, body, and breathe.

We have different schools of Meditation, although the most common and accepted practice is Transcendental Meditation. It offers deep relaxation. Transcendental Meditation was developed by Maharishi

Mahesh Yogi and was widely accepted by the western culture in the 1960s and 1970s.

Transcendental Meditation is a form of Meditation that uses a mantra or sound with a practice of 20 minutes, twice per day. Transcendental Meditation does not require any

Over 600 research has been carried out on Transcendental Meditation, and it has been proven that it improves, emotional, physical, and cognitive effects of stress and anxiety. Looking at these attributes, then it can agree that it's the best option to adopt. Meditation is cost-effective, does not have any side effects and it is easy to adjust when it comes to managing your anxiety. Looking at these qualities, it will be advisable to consider this option.

How do I Get Started?

There are ways to find a meditation center and instructors in the United States. You can as well check via the local YMCA for information, go through notepad, browse the internet, and ask for referrals from friends and family. You don't have to leave the comfort of your home to start the process of Meditation. There are books, audiobooks, videos, and internet resources that can guide you through the process.

To start your basic Meditation without a mantra, then you can start with the following process:

Ø Find the right time and place

Give yourself twenty minutes for this session, look go a quiet, solitary, and free form distraction. Turn off anything that can distract you, such as the radio, cell phone, and other devices that could easily distract you.

Ø Get into a comfortable position

People are fond of imagining the full-lotus position when Meditation comes to mind. While this is considered to be an effective position to meditate it is not necessary. You can perform different types of Meditation while lying down, sitting, standing, and walking. For lying down position. Lie on the floor; lay your back flat on the floor. You can use a yoga mat or blanket, depending on your choice. You might be asking why floor and not a bed? You can feel the sensation of your body against the floor. With this position, you can feel your breath entering and leaving your body. Lie with your palms up and your arms by the side. Your legs should be on the floor; if you don't feel comfortable, you can place a pillow under your knees to help balance your bank.

Ø Close your eyes and relax

This looks easy, right? There are lots of tips that can help you relax. Focus on your breath. Pay attention to your inhalation and exhalation. It is normal to feel distracted; however, you need to focus on your breathe and let the thoughts go. There's no need to blame you or beat yourself over little things. Just focus on your breath.

CHAPTER 91:

Cognitive Behavioral Therapy

Cognitive-behavioral therapy is a psychotherapy treatment that helps people manages their problems by understanding their thoughts and behaviors and ultimately changing them for the better. It enables you to absorb all beliefs and practices that are not useful to you and replace them with those that truly benefit you.

Many studies have shown that cognitive-behavioral therapy can be one of the most effective treatments for anxiety, depression, and other disorders. It is mainly used to treat various diseases, including addiction, phobias, depression, and anxiety. This particular therapy is known to outperform most other treatments in their effectiveness and ability to heal people. However, you might not appreciate the power of cognitive behavioral therapy until you see it in action. If you study it, you will see how reasonable this approach is. Therapists recognize that cognitive-behavioral therapy not only helps them treat patients but also benefits them personally in the long run.

Cognitive-behavioral therapy was initially developed specifically for the treatment of patients with depression. However, its use has increased over the years, and cognitive behavioral therapy has helped people with various types of problems. It functions as a psychosocial intervention that focuses on changing negative thoughts, beliefs, and behavior by increasing emotional processing capacity and developing better-coping strategies.

When people become more aware of their thoughts, it helps them make significant changes in their lives. This change may seem small at first, but the impact is substantial. Most people experience anxiety, mood swings, stress, etc. At some point in your life. If you think about it, cognitive behavioral therapy can help each of us. Cognitive-behavioral therapy tools help overcome many problems and prevent them from appearing again in the future.

People will have different reasons for seeking cognitive behavioral therapy. Many of them may have tried other treatments but have not found them useful enough. Some may have noticed that they still avoid uncomfortable situations when faced with them. Some people want to ward off the habit of self-criticism with self-love. Some people may not be fully aware of their problems or might be rejected. However, many people who want to start cognitive behavioral therapy are aware of their problems and are looking for tools or skills that will help them deal with them once and for all. They know that cognitive-behavioral therapy can be a type of therapy that can help them change existing knowledge into real change. I recommend cognitive behavioral therapy to as many people as possible, always because I know how it works, and I think it can help more.

Some people don't seek therapy or professional help because they think their problem is too big for a therapist to solve. You believe a psychologist or therapist can only help someone with insomnia to get a prescription for some sleeping pills. They do not trust anyone to understand or solve their problems that seem bigger. However, keeping these things locked will only damage you. Everyone can benefit from cognitive behavioral therapy in one way or another. This might not solve all your problems and make it go away entirely, but it will help you handle your struggle much better.

If you want to start cognitive behavioral therapy with some support, we recommend that you find a good therapist. Look around and ask for people's recommendations. Find someone who makes you comfortable, even if they challenge you to change for the better. A good therapist will listen to you and confirm your point of view. They gently challenge you when needed, but also offer you a safe place to talk openly without being judged. You will get to know a good therapist when you meet one.

If you look at the principles of cognitive-behavioral therapy, you will find that it is based on the principles of cognitive psychology and behavioral psychology. However, cognitive behavioral therapy is more focused on the problem and takes steps to solve it. This is a much newer concept than other older psychotherapy programs. It fell under the second wave of psychotherapy, and its popularity only grew.

For example, in the beginning, therapists focus on examining a person's behavior to find the hidden meaning behind it, and this helps them diagnose the problem. In cognitive-behavioral therapy, however, the therapist will focus more on treating problems that result specifically from the client's mental illness. Effective strategies help patients achieve specific goals. These cognitive-behavioral therapy strategies are known to reduce the symptoms of the disorders they encounter and improve their lifestyle. It contains mechanisms for overcoming training and some skills that will help reduce the effects of this diagnosed disease. Cognitive-behavioral therapy theory shows that most mental disorders are related to distortions of thought processes and wrong behavior. Therefore, therapy focuses entirely on changing mindset so that it does not lead to immoral behavior. So you see that cognitive-behavioral therapy focuses less on diagnosing certain diseases rather than finding solutions to patient pain.

Another advantage of cognitive behavioral therapy is that it is beneficial even when psychopharmaceutical is not used. This type of treatment can be used to recover from mild addictions, stress, anxiety, and personality disorders. Cognitive-behavioral therapy also helps adults or children who can show aggressive tendencies and help them improve their behavior. This drug is used in combination with cognitive-behavioral therapy if the client suffers from one of the more severe illnesses. For example, a therapist will most likely prescribe psychiatric drugs if a person has severe depression, bipolar disorder, or OCD. All psychiatric residents have mandatory training in cognitive behavioral therapy and interpersonal psychotherapy during their courses.

As mentioned, cognitive behavioral therapy is a combination of behavioral therapy and cognitive therapy. Let's try to understand what each one contains.

Cognitive Therapy

This therapy was first developed by an American psychiatrist named Aaron T. Beck. He said there was a relationship between a person's thoughts and feelings and their behavior. Therefore, it would be ineffective to examine each aspect individually, and it is far better to conduct a joint study. According to him, this will be a more effective way to understand the client and diagnose his condition. Adequate care can only be offered if the right condition is diagnosed. This is how cognitive therapy is developed. It focuses on current thinking and behavior, along with communication. It focuses less on a person's past experiences and instead helps them learn how to deal with current problems. Various problems can be treated with cognitive therapy. These include anxiety, depression, eating disorders, panic, and substance abuse.

Behavioral Therapy

Behavior therapy involves studying patterns of behavior to understand one's mental state. Instead, this is a general term for various types of mental health therapy. It focuses on identifying and changing someone's unhealthy and possibly self-destructive behavior. The basic principle is that he believes that every practice is learned, and every harmful behavior can be learned and changed. With this therapy, individual behavior is observed in response to different stimuli and various types of situations. The therapist will focus on multiple conditions that you usually experience in everyday life. This helps them identify the cause of the problem and then can help customers find ways to solve it and solve the problem.

The combination of behavioral therapy and cognitive therapy has become one of the best psychotherapy methods in recent years. Instead of trying one treatment, the therapist can combine two effective therapies to get better results. Cognitive-behavioral therapy emphasizes that logic and reason cannot be the only factors in the treatment of mental disorders. Other logical barriers must also be considered. Concerning mental health, it is also not possible to treat treatment problems individually. Several factors need to be addressed, and different treatments must be tried to reach the goal and solve the problem.

History Of Cognitive-Behavioral Therapy

Various aspects of cognitive-behavioral therapy come from many different philosophical traditions from ancient times. One of the primary sources of influence is courage, and many principles of cognitive-behavioral therapy are adopted from it. Stoicism teaches that logic can be used to get rid of all false beliefs that lead to destructive emotions in a person. Here you can see similarities with the identification and treatment of cognitive biases. This stoicism is often cited by Aaron T. Beck, who is known as the father of cognitive therapy. Some famous people who support cognitive behavioral therapy are Albert Ellis and John Stuart Mill. One of the first therapists to recommend cognitive therapy was Alfred Adler. His work then influenced Albert Ellis, who later developed therapies for rational, emotional behavior. This REBT treatment is the earliest knowledge-based psychotherapy. Beck is the person who said that Freud's theory did not apply on a case-by-case basis and that specific thought can cause emotional stress on everyone.

Cognitive therapy is developed through this way of thinking, and here refers to automatic thinking. The conditioning study was conducted by Rayner and Watson around 1920. This is an integral part of the study of behavior. Cognitive therapy was established around 1960, but behavioral therapy has existed since the early 1900s. More prolonged research has been used to develop Joseph Wolfe's behavioral therapy further.

In the years that followed, much research was done with people like Glen Wilson and Arnold Lazarus. Watson, Pavlov, and Hull presented theories that inspired much research into cognitive and behavioral therapy—immediately used in various countries around the world. Joseph Wolfe's work is significant in laying the foundation for the anxiety reduction techniques used today. Rother and Bandura are other participants whose primary task is in social learning theory. They show how learning and behavior change are influenced by knowledge. Emphasis on behavioral factors contributes to the first wave of cognitive-behavioral therapy. REMT and cognitive therapy trigger a second wave. The third wave is the result of mixing technical applications with behavioral and cognitive theories. This should give you a brief description of the origin of cognitive-behavioral therapy.

CHAPTER 92:

Power Of Cognitive Therapy

The reason cognitive behavioral therapy is so effective is that it is unlike other forms of treatment. It takes your issue and dissects it to the root so that you can resolve the situation, not just find a quick fix. There are ways that you can make yourself feel better in the moment, but when you have continual negative perspectives, it is going to make it challenging to find a solution that sticks and not just something that provides temporary relief.

Cognitive behavioral therapy uses your own mind to try to help you change the way that you think. When you can find that sort of resolution and not just one that will make you feel better in the moment, there will be long-term results.

It looks at the root cause without just giving you a quick solution. Think of it like any other sickness. You can take Tylenol, ibuprofen, or another pain medicine to make you feel better, but if you don't actually treat what's making you sick, then it is only going to make things worse.

You can put a Band-Aid on a cut, but if it is not treated correctly, it might still get infected and cause more health issues. CBT aims to help you find a resolve, a solution, and something useful you can use, without any other negative side effects. It is not going to be easy for everyone, but it is something that could potentially help everyone if they put the effort in to change.

Remember that cognitive behavioral therapy isn't something you are always going to be able to do on your own. There might be moments you need to ask for help, and never be afraid to do so. Even if you have a small social circle, inability to access professional help, or are scared to reach out, there are always online tools you can use to help you overcome your depression and anxiety.

The most important part of cognitive behavioral therapy is having compassion. You have to make sure that you are not being too hard on yourself. If you spend too much time trying to be perfect or making sure that you is going about recovery in the right way, you are only going to set yourself further back. It is essential to make sure you are treating yourself with as much kindness as you would anyone else going through the same process.

By picking this book (or your reading device) up in the first place, you have already shown yourself love. You have admitted that there is an issue and that it needs to be fixed. You have accepted that it is time for a change and that you are ready for a new perspective and chance at life.

Hating yourself can be exhausting. If you spend so much time always breaking yourself up, you get to a point where you don't even see your own purpose anymore. When we make ourselves feel like we don't deserve anything, we'll start to believe it, and it can cause serious issues, both mentally and physically, down the line.

Make sure that you are giving yourself realistic time periods to achieve specific goals. Don't put too much pressure to change right away. After reading this book, you might already feel enlightened. You might have gotten to a point where you instantly feel better and now have the tools to go toward a path to improvement. There are still going to be moments that challenge you and times where you have the same thoughts that caused you to pick up the book in the first place. Always remind yourself you are stronger than your most negative thought and continue toward a path of recovery. Be patient to yourself so that you don't expect too much and get disappointed when you don't meet unrealistic goals.

If you do have a moment where you fall back into old habits, don't beat yourself up over it. In fact, expect these moments so that you don't feel completely lost if they do happen. Having too low of expectations can sound depressing to some people, but if you set them too high, you are putting yourself at risk for just as much failure if you don't push yourself at all in the first place.

Look at it like you are losing weight. If you try to force yourself to lose twenty pounds in a week, you are either going to have to go through something seriously unhealthy or not meet those goals and get an overall feeling of discouragement. Be realistic with your goals, and then give yourself some more time as well. There is no rush. Life is short, but it can be easily wasted by spending too much time ruminating and thinking of all your regrets.

Using Your Mental Illness

"Why is this happening" or "Why me?" is going to be a common thought among depressed individuals. It is hard not to question why you might be experiencing something that others will never understand. If you don't know others that are depressed, it can be a very lonely feeling. You might often ask yourself why you can't just be normal, as you would with any other ailment or quality that sets you apart from the norm.

Though you might wish you could go back in time and never have to experience it in the first place, find a way to be happy that you have it now. Depression and anxiety aren't fun. No one is excited that they continuously have anxious thoughts or depressive feelings. However, we can find meaning within them and see these illnesses as something part of our character. What doesn't destroy us can make us stronger, but it also gives us perspectives that no one else will have.

Of course, we are never going to be excited that we were depressed or anxious. But looking at something only in a negative light can make you feel very hopeless. Until we find a way to travel time, there's no going back. You can't reverse time and fix things you regret. Instead, we can look at the experiences we've had and relate them to ways that we've grown.

If you live in a warm place all your life, you might never fully appreciate the sunshine if you don't know what it means to be caught in a snowstorm. Be grateful that you have a newfound perspective on life.

Sometimes, those with anxiety will have better thought processes. You might have anxiety all the time, but you have also gained the ability to be prepared for specific situations. You have also been taught now how to manage some of your feelings. By being forced to express yourself through journaling, you have even gotten the chance to understand how to properly discuss emotions, rather than keeping them in for so long.

Even people without chronic depression and anxiety could use some cognitive behavioral therapy tools just to help them navigate easier in general. If you are not depressed at all and you take Prozac, it might end up making you feel bad. But if you use CBT or see a therapist and you are not depressed, you might find that you had automatic thoughts that were still affecting your life.

Creativity

Those who experience mental illness will often be more creative than others as well. They've gained the ability to look at things from a completely different perspective. That doesn't mean that someone who hasn't struggled with mental illness can't be creative, but it is still a beautiful result of something so tragic.

When you are depressed, you are more likely to question the meaning and ask yourself why something is the way it is. If you go through life moderately satisfied, you might never have to face reality or question some of your thoughts. When you are depressed and anxious, you have to look yourself in the mirror and reflect on the past, helping you grows as an individual from within.

Those who are anxious have creative abilities to make predictions. There's no one better coming up with wild stories than an anxious person that's thought of all possible outcomes! Find a way to use your mental illness to express yourself, whether it is through music, painting, drawing, writing, or some other creative medium.

If you feel like you don't have a creative bone in your body, that doesn't mean you can't still express yourself. Perhaps you can be a mentor, a teacher, a cook, an organizer, a manager, or a coordinator.

When you use cognitive behavioral therapy to overcome your thoughts, you become your own leader. When you do this, you'll be able to have better the tools it takes also to lead others.

CHAPTER 93:

Benefits Of Cognitive Behavioral Therapy

Firstly, it is amazing that this one therapy can help in the treatment of so many different conditions. As you gain a better understanding of Cognitive Behavioral Therapy, you may be curious about what exactly it can be used for treating. Unlike what some believe, Cognitive Behavioral Therapy can actually be used to treat quite a few different conditions. It is also used as additional therapy, alongside other treatments, for various diseases to help the client heal faster and improve their way of life. The following are some of the common conditions that can be treated with the help of Cognitive Behavioral Therapy.

Amongst many other conditions are the following:

- Depression
- Dysthymia
- Eating disorders
- Substance abuse
- Anxiety disorders
- Sleep disorders
- Personality disorders
- Aggression issues
- Anger issues

- Chronic fatigue
- Psychotic disorders like schizophrenia
- Muscle pain
- Somatotropin disorders
- Criminal behavior

Research has shown that Cognitive Behavioral Therapy has been used in the treatment of these conditions successfully for a very long time. Cognitive Behavioral Therapy has shown higher response rates when compared with other therapies used for the treatment of these conditions. This is why psychotherapists all around the world tend to recommend Cognitive Behavioral Therapy in particular to their patients.

The following are some of the significant benefits of Cognitive Behavioral Therapy:

Helps Reduce Symptoms Of Depression And Anxiety

The best-known treatment for depression is Cognitive Behavioral Therapy. Numerous studies show how beneficial this particular therapy is in alleviating symptoms of depression, such as lack of motivation or a sense of hopelessness. This treatment is short term, but it lowers the risk of relapses happening in the future. It helps relieve depression because it facilitates the changes in thought process for the person. Hence they stop thinking negatively and try being more positive. Cognitive Behavioral Therapy is also used along with anti-depression medication. Some therapists also recommend it for patients suffering from postpartum depression. These are the conditions that are most commonly treated with Cognitive Behavioral Therapy. Depression has become a prevalent condition in the last decade. As the stigma on mental health is slowly being removed, more and more people are being diagnosed with this condition. The causes may vary, but the symptoms are mostly the same, and it has a very negative impact on the person suffering from it. Depression can be a result of some traumatic event or even arise out of the blue. Nonetheless, it is essential to validate this condition and give the person the help they need. The hectic lives we lead also cause a lot of people to have anxiety issues. Cognitive Behavioral Therapy is beneficial in overcoming both these conditions and helping the client lead a happier life.

It Helps Treat Personality Disorders

Each person has their own unique personality that others identify them with. How you come across to others is reflective of your personality. Every person will have specific good characteristics, along with a few bad features. When a person displays more of the bad, their personality appears non-appealing to those around them. This can cause them to face various social issues. Cognitive Behavioral Therapy can help improve such negative traits and develop more of the positive characteristics that will benefit the person and make them well-liked amongst others. Cognitive Behavioral Therapy can be used to improve someone's general personality as well as for treating various personality disorders that are a much more serious concern.

Helps In The Treatment Of Eating Disorders

According to research, eating disorders have some of the strongest indicators of Cognitive Behavioral Therapy. Today's society lays a lot of emphasis on appearance, and this causes people to be much more conscious of how they look and how much they eat. This is especially so for women all over the world. This is one factor that causes eating disorders. They may develop these disorders to lose weight or even as a reaction to stress and anxiety.

Most eating disorders will fall under two types. One type is when a person develops a tendency to eat much more than required. This can cause the person to become overweight and develop weight issues. The second type is where the person may do the opposite and eat too little. This can cause problems like anorexia or bulimia. It is to be noted that these conditions can also be fatal, in the long run, if they are not treated soon. The human body requires a healthy amount of food to sustain itself well, and too much or too little can cause various illnesses in the body.

There are also people who have body Dysmorphic disorder. This makes them see their body different from what it really is. Even when they are thin, they might look at their reflection and criticize themselves for being fat. You have to understand that these kinds of eating disorders are quite strongly related to mental health issues. Your brain plays an essential role in the development or prevention of such problems. When there is an imbalance in certain hormones in your body, either type of eating disorder may develop. It is essential to identify the cause of any eating disorder and help the person overcome it. This is where Cognitive Behavioral Therapy comes in by finding the cause of any problem and then implementing techniques that help them overcome the issue.

Cognitive Behavioral Therapy has been incredibly effective in helping people with eating disorders. It has helped identify the underlying cause of such issues and questions about why shape and weight are over evaluated. Cognitive Behavioral Therapy will help a person maintain a healthier body weight. They will learn to control the impulse for binge eating or the tendency to purge after eating. They will learn to feel less isolated and will also learn to be more comfortable around food that might usually trigger their unhealthy behavior. Exposure therapy teaches them to avoid overeating even when their favorite food is right in front of them. Cognitive Behavioral Therapy is especially useful in helping patients suffering from bulimia nervosa. It also helps in treating conditions that are not specified either. Anorexia is one of the most challenging eating disorders to cure, but research shows that Cognitive Behavioral Therapy is quite helpful in dealing with this condition.

Helps Control Substance Abuse And Addictive Behavior

At this point, no one is unfamiliar with substance abuse. Drugs are obtained too easily, and people tend to abuse their privilege as and when they can. No matter where you look, you will see that marijuana, meth, cocaine, and other such drugs can be obtained quite easily with a little bit of money. Going to the clubs or even walking down some streets will bring you across dealers who will coax you into trying these harmful substances. A lot of people think that it's okay to try it once, but it is a well-known fact that once is just the start of a habit.

Most people fail to exercise control over the consumption of such substances and deaths from overdose have taken an alarmingly high number over the years. Taking such drugs can cause a person's life to take a very drastic change for the worse. It changes them, affects their body, and has nothing but a negative impact on their entire life. But for those who seek help on time, there is a way to turn things around. Cognitive Behavioral Therapy is one of the well-known therapies used to help addicts get over their addiction. The exact approach can be different for each person, but Cognitive Behavioral Therapy can help most addicts in getting over such harmful addictions. They will learn to control their thoughts, resist urges and take actions that will only benefit their well-being. And although it will take time, Cognitive Behavioral Therapy will train them in a way that they learn how to say no to drugs.

There is a lot of research that indicates Cognitive Behavioral Therapy is effective in helping people who are addicted to cannabis, opioids, alcohol, smoking, and even gambling. The skills taught during Cognitive Behavioral Therapy will help these people control impulses and prevent a relapse after the treatment is over. The behavioral approach is more effective than any control treatment.

Helps Reduce Mood Swings

Have you ever felt really low or really high? The answer will be yes for most people. Everyone has mood swings, but for some, it is not in the usual manner that others experience them. Your mood is your current state of mind and it will affect how you feel and what you do. Sometimes your mood may be good and you feel very happy. The next moment you may feel very low and feel quite unhappy. If you have mood swing issues, your emotional state may be quite unstable. The smallest provocation may cause you to get angry and react in a bad way. It is essential for people to have better control over their emotions and to learn to process them well. This way, they will know how to react appropriately in different situations. Cognitive Behavioral Therapy is one of the treatments that can help people deal with such mood swings and be more emotionally stable.

Helps in the Treatment of Psychosis

Are you familiar with this term? It is a state where your mind starts losing touch with reality. In this state, your mind is pushed into a different world that is filled with sounds or images that lull you into a momentary but deep mental slumber. People with psychosis issues tend to have minimal contact with other people. Their mind tricks them into thinking in a very different way that is not healthy for them. Psychosis is more severe than you may realize. It can push the person in a way that their usual way of life is completely changed and not for the better. Cognitive Behavioral Therapy is used for helping people who face psychosis issues. It is used to identify the root cause behind this state of mind and then the therapy helps them to establish a routine that is more of a usual way of life. The client will slowly move back towards reality and out of their virtual world.

Helps Improve Self-Esteem And Increases Self-Confidence

There are a lot of people who suffer from low self-esteem or self-confidence. They may not have any serious mental conditions but having low self-esteem can cause the person to have very negative and destructive thoughts. These thoughts are replaced with positive affirmations with the help of Cognitive Behavioral Therapy. This therapy helps the person learn better ways of dealing with stressful situations and how to improve their relationships with others. Cognitive Behavioral Therapy also gives them the motivation to try new things. Cognitive Behavioral Therapy techniques will help the person improve their communication skills and develop better relationships. Many other conditions can be treated with the help of Cognitive Behavioral Therapy as well. There are even people who suffer from more than one condition at the same time. You also have to know that this therapy may not be effective in completely treating someone at times, but it makes a huge positive difference, nonetheless. This is why it is always worth it to give Cognitive Behavioral Therapy a try. Cognitive Behavioral Therapy is a much more practical solution than most other treatments and it has helped a lot of people over the years.

CHAPTER 94:

Identify Your Negative Thinking

Social anxiety comes with harmful and destructive thoughts. You need to be observant so that you do not hurt yourself in the process. The negative thinking is likely to rob you of the confidence that you have and feel that you cannot stand going before people. The thoughts instill fear in you, and you end up avoiding social gatherings. When you subject your opinions to negative thinking always, it will result in negative emotions. It can end up making you feel bad and can even lead to depression. The thoughts that you have will determine your mood for your entire day. Positive thinking will make you happy, and you will have a good feeling. Finding a way to suppress negative thoughts will be of importance to you. Replace them with positive ones so that they will not torment you. Some of the negative thoughts that come along with social anxiety are;

Thinking That People as bad

When you are in a social setting, each person tends to be busy with their issues. You can meet a calm person and begin a friendship if you can build a good rapport. When you have social anxiety, you are likely to avoid people and think that people do not care about you. You may feel that you don't see your importance being there while as you is the one who is avoiding them. When you find yourself in such a situation in a social setting, its time you know that you have to handle your social anxiety. That will make you feel like the people around you do not like you and hate you for no reason.

Unnecessary Worry

It is obvious to have unnecessary worry when you have social anxiety. Even when you are on time, you are always worried that you will get late. You will portray a bad image for getting there late. When in a family setting, you are worried that your partner will scold you for lateness. You do things in a hurry so that you do not get late even when you have enough time to go to your thoughts. When you have someone else to accompany you, you will make them do things in a rush. You think like they are consuming all the time and they will be your reason for being late. You will even, at times, threaten to leave them if they don't do things quicker than they are already doing.

When you are needed to make a presentation, you get worried about whether the audience is going to like it or not. You are not sure whether there is someone else who will do better than you. The worry will make you start thinking that your boss will not like what you have even before they make their remarks. You believe that you have nothing exciting to say. The worry will make you lose strength, and any change you feel makes you think that you are not ready. When you start experiencing such concern, you need to know that you have social anxiety. You require having an approach that is appropriate to handle your social stress so that it will not escalate into something worse.

Judging Yourself

Judging yourself is the worst thing you will ever go, and that will make you fear. Deciding whether you will find pleas people or not will make you have much social anxiety. You will be nervous when you start thinking about how people are thinking about your physical appearance. Judging how you other people will view you will make your self-esteem go down. The people you are worried about will think you do not look outstanding may not have an interest in how you look but what you have to deliver. At times, people ignore the minor details that are making you judge yourself unnecessarily.

Criticism

Anytime you know that you will intermingle with people, you fear they will criticize you. You do not even have a valid reason why they will criticize you, but you think it is not wise for you to join them. It is a negative thought, and you need to stop thinking in that direction. No one is going to critic you for any reason, and that should not be considered close to you. You will fear to go to social gatherings, and you will deny yourself an opportunity to learn from your fellow partners. Fearing criticism will make you be an introvert, and this will make you mean that you will miss much when you choose to stay indoors. When you say something, and you get someone to challenge you, it will help you be more creative. Criticism is not evil even though that people with social anxiety do not like a situation where they are subject to critics. They will avoid such problems at all cost for fear of humiliation.

Negative thoughts will only escalate worry as well as fear in you. For you to avoid negative thoughts, you can practice cognitive-behavioral therapy. That will replace your negative thoughts with those that are accurate as well as encouraging. Though it will take quite some time to replace the negative thoughts, practicing healthy thinking daily will make that natural to you. If you have social anxiety and you feel that all the approaches you use are not helpful, seek the help of a therapist. Positive thinking will help you in coping with social anxiety. When you notice that you have negative thoughts disturbing you, you should try to drop that with immediate effect. Filter the bad and focus on the good. To change how you are thinking, the first thing you need to do is try and understand how your thinking pattern is at the moment. Do not view yourself as failure always because that will never change your thinking patterns. When you avoid negative thoughts, you will be in an excellent position to fight social anxiety.

Prejudgment

Although it is wise to think about the future, the art of judging tends to be detrimental. In other words, the aspect of prejudging situations tends to be worse, especially when the opposite of the expectations is meant. In most cases, social anxiety disorder tends to cause people to decide the results of a particular situation. It is worth noting that prejudgment is done in regards to the history or rather a forecast of what might happen in the future.

In most cases, the people who are involved in this practice are always negative. Thus, they will think evil of someone and need up piling unnecessary pressure on someone. There are cases where the prejudgment causes one to overthink rather than spend quality time improving on their lives. In other words, prejudgment may force one to change all her aspects in an attempt of meeting the expectations of the peers. The peers will, in most cases, demand or look for everyday things. However, there are cases where the prejudgment predicts a near future that is yet to be accomplished. The aspect causes one to fear as they loom for an alternative as well as ways of meeting the expectations.

However, when the expectations aren't satisfied, the anxiety of what the society will say pile up, and one may quickly lose the focus.

Blame Transfer

When a society or the peers piles up unnecessary pressure on someone, the chances of missing the mark is relatively easy. In other words, one quickly loses focus and misses the point. As a means of evading the shame or rather the punishment, the victims, in most cases, transfer the blame. For instance, if it issues to deal with academics, the victim may start claiming that time wasn't enough to deliberate on all issues. Others may associate their failures with the climatic changes or the lack of favorable conditions for working. There are cases where the victims tend to be genuine and claim diseases as the cause of their failure. However, the art of transferring blames from one point to another tends to be detrimental and shows signs of being irresponsible.

Procrastination

One of the significant effects of social anxiety is that is causes one to fail in deliberating on duties and transfer them to a later day. In other words, procrastination becomes the order of the day. However, it is worth noting that with procrastination, the expectations are never met. The fear, as well as the sensation of being anxious sets in. In other words, the victim starts feeling as if they are a failure in the collective and loses focus. More time is wasted as they try to recollect themselves up. More fear sets in and the victim may end up being restless. Improper management of time is the primary cause of procrastination. In other words, the lack of planning causes individuals to keep working over the same issues and forget about others. For instance, scholars may spend more time with the subjects they like and forget about the others. In other words, they may end up forgetting that all the items will be examined in the long run. The sensations bring more fear and restless.

CHAPTER 95:

Regenerate Your Brain With Relaxation And Cognitive Therapy

There are many pills that you can take for anxiety. Unfortunately, these are just temporarily reliefs that won't fix the root problem.

There's a chance that by this point, you are already feeling a little bit better. Sometimes, all it takes is sharing and confronting your feelings for you to have a better sense of yourself and the things that trigger you. Telling just one person about a trauma, a secret, or something that you fear can make you feel better. Having to keep that all in yourself can cause a lot of strain. That's added responsibility that you might not always have the mental capacity for.

It all starts with identifying your root causes. Is it body image issues from society or your parents? Did you have abusive family members that hurt you physically and mentally? Perhaps it was a significant trauma that has led to depression and anxiety in your adult life.

Make sure that you are taking notes on what those are. If something pops into your head, jot it down so you can look back at it later. Remember that you aren't going to be able to unravel it all right away. At first, you might just be able to look back at your life and see the loss of a parent as the root cause for some issues. Once you explore that, you will discover other small instances that experience affected and how that has resulted in the way that you think today.

After you have identified the root cause, you can start to see how it has manifested itself into your life. Do you experience obsessive-compulsive disorder after parental abandonment? Do you have social anxiety because of your body image issues? When we look at the 'what' of our mental illness, we can look at the 'why', and vice versa. Sometimes, you might first notice your habits before the cause, or maybe you already know the reason and can now use that to explain your behavior now. Either way, it is essential to make sure that you are aware of this so you can come up with specific cognitive behavioral therapy tools to overcome that.

Now, it is time to start to see what we can do to turn these thoughts and feelings around. Instead of finding a quick fix, we have to look at long-term solutions to help us overcome fear and anxiety. These are all methods that aren't going to solve all your problems right away. They are things that you need to practice, but with time, they become easier. It took a while to develop the automatic thoughts you have now, so you can't expect that they'll stop after reading this book. After a while, however, you will find that they come naturally to you.

Mindfulness

Mindfulness is going to be one of the most helpful tools of Cognitive Behavioral Therapy for all forms of mental illness. Whether you are anxious all the time, or just experience social anxiety in situations with other people around, you can benefit from mindfulness. It can be helpful for those that have constant depressive thoughts as well.

It can be easy to run away with your thoughts when you are feeling stressed and overwhelmed. You might escape to a place in your mind, real or not, where you don't have to think about the present. It might be a fantasy of a life that's better than the one you are living, or perhaps you often reminisce on times that were better than now.

Mindfulness is the attempt to ground you and keep you in reality, rather than lost in your anxious and stressful thoughts. When you think of one anxious thing, it can easily lead to many other worries. For example, you might think, "Oh no, I have to pay the cable bill because it is already late." Then you might start thinking of all the other bills stacking up, not to mention the debt that's increasing in interest. Then you begin to think about how you aren't making enough money at your current job, or that you should have gone to school for a different subject, or you shouldn't have broken up with that guy who's now a doctor. One thought about a late cable bill can lead to an existential crisis if you don't stop anxiety right in its tracks. Mindfulness will help you through this. Instead of ruminating on what could have, would have, or should have been, you instead focus on the now, which is really what is most important.

It is something that needs to be practiced. You might only be able to be comfortably mindful for a few seconds at a time before your thoughts go straight back to something else. When you find that you are losing yourself in your own thoughts, you have to make sure that you redirect. Back to the example of the cable bill, you might notice your anxiety rising when you think of the other bills stacking up. That is when you would stop your thoughts and instead use a mindful tactic to bring you back to the present and instead focus on the cable bill.

Then, you might find that your mind is wandering again, and before you know it, you are thinking about that wealthy guy and what could have been. You have not failed, but you just need to redirect again, either using a new mindful technique or just practicing the other one again. Don't punish yourself for drifting back into those thoughts.

Picture yourself driving down a highway. The present is the road in front of you, the past the grass to the right, and the future the other road on the left. If you drift too far right or left, then you'll get lost and off track. Mindfulness is what's going to keep you moving forward in the right direction. At first, you are going to be driving down the road very wobbly. Every time you feel yourself drift off to the right or the left, redirect and focus back on going straight. Eventually, you'll find it very easy to avoid drifting at all.

Mindfulness should be practiced whenever you find that you are experiencing symptoms of anxiety. You should practice mindfulness even when you aren't in a negative mindset. Maybe you are just bored or restless, which are also forms of anxiety. We don't always view being bored as a bad thing, and many people feel lucky if they get the chance to have nothing to occupy their time. If you are bored for too long, however, you might find your thoughts drifting somewhere dangerous, so practice mindfulness even if you don't feel like you are that obviously anxious.

Some activities would be considered mindful, such as playing a game or reading a book. Anything that keeps your mind on the present is going to be mindful. We might not always have the time to play a game or read a book to pull us out of our thoughts, however, so it is important to remember to use these tools for when you are feeling anxious.

CHAPTER 96:

Self Esteem

What is self-esteem? A simple definition would be having confidence in your ability and worth. It is the way you perceive the kind of person you think you are, the negative and positive thing about you, your abilities, and the things you want for your future. If your self-esteem is healthy, you will believe positive things about yourself. You might experience some hard times during your lifetime but you can handle these times without causing any long term negativity. If you do have low self-esteem, you will always think negatively about yourself. You are only going to focus on any mistakes or weaknesses that you may have. You might not be able to see that your personality does hold some very good things. If you have any failures or hardships, you will continuously blame yourself.

Causes Of Low Self-Esteem

There isn't any way to find one single thing that causes low self-esteem that will work for everybody. You have believed these things about yourself for a long time and this process can be affected by many things.

Here are some factors that might cause low self-esteem:

- Mental health problems
- Bullying, abuse, or trauma – things like psychological, sexual, or physical abuse, bullying, and trauma could lead to feeling worthless and guilty.
- Loneliness and social isolation – if you haven't had a lot of contact with others or you can't keep a relationship with another person; this could cause you to have a bad self-image.

- Stigma and discrimination – if you have been discriminated against for any reason, this could change how you look at yourself.

- Thinking patterns that are negative – you might develop or learn thinking ways that reinforce your low self-esteem like creating impossible goals that you won't be able to achieve or constantly comparing yourself to other people.

- Excessive pressure and stress – if you are under a tremendous amount of stress and you find it had to cope; this might cause the feeling of low self-worth.

- Relationships – there might be other people who feed your low self-esteem if they make you feel like you don't have any worth or they are negative about you. You might feel as if you can't live up to other's expectations.

- You feel different – you feel like you are the odd duck or are pressured to fit into what others think is the social norms. These can change how you view yourself.

- Temperament and personality –personality elements like having the tendency to think negatively or making it impossible to relate with others might cause you to have low self-esteem.

- Life events – if you have had difficult experiences during your adulthood like ending a relationship, illness, the death of a loved one or losing your job could affect your self-esteem especially when you experience many horrible events in a short time span.

- Childhood experiences – negative experiences during childhood like having hard times in school, hard family relationships, and bullying could damage your self-esteem.

Self-Esteem And Mental Health

Having low self-esteem isn't a recognized mental health condition but they are similar.

Having low self-esteem could cause you to develop mental health problems:

- You won't try to do things that aren't familiar to you and you don't even finish a task, like beginning a new art project. This might make it hard for you to have the life that you've always dreamed of. It could cause depression and frustration.
- The thinking patterns that are negative that have been associated to people who suffer from low self-esteem like automatically thinking you are going to fail could happen with time and cause you to develop mental health problems like anxiety or depression.
- If specific situations seem hard due to low self-esteem, you might begin avoiding them and this causes you to become even more socially isolated. All of this could cause you to feel depressed or anxious and could cause mental health problems with time.
- You could develop bad behaviors to help you cope like drinking to excess, taking drugs, and getting into a damaging relationship. This can cause problems and make your life harder and this might cause mental health problems.

Once you have developed mental health problems, this, in turn, could cause low self-esteem:

- Discrimination about your mental health problems could cause you to develop negative opinions about yourself.

- Mental health problems may cause you to withdraw from society if you worry about how others see you. This can make you feel lonely or isolated and could cause low self-esteem. Moreover this problem could influence daily activities like keeping a job or using public transport, and this has a negative impact on how you look at yourself.

- Specific mental health problems like social phobia, depression, and eating disorders involve negative thinking patterns about you.

Things You Can Do To Improve Your Self-Esteem

In order for your self-esteem to improve, you are going to have to change and challenge all those negative beliefs. This may seem like a daunting task but there are various techniques that could help.

Do Things You Like to Do

Doing things you like to do and can do well could help give a boost to your self-esteem and build your confidence. This might be anything like your favorite hobby, caring, working a soup kitchen for the homeless, and working.

Working

Working a steady job could give you a salary, routine, friendship, and identity. Some people like to work on ambitious targets and they love being in the middle of a busy environment. Others will view their career as just a way to have things they want or they only do volunteer work. It does not matter what you do but it is essential that you feel supported and confident in whatever you do. You need to feel a balance between your home life and work-life that is right for you.

Hobbies

This might be something from painting, singing, dancing, or learning a new language. Think about things that you have a natural ability at doing or something you have wanted to try. Choose activities that are not too challenging so you will feel that you have achieved something and that will help you build your confidence. You should be able to find some classes or activities at your local library, adult education center, or the internet.

Building Positive Relationships

Try associating with people who do not always criticize you. These need to be people that you feel comfortable talking to. If you spend more time around people who are supportive and positive you will start seeing yourself in a better light and will start having more confidence.

If you support and care for others, they will give you positive responses. This could give your self-esteem a boost and change the way others see you. If you suffer from low self-esteem, people who are close to you might encourage your negative opinions and beliefs that you have about yourself. It is imperative that you find these people and stop them from doing this. You need to become more assertive or limit the amount of time you spend with them.

Be Assertive

When you are assertive, you value others and yourself. You are able to talk to others with respect. It can help you get boundaries set. Try some of these to help you be more assertive:

- Use "I" statements if possible: "When you talk to me like this, I feel…" This lets you tell them what you want to happen without being scared or aggressive.
- Tell others if you need support or more time on tasks that are challenging for you.
- It is okay to say "no" to unreasonable requests.
- If you have been upset, try to express your feelings or wait until you have calmed down to try to explain the way you are feeling.

- Pay attention to body language and the words you are saying. Always be confident and open.

Assertiveness is a hard skill to learn and you might have to practice with a friend or while looking in a mirror. Most adult education centers have assertiveness classes that you can take. There are many books that have practical tips and exercises that you can use. These can be found online or in stores.

Take Care Of Yourself

Taking Care of yourself physically could make you feel healthier and happier and might give your self-image a boost. Here are some things you can do:

Diet

Eating a diet that is well balanced at standard meal times that is full of vegetables and water is going to make you feel happier and healthier. If you can lower or stop drinking alcohol and stay away from illegal drugs and tobacco, it could help boost you are well-being.

Sleep

Not getting enough sleep could increase negative feelings and this can lower your confidence. It is essential that you get enough sleep.

Physical Activity

Getting exercise can help anyone's sense of well-being and how they view themselves. When you exercise, your body will release endorphins or "feel good" hormones. These hormones can help boost your mood especially if you exercise outside.

Challenge Yourself

You need to set goals for yourself and then work to achieve them. When you are able to do this, you are going to feel proud and satisfied with yourself. You will feel more positive about yourself because of this.

Be sure your challenge is something that you know you will be able to do. It does not need to be anything spectacular but it needs to mean something to you. You could decide to write to your pen pal or begin going to the gym regularly.

Identifying And Challenging Negative Beliefs

In order to boost your self-esteem, you need to realize your negative beliefs and how you got them. This process could get painful so you need to take the time and ask for help from a friend or partner. If you begin feeling extremely distressed, it could be better to find a therapist you can talk to. You can write down some questions and notes to help structure your thoughts:

- Do you have specific negative thoughts regularly?
- Can you think of a particular event or experience that could have caused these feelings?
- When did you begin feeling this way?
- What word could you use to summarize yourself? – "I am…?"
- What do you think other people think negatively about you?
- What do you think are your failings or weaknesses?

It could be helpful if you kept a thought diary or journal to write down situations or details about how you feel and what you think these underlying beliefs are.

As you are able to identify your beliefs about yourself and where they originated from, you will then be able to change them. This can be done by writing down some evidence that will challenge every belief so you can start to explore other situations.

CHAPTER 97:

Believe In Your Self-Esteem

Be Happy Now

One of the ways to start working with your self-esteem is to be happy. Happiness is a journey, not a destination. Happiness is not something that happens to you from the outside. Happiness is a habit, a state of mind. Happiness is so many things. But the decisive and most important thing is: What is happiness to YOU?

The latest studies have found that happiness is not a thing that happens to you from the outside. It's a choice you make, but it requires effort. The good news is: It can be taught. It's those small habits like gratitude, exercise, meditation, smiling and asking yourself "What can I do to be happier in the present moment."

You can be happy right now! Don't you believe me? Okay. Close your eyes for a moment. Think of a situation that made you really, really happy. Relive this situation in your mind. Feel it, smell it, hear it! Remember the excitement and joy! So? How did it feel? Did it work? How are you feeling now? Happiness doesn't depend on your car, your house, or anything in the outside world. You can be happy right here, right now!

Science has found that your external circumstances make only 10% of your happiness. Surprisingly where you are born, how much you earn, where you live, where you work has a remarkably small impact on your happiness.

50% is genetical. Yes, some people are born happier than others. A whopping 40% of your happiness can be influenced by intentional activities. This is where the gratitude, the long walks, the meditation comes in. This also means if you are born less happy you can improve your happiness by doing these intentional activities.

Don't postpone your happiness to the future, the new apartment, the new car, the promotion. Happiness is right here, right now. In a sunrise, in the smile of your children, in a beautiful piece of music, you are listening to. Sometimes - when you stop chasing happiness and just stand still, you might notice that happiness has been on your heels all along.

Your happiness - same as your self-esteem - depends solely on you. Other people may influence it in specific ways, but ultimately, it's always you who decides, who chooses how happy you want to be.

Be Nice

How you treat others is very closely related to how you treat yourself. So, be nice! It will pay you dividends in the long-term.

Emotions are contagious. Scientists have found that if you put three people in a room together, the most emotionally expressive one "infects" the other two with his or her emotions - this works both ways: positive or negative.

Choose to infect others with positive energy. It will only be beneficial, because as they say "what goes around comes around."

You already learned the power of your spoken words. Use your words positively, use them to empower people. Words have such a significant impact. It's scientifically proven that our words can influence the performance of others. They can change the mentality of a person which in turn changes their achievements. For example, when researchers remind older people that memory usually decreases with age, they perform worse on memory tests than those who weren't reminded of that detail.

See the greatness in others. If you can see their greatness, you are actually contributing to that greatness. The Pygmalion Effect teaches us that our belief in the potential of a person awakens this potential. When we believe that our colleagues, friends and family members can do more and achieve more, this is very often the exact reason for why they do. Unfortunately, this also works the other way round - which very often is the case.

Every time you meet someone, try and see the greatness that lies in this person. Ask yourself "What makes them special? What's their gift?" As you focus on it, you will discover it. It also makes you more tolerant of not-so-friendly people. You can say "I'm sure they have great qualities, and today they only have a bad day..."

Be nice! And let me know how it goes for you.

P.S.: Being nice doesn't mean that you have to let other people fool you, or say yes to everything. Nice people also say no or enough is enough.

Be Prepared

Even if we are super prepared, there's always something that we can't know. Be ready to admit it. You don't have to know everything. My former University professor Angel Miro once told me: "Marc, you don't always have to have all the information, but you have to know where to find it."

Keep working on your personal and professional development. Commit yourself to becoming the best person you can be. Stay hungry! When researchers studied extraordinarily successful people, those had two characteristics in common that separated them from the rest. First of all, they believed in themselves. They thought they could do it and second: They always wanted to learn more. They kept asking questions. They kept on learning. Stay curious and eager to learn new things and better yourself. The wiser you become, the more valuable you become for your company.

Read books, take a workshop. Today, you can learn the best tricks of management, leadership, time management or financial planning in a two or four-hour workshop that will benefit you for the rest of your life.

I make it a habit to read at least one book a week, buy a new course every two months, and sign up for at least two seminars or workshops a year. What are you going to do?

Make A Difference - The Power Of One

In a world full of problems, wars, scandals, corruption, terrorism, climate change and much more… what can YOU do to make a difference? Is there anything? Yes. I have good news for you. You are more powerful than you think.

We generally underestimate our power to bring change. Yes. One person can really make a significant difference. Why? Because every change begins in the mind of one single person and then it expands. And my friend…It grows exponentially.

We underestimate our capacity to bring change because we underestimate the potential of the exponential function. Think for example about the exponential nature of social networks. It's said that within six degrees of separation we are connected to everyone on this planet.

You are much more powerful than you think. Although there are many things in this world that you can't control, there are also things you can handle. You will not stop world pollution, but you can walk, go by bike or public transport or separate your trash. You can choose healthier, non-processed food. If you're not happy with the policy of a particular company, you can stop buying their products. Yes, you are only one. But if a thousand people do the same thing surely someone will notice.

In these rough times, you can decide to be polite to everyone you meet, no matter their color or religion. You can choose to affect the 4 square meters around you positively. What would happen if everybody did this?

You can gift five smiles a day. Smiles are contagious and if everybody that you are smiling at gifts smiles to five other people, in no time the whole world will be smiling :-) The same happens when you compliment people or make people feel good. We are influencing people every minute of our life with our actions and emotions. The only question is: in which direction are we going to do it?

Embrace the power of ONE. It will be very beneficial for your self-esteem.

Let Go Of The Past

Letting go of the past and learning from past behavior is crucial to developing a healthy self-esteem. Feeling guilty about things you have done, or staying stuck in situations that have already passed is not learning from the past. Every moment you spend in your past is a moment you steal from your present and future. You cannot function in the present, while you live in the past. No mind in the world can cope with two realities at a time.

Don't hang on to your drama by reliving it indefinitely. LET GO OF IT! It's over. Focus on what you want instead. You can't change it anymore. What you can do is live your present with higher awareness, knowing that this is what will shape your future.

Only when you dare to let go of the old, can you be open to new things entering your life. Don't waste your time thinking of things that could or should have happened or that didn't work out as you wanted in the past. It doesn't make sense! Your life reflects whatever you dominantly focus on. If your focus is mostly on your past, on the "could, should, would," you will be constantly frustrated, anxious and confused in the present. This is far too high of a price to pay.

Learn from your past experiences and move on. That's all you have to do from now on. Easy, isn't it? This means recognizing your mistakes and - to the best of your ability and awareness - not repeating them.

Concentrate on what you want to do well in the future and not what went wrong in the past. You need to let go of the past so that you are free and new things can come into your life. Let go of old baggage, finish unfinished business, and get closure with people. Torturing yourself over what you have done, feeling guilty, ashamed or even unworthy is a pure waste of valuable time and energy. These negative emotions will only prevent you from enjoying the present. Use your memories, but don't allow your memories to use you. Complete the past so that you can be free to enjoy the present.

Forgive Everyone

I know. Why forgive someone who did you wrong? Because it's not about being right or wrong, it's about you being well and not wasting energy. Being resentful or angry with people - or even worse - reliving hate and anger over and over again is toxic. It's bad for your energy; it's bad for your health, it's bad for your relationships so, do yourself a favor and forgive. It might be challenging to accept, but you're not doing it for the other person, you are doing it for yourself. Once you forgive and let go, you will sleep better, you will enjoy your present moments more, and a huge weight will be lifted off your shoulders.

CHAPTER 98:

Overcome Anxiety

Embrace Your Character

Personal stories that include victories over panicked heroism as depicted in films, dramas and novels. Over the years, I have had the privilege to witness many such journeys and transformations from fear to perseverance. Just like sturdy plants that cannot be blown by strong winds, strong individuals withstand whatever pressure they face. Endurance, which is goads fear and panic, can be maintained.

To build resilience, access your strengths and virtues. This includes targeted positive psychology, areas of study that relate to optimal functioning and factors that contribute to well-being. Positive psychology was developed by Martin Seligman and Michal Csikszentmihalyi and focuses on what is best for people and what their best strengths and virtues are. I want you to look after your best and not try to fix what is broken. Step 10 is about using your strength.

You may have heard about the Diagnostic and Statistical Guide to Mental Disorders, a psychiatric guide that classifies what is wrong with you. As you have learned so far, I believe that there are better ways to deal with the symptoms. For this reason, in step 10, we turn to the strengths and virtues of Chris Peterson and Martin Seligman (2004), the "Technician" manual that classifies what is right with you. Instead of "happiness", positive psychology integrates all aspects of human experience, both positive and negative.

The strength of the signature is the characteristics of individuals who are stable and universal, which forms the basis of building the goodness and prosperity of humans. They are the essence of well-being. Through empirical and historical analysis, 24 forces are identified and divided into six virtues: wisdom, courage, humanity, justice, moderation and transcendence. You have each of these strengths, but some may stand out as your main strengths. If you are interested, visit www.viacharacter.org and do an online survey about the strength of character.

Here are some suggestions for how you can identify and use your strengths:

Start a power journal

Select a period, e.g. B. 7 days, and for each of these 7 days illustrate examples of how you have used your strength to overcome fear, fear, panic or negative thinking.

Check the strength protocol in Appendix III.

POWERS OF YOUR CHARACTERISTICS

The virtue of wisdom

The power that is grouped based on wisdom leads to the acquisition and use of knowledge. Just imagining that you can trust your inner wisdom will evoke a convincing response.

Creativity

Think about how you use intelligence, or think about creative solutions to problems. Every time you practice attention, you open up opportunities to find creativity. Do you remember the past when you used creativity to deal with difficult situations?

Curiosity

If you have ever been interested in expanding your knowledge, learning something new, or actively interested in an ongoing experience, you will be curious. When you read this book, you want to know. Can you practice experiencing the world around you with a beginner's mind and innovative meaningless patterns and combinations? Attention and curiosity go hand in hand and together they reduce fear.

Solution

When you practice thinking as a scientist, you use judgment. This means using logic, reason, rationality, and openness, not "destroying" and arriving at illogical conclusions.

If you are worried or panicked, can you play as a devil lawyer with yourself or realize that you are using tunnel vision?

Practice with your judgment and notice how far you feel calmer.

Love to learn

Do you like studying because you study? Do you feel positive for 10 steps when you learn a new skill? Do you want to relive a hobby or interest that you like or learn something new so that you are no longer afraid? If so, then you have access to your love of learning.

Brochure

If you continue to practice panic and fear, you will get access to the power of perspective which is part of thinking as a scientist. You have the opportunity to make judgments based on information about what is essential in life. If you are frustrated, how about remembering the wise choices you made in the past? Access your inner wisdom to put fear in the right light.

Virtue of courage

If you are familiar with the classic film The Wizard of Oz, then you know that cowardly lions looking for courage always have this quality. I often tell my customers to find their inner courage and to "be brave".

Courage

Overcoming panic and fear requires courage. Think back to the times when you voluntarily showed your courage. Can you be brave enough not to be afraid?

Durability

Perseverance is one of the strengths that are most associated with good health and authentic happiness.

How many times have you continued despite challenges?

Can you exercise stamina by turning on optimism and imagining yourself as a calm and panic person instead of giving up and telling yourself that you will never improve?

Honest

If you accept "as is", a change occurs. When you accept fear, you are honest with yourself and accept yourself as you are. As a result, you understand your authenticity and truth and act so honestly and on your way to prosperity. Can you combine courage and perseverance with your honesty? In other words, admit fear, but keep doing it.

Spirit

Vitality, strength, excitement, energy, and enthusiasm are words that describe passion. The strength of this character is most closely related to a life full of joy, commitment and meaning. I encourage customers not to be too afraid. Watch when you feel full and full of energy. Often this is before a big event that can also cause fear. Instead of calling this feeling fear, experiment by changing the label of passion and see how this change changes meaning. Find many ways to live with heat. Whenever you are fully consciously active, exchanging positive experiences or just enjoying slow breathing, you have the opportunity to feel heat.

Virtue of humanity

While the power of humanity is interpersonal and includes compassion and friendship with others, I will ask you to consider this power as well.

Love

Love focuses on caring for others (significant others, parents, children, friends, pets) and living a life that emphasizes commitment. But excessive love and friendship can be stressful. Can you get involved in these 10 steps to love yourself?

Kind

Kindness means showing generosity, caring, caring, compassion, or just being pleasant. What about self-esteem? Can you direct the power of your goodness inward? Can you calm down Can you practice saying no?

Social intelligence

Social intelligence is about recognizing feelings, signals, motives, and signals both to you and others. How quickly can you recognize the uncomfortable feelings associated with fear? How fast can you use 10 steps to delete it after you recognize it? If thoughts lead to negative emotions, ask whether those feelings are beneficial. Do they motivate or limit you?

The virtue of justice

The power of justice is at the heart of a healthy community life. Being part of a community promotes your well-being and ability to develop.

Justice

Remember that your strength can be against you. If you have a strong sense of social justice and caring, loving nature, you may have difficulty accepting what is unfair. Can you practice letting go of "must" as in "that should not have happened" or "he shouldn't act like that"?

Leader

If you follow these 10 steps to overcome panic, your behavior will change. You will show gentle strength and other people will pay attention. If you are calm, grateful, and very aware of the world around you, the electromagnetic signals coming from your heart will change and the people around you will feel it. You will use your manager to make calm reactions to other people. If you share stories about how to overcome panic, you will inspire others to follow your example. This is another reason why I encourage you to join our Facebook group through Feed Your Mind Wellness.

Temperament

Anxiety can take the form of thinking that is too worrying and disastrous. This can also cause overeating or emotional eating. Use your strong power to protect yourself from this excess.

Forgiveness

Practice self-forgiveness first. In addition to accepting other people's disabilities and releasing anger, accusations, and hatred, use your forgiveness to apologize for feeling nervous, overreacting, acting horribly, or working towards a total panic attack. Practice forgiveness when you make bad decisions or overeating.

Modesty

Can you accurately assess your skills and identify your limits? Practicing humility leads to openness to new ideas and this guarantees success in overcoming panic. In other words, get rid of the irrational statement "I must" as in "I must do it perfectly" or "I must stop panicking completely".

Be careful

Considered the mother of all forces, caution involves using reason to control behavior, weigh risks to benefits, and avoid excess. If there is a silver line that suffers from anxiety, you might find that you are taking excessive care. It makes sense to act cautiously and avoid risky behavior. If you spend a lot of time thinking about the consequences of certain behaviors or decisions due to fear, accept your fear and label yourself carefully. I must have had a lot of fear in my childhood and adolescence, but what's the point of not being a brave Devil?

Self-control

Think of self-regulation as the ability to control your thoughts, actions, and feelings. He is considered one of the most crucial character strengths because self-control exercises reduce the harmful effects of stress and trauma. The more you practice 10 steps, the better you can manage yourself, how you think, feel and react to challenging people and situations. The mind and body techniques that you have learned are known as self-regulation strategies.

The virtue of transcendence

To genuinely develop, you need meaning or deeper meaning in your life. The power of transcendence is related to finding meaning or building relationships that go beyond interpersonal relationships. When you access these forces, you will feel deep inner peace.

Beauty eventuation and advantages

Find hospitality, beauty and excellence in your area. When practicing mindfulness, the opportunity waits for admiration, admiration, admiration, wonder, and exaltation. For example, if you adjust to your body's natural ability to create a state of calm, you evaluate the strength of your extraordinary relaxed response.

Thanksgiving

A grateful experience creates a robust healing response. It significantly reduces anxiety, improves mood, reduces physical symptoms and leads to better sleep. How many ways can you say thank you and make sure life is good? Can you turn stress into a blessing?

Hope

Hope is closely correlated with happiness. Even if you are still panicked and afraid, if you hold on to the hope that you will be better than you have already started a healing reaction. If you fail, remember that when you close one door, another door will open. Use the power of hope by practicing realistic optimism: forgiveness for the past, appreciation for the moment and opportunity to look for the future.

Humor

Humor is one of the strengths of character most associated with the path of excitement to happiness. Laughter helps you overcome despair. As mentioned above, it is not possible to laugh and feel anxious at the same time. Grab the power of your humor during the height of the anxiety attack and ask yourself, "What will the comedian say?"

CHAPTER 99:

Panic Attacks

Paula suffered from her first panic attack two months ago. She was in her office getting ready for an important presentation when, all of a sudden, an intense wave of fear hit her. A chill ran down her spine and before she could understand what was happening to her, the room started spinning. Paula felt like she was just about to throw up. Her body was badly shaking, her heart was literally pounding out of her chest, and she was unable to catch her breath no matter how hard she tried. She held onto her desk tightly until the episode passed, but it left her devastated.

Three weeks later, Paula found herself in the exact same situation, and since then, these attacks have been hitting her every now and then. She never knows when and where they will hit her, but she is terrified of having one in public. To avoid this, she has been spending most of her time at home. She has stopped going out with friends and is afraid to ride the elevator to her office on the 12th floor because she fears that she will get trapped if she suffers from another panic attack.

What Causes Panic Attacks?

As with depression, paranoia, anxiety, and other clinical terms that you use every day, a panic attack may carry a different meaning for different people. For this reason, it is useful to decide on a working definition before going into any further detail. A panic attack can be described as a sudden episode comprising of intense fear that may trigger severe physical reactions even though there is no apparent reason or any actual danger. Panic attacks are extremely frightening. When these attacks occur, you may feel like you are suffering from a heart attack, losing control, or even dying. While unraveling the brain chemistry that underpins panic attacks, it is essential to think of these panic attacks as short spells of extreme, visceral fear. The type of fear that tends to keep you violently alive right in the face of danger.

How Do You Get Panic Attacks

Amygdala is vital for your brain. It comprises of compact neuron clusters and is said to be the integrative center for motivation, emotions, and emotional behaviors in general. However, it is best known for the part it plays in aggression and fear. Panic attacks also stem from the abnormal activity going on in this cluster of nerves. In a review of studies published in 2012, scientists have successfully linked the stimulation of amygdala to the behavior analogs of panic attacks in humans.

Another possible culprit for a panic attack is a section in your midbrain known as the periaqueductal gray responsible for regulating defense mechanisms such as freezing or running. Scientists used functional MRI scans to establish how this area lights up with activity as soon as an imminent threat is perceived. When your defense mechanisms start malfunctioning, this may lead to an over-exaggeration of threat, causing the development of anxiety, and in extreme cases, panic. But what makes people more prone to these attacks?

Genetic Predisposition

Panic attacks are almost always genetic. There are some people who will get panic attacks even with a slightest trigger, while others won't have them even if you paid those millions. All of us are born with certain innate tendencies. If you keep having panic attacks, this may be one of yours.

Anxious Patterns During Childhood

People who have a history of a depressing or stressful childhood are often the ones to suffer panic attacks in later stages of life. Such people fail to consider the world as their oyster-a place where they can enjoy their lives. Problems such as long-term illness, the death of a parent or a sibling, or an early exposure to issues like domestic violence or divorce of the parents may play a role in this. In some cases, the reason is a parent exhibiting anxiety or over protectiveness in response to their own anxiety. In other cases, the child is made to believe that it is his responsibility to make others happy and take care of them, forcing him to spend a large part of his life trying to please others.

A Reaction To Challenges Faced In Emerging Adulthood

Another reason why some people get panic attacks is due to the stressful experience they faced shortly before the onset of panic attacks. It can be anything, from feeling suffocated in a detestable job or relationship to losing a loved one. Alternatively, it is possible they had to face different changes which were not bad in themselves- such as completing school, switching to multiple jobs, getting married, having kids, moving to an other city, etc.- all of which imposed tensions to the point that they found it impossible to deal with collectively.

It is quite interesting that for the majority of people, who suffer from panic attacks, the problem usually starts in their twenties or thirties- the years where they are struggling to establish a self-supporting life for themselves.

What Is Common In All These Factors?

What causes a person to develop panic attacks? A genetic predisposition, early childhood trauma, and challenging changes in adult life. The factor common in all of these things is that these are not under your control. All three elements are growth and developmental events that can happen to some people and they're not something that is out of your jurisdiction.

Therefore, there is no reason to feel ashamed, apologetic, or guilty about experiencing panic attacks. They are not a consequence of living badly; or of making poor decisions; or of being cowardly or dumb.

Symptoms And Sequel

When a person experiences a panic attack, they feel like they are stuck in an escalating cycle of catastrophe that something bad is going to hit them at the very moment. Others feel like they are about to have a heart attack due to a pounding heart and continuous palpitations. A few people may think that they are "losing control" and will end up doing something that embarrasses them in front of other people. Sometimes, they start breathing so quickly that they hyperventilate and feel like they're going to suffocate from a lack of oxygen.

The symptoms of a panic attack can be divided into two categories: emotional and physical

Psychological Symptoms

Suffering from panic attacks can badly affect the health of an individual. Besides the inexplicable anxiety and fear, some common symptoms of a panic attack include:

- Failure to relax
- Absent-mindedness
- Expecting danger
- Inability to focus
- Feelings of tension
- Getting annoyed easily

Physical Symptoms

The physical symptoms of panic attacks include the following:

- Inability to sleep
- Exhaustion
- Muscle contractions
- Shakiness
- Excessive sweating
- Nausea
- Increased heart rate
- Difficulty in breathing
- Increased blood pressure
- Chest pain

Sequel Of Panic Attack

Panic attacks have an ability to disturb the normal functioning of your body. Mentioned below are the consequences of this disorder on various body systems.

Cardiovascular System

A panic attack can lead to palpitations, chest pain, and a rapid heart rate. The risk of developing high blood pressure or hypertension increases several folds. If you are already a heart patient, panic attacks can also raise the risk of coronary problems.

Digestive And System

A panic attack can also affect your digestive system. You may suffer from nausea, vomiting, diarrhea, and stomach aches. Sometimes, loss of appetite may occur leading to weight fluctuations in the long run. A connection between anxiety disorders including panic attacks and the development of irritable bowel syndrome has also been established.

Immune System

Panic attack triggers the fight-or-flight response in your body by releasing certain hormones like adrenaline. In the short term, this causes an increase in your breathing rate and pulse so that your brain can receive more oxygen. This helps you respond better to the stressful situation. Panic attacks may also boost your immune system. When this happens occasionally, your body returns to normal functioning as soon as the attack is over. But if you repeatedly suffer from panic attacks, your body never receives the signal to resume normal functioning. This may weaken your immune system, making you vulnerable to infections.

Respiratory System

Panic attacks usually trigger rapid, shallow breathing, if you are a patient of chronic obstructive pulmonary disease (COPD), you may even require hospitalization. Panic attacks can also make asthma worse.

Summary

A panic attack may happen anywhere, at any time. You may feel overwhelmed and terrified, even though there is no danger. You should remember that it is not something to feel guilty for. You were never in control of the factors responsible for this disorder so there is nothing that could have been done to prevent this. However, you can follow these tips to reduce the frequency of attacks:

- Exercise every day
- Get enough sleep
- Follow a regular schedule
- Avoid using stimulants like caffeine

CHAPTER 100:

Post-Traumatic Stress Disorder

PTSD is a mental condition usually triggered by witnessing or experiencing an emotionally terrifying event. You do not have to have experienced the event firsthand for it to lead to post-traumatic depression. Perhaps you saw someone close to you experiencing something horrifying.

For instance, war veterans who have seen their friends and enemies die in front of them may have those images and sounds of guns and mortar tormenting their minds for the rest of their lives—in the recent past, PTSD has associated very significantly with combat veterans. However, the condition does not discriminate; all people, regardless of career, ethnicity, or age, can experience PTSD.

PTSD: The Symptoms

It is normal—expected, even—that after a distressing event, one would have a hard time coping with the aftermath of the occurrence or returning to normalcy.

The human mind and body are versatile and capable of handling a lot of pressure; as such, difficulty adjusting to a distressing event should have a short-lived life span.

However, if the symptoms stretch beyond one month, the chances of PTSD having taken root are exceptionally high. Worth noting is that sometimes a person may seem okay after a traumatic event, only for the symptoms to manifest years later.

Common PTSD symptoms fall into four categories: intrusive memories, adverse changes in thinking and mood, avoidance, and changes in emotional and physical reactions.

Let us examine the specific symptoms in each group.

Intrusive memories
These normally include:

- Spontaneous, involuntary, invasive, and distressing recollections of the event.
- Intense and disturbing thoughts and visions about the event. These visions last long after the event has passed.
- Flashbacks that cause you to relive the event as if it were happening in real-time.
- Dreams and nightmares related to the event.
- Emotional distress or adverse physical reactions to anything that reminds you of the traumatic event.

Avoidance

These include:

- Avoiding reminders of the event: you may not want to associate with similar events; for instance, witnessing an accident may cause a desire to avoid travel by road.
- Avoiding people or places that trigger memories of the event.
- Resisting talking about what happened or discussing any sentiment related to the event, even when it becomes evident that the occurrence was bothersome.

Changes in physical and emotional reactions (also called reactive or arousal symptoms)
These include:

- Severe anxiety.
- A tendency to stay on high alert for danger, which can come off as paranoia.
- Reactive symptoms, such as angry outbursts or being easy to startle, especially as a reaction to ordinary occurrences such as loud noises or accidental touches.

- An overwhelming sense of guilt or shame over what happened.
- Destructive behaviors such as excessive drinking or substance abuse, behaviors the person engages in as a way to dampen negative feelings.
- Aggression in the way the person acts or reacts. For instance, a careful driver becomes super aggressive on the road.

Negative changes in thinking and mood
These include:

- An overwhelming sense of numbness.
- Negative thoughts about oneself and others, usually bearing the blame or blaming others for the events that occurred.
- Mood swings.

These are the commonly displayed symptoms of post-traumatic depression, but they may vary from one individual to another. Have you been experiencing this, or witnessed a loved one showing such signs?

CHAPTER 101:

How To Recognize Depression

If you think you may be battling depression, or if someone else around you might be depressed, there are some signs and symptoms to recognize this condition by.

- There should be a whole range of emotions that you or someone else is feeling, if depressed, such as -
- Overwhelmed with everything that's happening
- Easily irritated
- Lacking confidence
- Easily disappointed
- Generally unhappy with their life
- Miserable and sad over nothing
- Indecisive about choices
- Quickly frustrated
- Guilty over their emotions, etc.
- Due to all these emotions, there might be some behavioral changes as well that will be mostly visible to those closest to the person affected -
- Lagging behind in work or studies
- Not being able to concentrate on anything
- Not wanting to go out of the room or out of the house
- Becoming distant from friends and family members

- Not engaging in activities that they liked doing
- Getting dependent on alcohol
- Starting to smoke
- Relying on sedatives and sleeping pills without
- Becoming non-responsive to queries and conversations
- Although few, there are some physical changes that can occur in depression, namely -
- Exhausted all the time
- Having trouble sleeping
- Loss of appetite
- Occasional headaches and pains
- Sudden weight loss
- Vomiting tendency
- It's not only the emotions that change in depression, but the thoughts of a person. In a depressed state, their central thoughts feature ideas like these -
- "I don't deserve anything good in my life!"
- "It's my fault that I am a failure!"
- "Nobody cares if I'm dead or alive!"
- "I'm worth nothing!"

Depression may be an emotional state, but its effect on a person is manifold. A person could go further into depression and think about hurting themselves or committing suicide. In some rare cases, they can, uncharacteristically, start engaging in some risky activities that could lead to an accident or even death, such as speeding, running a red light or taking up an extreme sport.

In short, a person changes completely when they are depressed. Someone who was always extremely amicable and kind can turn spiteful overnight; a previously calm and content person can show signs of extreme mood swings. In almost all of the cases, people are not aware of what's happening to them and get confused about everything they are feeling.

What Are The Types Of Depression?

The types of depression differ not on the basis of their symptoms, but actually on the causes behind the problem. Based on what conditions can cause depression, there are three main types:

Seasonal Depression

Astounding it may seem many people start getting depressed at the end of fall or at the beginning of winter, especially in the colder countries. This kind of depression is also called the 'winter blues' or 'winter depression', or 'Seasonal Affective Disorder (SAD)'. Alternatively, it is also called 'Summer Depression' or 'Summertime Sadness', in which the depression starts at the beginning of the summer, which is rarer. People who have a healthy and balanced emotional state throughout the rest of the year start feeling depressed when winter or summer start. It is the change of the weather that brings on an onset of mood swings; people start feeling completely different when the weather changes drastically.

Especially when winter starts and the evenings come sooner, some people start feeling tired and lethargic. They show no interest in work or social activities, but prefer to keep to themselves. These mood swings decrease in severity and frequency once the spring approaches again.

Postpartum Depression

Depression after a pregnancy is extremely common in women; this happens to at least 1 out of 5 women after childbirth. Postpartum depression - or depression that comes immediately after or during the last months of pregnancy - can be severe or mild depending on the person.

Postpartum depression can last for days or weeks, or even months in some cases. The new mother can feel unnecessary sadness during the days that are supposed to be the happiest in their lives. This kind of depression can go away on its own after some time.

Situational Depression

Situational depression, as is evident from the name, happens during a particular circumstance in a person's life - the passing of a loved one, a divorce or a break-up, losing a job or failing a class, etc.

This kind of depression is actually a reaction to a particularly stressful time in someone's life. It usually passes with time or with the passing of the crisis; however, in some cases, therapy may be needed.

Premenstrual Depression

About 85% of the female population around the world feels melancholy and depressed during their menstruation every month. Among 5% of them, depression reaches a severe level. Women may feel moody and anxious the whole time, and become easily irritated with everyone around them. They also have trouble concentrating on work and become unwilling to socialize with others.

Manic Depression

People with manic depression mostly show alternate periods of melancholy and joy. When they are feeling good, they are extremely energetic; they talk a lot and quickly, engage in a number of activities, sleep little and prefer to remain active most of the time.

On the other hand, they become depressed at other times and their personalities change overnight. They start to lose focus in their work, prefer solitude to company and become irritable.

Besides these common types of depression, some people suffer from atypical depression - a kind of depression where a person tends to overeat and oversleep; dysthymia - where a person can have a low mood for months or even years; psychotic depression - in which a person loses touch with reality and suffers from delusions; or melancholia - where the depressed person tends to move and react slowly.

Depression is known by a lot of names, especially among the population who are not aware of its severity and consequences - the blues, feeling under the weather, feeling down, or melancholia. In truth, depression is a real illness of the mind that can make a person feel completely numb and unhappy. Without any apparent reason, they feel helpless and meaningless; they can change completely overnight.

CHAPTER 102:

Causes Of Depression

There are many personality disorders experienced by many people. Some people may experience a change of mood, especially feeling angry and feeling isolated. You may have come across and individual with stress, think of how that person usually behaves. Such people are emotional and upset. You cannot even establish the real cause of their sorrow, but they mostly attribute the causes from many weird episodes happening to their life. At that moment you can wonder what to therapize them. However, it is good to visit a physician if you find your relative suffering from that menace. Such disorder is generally referred to as a depression.

Depression can mostly be associated with stress and sadness; however, is a broader concept. Some scientist defines this disorder as the long-term effect of experiencing anxiety and grief. Yes, it is accurate; however, in some other instances, depression is not stress or sadness. Take an example of the emotional reaction you feel after seeing your beloved or relative dead. Then it would be exceptional distress or sorrow. However, after sometimes, you will recover from that feeling and feel whole again. In depression, one typically loses their self-esteem, but when stressful, the self-worthiness is not depreciated. Your health is not deteriorated when distressed but you experience medical issues like mental illnesses.

Depression has mostly been experienced by many people in society. These people usually show some mood disorder. It is a feeling of being sad, angry, low esteem, and a sense of disliking things. A person may be your close relative, friends, or loved ones, and the worst results that happen to them are by committing suicide. Psychologists have tried to explain depression even in scientific theories. Some sites the menace is caused by an imbalance of hormonal nerve transmission in the brain; however, this cause is not proved. To know what is depression one have to recognize the different causes of this epidemic which are listed below

The Causes Of Depressions

The faulty mood causes this disorder. The emotion is about the general feeling one has about a specific event or a person. This aspect can also be called as an attitude or the public emotions people have. Moods can be positive or negative. Some positive moods include joy, happiness, a feeling of self-worthiness, and many others. However, when considering depression, it is associated with bad attitudes, which involves one being sad, anxious, low self-esteem, anger, narcissistic, arrogance, and many other emotions.

Other specialists believe this menace originates from the genes. They say that like 'father like son. Therefore, if the parent experiences some personality disorder, the offspring is likely to experience the same jeopardy. Some studies show that if your twin is depressed, there is a high percentage that you will experience the same illness. People of similar genetic makeup show similar vulnerabilities if especially the people are angry. They are likely to experience some mood swings and unstable emotion that lead to stress.

Another cause is drug abuse in some people. Some individuals take hard drugs that affect their mentality. Such personnel may change from a happy individual to a stressed character. Some medications have impacts like hallucinations, sedation, stimulation, and other effects that increase their anxiety and personality disorders. Hence such fellows will anticipate of an illusion and unrealistic goals if otherwise they are not attained thy quickly get depressed. Intoxication affects your rational beliefs and values and makes you moody, such that you are stressed due to less irritation.

Some medical conditions, like chronic diseases, put one under trauma. Such illness has painful ordeals and events. Moreover, they cause stigma to your beloved ones. Therefore, the victims view themselves as a deadweight to the family where the end-results being losing their self-esteem and feeling miserable. If you have a chronic disease like stroke, you will think others are doing more for you, which is unnecessary. Therefore, you will likely feel heartened and stressed. Consequently, one will be miserable and hopelessness.

The environmental conditions that surround you may lead to distress. Imagine being surrounded by assaultive parents, poverty conditions, or your family is regarded as an outcast. That atmosphere you are living makes one feel unappreciated. Some instances of early childhood trauma can cause depression. Take a situation a child being assaulted, that scar will probably remain with him or her rest of their lives.

Types of Depressions

Depression occurs in diverse, which means it occurs in very many ways. Being anxious or having unstable emotions is some of the renowned personality disorders. Imagine the emotional impact one gets after watching your loved one feeling depressed.it is, therefore, a hurtful situation. The consequences are destructive, which can even result in the individual committing suicide.

For a proper treatment of such victims, it is wise to know the type of depression one is experiencing. It is advisable to visit a clinic as soon as you realize you are experiencing some sorrows disorders. Visit the therapist as early as possible because depression is manageable at an early stage. When it is at its late-stage, it develops to become a chronic condition that puts one under pressure of health complication. Recognizing the type of depression, you are suffering from is essential because the psychiatrist will know the kind of therapeutically program to put you through. The following are some of the types of this illness.

The first type is major depressive disorder. It is the commonly known type of depression because it affects many people in society. It exhibits the typical symptoms that are accustomed to stressful reactions. These emotions involve the feeling of sadness, hopelessness, emptiness, low self-esteem, and loss of interest in major recreational activities. These symptoms are easily recognizable to a patient, and one should seek medical attention as early as possible. This disorder falls under two main categories, which are atypical depression and, melancholic type. The atypical ones always anxious therefore they eat and sleep a lot, and the melancholic beings tend to suffer from insomnia and guiltiness.

There is this type of stress which is resistant the antidepressant drugs. This kind of condition is referred to as the treatment of resistant depression. You can administer any medications to this patient, but still, they are not working. They always have unknown causes where their most prominent suspects are the genetic, or environmental causatives. For one to treat those victims, psychotherapy is recommended for one to assess the reason for that moody feeling. You may also administer different types of antidepressants to establish the medicine that heals that person.

Other people who are not healed quickly from significant depression experience the subsyndromal condition. This menace involves one experiencing many melancholic disorders. In simple one is engrossed with different symptoms showing varying characteristics. You may experience melancholic and at the same time, atypical illness. The physician must be quick to detect this condition because if the patient is affected by many sicknesses, chances of healing are less.

The persistent depressive disorder involves a state of stubborn symptoms. What does it mean by them being stubborn? It means that the syndrome is continuous over time. The syndromes seem to restructure themselves where if one sign is treated, it changes and reforms to terminal disorder. Such complications involve sleeping problems, fatigue, loss of appetite, and many other conditions. The best thing for a psychiatrist to diagnose such a patient is by combining both psychotherapy and medicinal diagnosis.

Depression due to diseases is another type. Some of the chronic illnesses cause stigma to the victims. Think of how you would react if tested HIV positive, cancer, or any other fatal diseases. 'I will kill myself,' 'everybody will laugh at me and despise me.' These are your probable thoughts you would experience if told the bad news. Feelings of loneliness, regret, and guiltiness will eat you strike on you like a hungry lion.

Substance intake depression is a major one attributed to the intoxicants. Intoxication comes from people indulging in drugs and alcohol. The results were those people hallucinating or do unusual things. They will, therefore, find other people do not agree to those deeds, hence feel emotionally discouraged. Some end up in crimes and theft to buy those drugs. These substances change your mood, loss of concern on pleasurable practices, and feeling empty always. Specialized rehabs centers are useful in healing those patients. It is every parent's joy to have a baby. However, did you know that for some people, child giving can be stressful? Probably this amazes you, but it is a fact. Some mothers change their attitude after giving birth because there is a change in hormones, fatigue, or fear of raising a child. Fathers, in their part, can change their mood when they feel their workload will be increased. Consequently, some folks become stressful.

Principal Frequently Depressive Disorder

It is sometimes so hurtful to take a depressed patient into a clinic where there are no guiding principles to treat such a person. Therefore, facilities for rehabilitation must operate under the given guidelines in treating depressed people. Remember that this condition affects you passively because you are suffering when seeing your beloved one acting abnormally. Therefore, when you are promised of quality recovery from a recognized clinic, you feel contented that you're beloved is going to heal.

In any organization, there must be rules and regulation and policies in their operations. Treating a depressed patient is not a one-day affair; however, it involves the therapist to assess the health status of that particular person continually. This treatment is applicable, especially for frequent depressive disorder. Others may term this as the persistent depressive disorder because it is not curable at the first instance. This condition warrants the physician to apply all types of treatment methods. They may use the medical and psychotherapy ways which must be combined to nurse the victim.

Treatment of such patients involves establishing correct guidelines of principles which should be goal-oriented ones. The most particulate goals are to curb the effects of the disorder. To improve the quality of life belonging to that individual and to avoid any recurrence of stressful emotions that should not happen at the person in the later stage of life. Remember that for this sickness, it is hard to detect the symptoms, as those signs correlate and masquerade as different disorders. The following are some of the principles to support in healing those patients.

Your first objective as the psychologist is to set goals for the treatment. Even Rome was built from a single step. In simple everything that seems complex must start with one identifying your targets and goals, which is the primary step. Your goals must be oriented to the healing of that particular victim. The treatment must be practical and satisfiable to the person. Theorizing of this disorder mostly occurs in two-phase, which includes acute and maintenance. The acute phase should involve reductions of symptoms and returning the patients the state they were before suffering the illness. In the maintenance protocol, it includes the prevention of recurrence of the situation.

The second step is to treat the general medical conditions associated with the patient. You will realize most of them complains of disorders such as headaches, blurred visions, vomiting, breathing complications, and any other pain. Therefore, recognize and investigate those pains. Examine them appropriately and do not rush to conclude of depressions related diseases. Give the fatalities the right medications like appropriate painkillers.

CHAPTER 103:

Managing And Overcoming Depression

Treatment

And now, after providing you with the necessary information that I hope will have helped you obtain a clear, holistic and hopefully not negative understanding of depression, we come to the crucial section on how to treat it. As has been mentioned in the beginning of the book, throughout time, there have been many schools of thoughts as to how depression should be treated. Many forms of treatment have been proven less beneficial and dropped by practitioners and clinicians over time, and many more have cropped up in their place. In this book, we are going to look specifically at treatment options that encompass the following characteristics:

- Does not hold an overly negative view of the condition
- Is optimistic about working through the condition
- Functions as a part of a model and not as a 'miracle cure'
- Is concerned with improving quality of life of the individual as a whole

Key Factors to Dealing with Depression.

- Engage in Self Help Techniques
- Challenge Negative Thinking
- Practice Good Nutrition
- Exercise Regularly
- Build Support Systems
- Set Goals (Realistic)

- Manage your Stress
- Journal
- Get a Pet
- Seeing a therapist
- Engage in Mindfulness
- Educate yourself

Self Help

Working through curing depression is by no means an easy or quick feat, but it is definitely possible. One of the first recommendations practitioners make for those who are diagnosed with depression is to engage in some form of self-help. It is pertinent that the individual in question has to make a commitment to them and is ready to put in the effort to improve their condition. An important fact to remember when getting started on the path to overcoming depression is that regardless of how bad depression makes you feel, you do have control over yourself and your condition.

Challenge Negative Thinking

One of the symptoms of depression is the pervasiveness of negative thoughts and thought processes. These thoughts reinforce depression and depression reinforces the negative thoughts-it's an unhealthy cycle. One technique that we can employ to counter these is to challenge all the negative thoughts that occur in our heads. Particularly when depressed, the kinds of negative talk that occurs need to be monitored, and the cycle needs to be broken.

This is understandably hard to do. Depression allows for these types of negative self-talk and criticism to run wild and attempting to rein them in and question them may seem impossible at first, but it is not. It is just like every other healthy habit that you want to cultivate-it needs practice. Firstly, understand that the reason it seems so difficult is probably because you have been doing this for so long-your brain is used to it. Introducing a new way of thinking will take time, and that's fine. As long as you keep at it, it can be done!

In addition to the specific CBT techniques addressing negative thoughts that we will discuss below, some helpful tips when working on this aspect are:

- Give yourself allowance-you will find yourself forgetting that you are trying to change your thought patterns and going back to your usual way of thinking. That is ok. It is a process and this is normal. Try again. It will eventually become easy.
- Be around people who are positive-people who have a generally positive outlook on life tend to have many positive characteristics that you will be able to learn from.
- Put yourself in another's shoes-think of what you would say to your friend if they were in your situation, or think of what a caring friend would say to you. We are often much harder on ourselves than we would be on others. This is a good technique to notice that.

After several CBT based sessions of therapy, Leela has now learnt to do this. Leela's homework for the week was to write down any negative thoughts that she had and then to follow these with challenging statements. Below is some of what Leela shared with me of her homework:

"I convinced myself to go to dinner alone last Wednesday. Since Tim and I separated, I've been finding it hard to do things alone. When I arrived at the restaurant I thought that all the people there were thinking that I was pathetic to be there all alone. I remembered what I was supposed to do and tried to ask myself if there is any evidence for this thought. There certainly was no one staring at me. Then I asked myself if I had been in a restaurant with someone and had seen someone alone if I would think such things and the answer was No, I wouldn't. This helped calm me down a little."

This is just one example of challenging a thought, but you get the idea. Often, the negative thoughts that pop into our heads are not true and have no basis, but due to habit and practice, we listen to them and believe them anyway, resulting in the worsening of our mood. Once we began to question and challenge these, it will begin to get harder and harder for us to believe these thoughts as truths.

Nutrition

Before we begin this section, can you honestly say that you didn't know this? I'm pretty sure that your answer is no. Most of us know how important nutrition is to us. We know of the abundant benefits of good nutrition and the detrimental effects of not practicing it, and yet health complications always result from bad nutrition.

Our bodies receive nutrients from healthy food that aids it to perform various functions, all of which contribute to our wellness. When we don't obtain sufficient nutrients, our bodies are unable to work as well as they could, and this could lead to illnesses.

Similarly, in order to work on treating depression, we want our bodies to be in the best form possible. One simple way we can aid that goal is to feed it what is good for it: good food!

There is of course no specific food that you can eat to cure depression. That would be too easy. But there are recommendations of what you should and shouldn't eat to ensure your body is working at an optimum level. Having your body working at its best is a big part of the treatment plan for depression.

Suggestions for keeping your body healthy:

- Eat foods rich in antioxidants-Antioxidants help reduce the effects of free radicals that cause aging and damage cells. Vitamin C, E and beta-carotene are antioxidants that can be found easily in common food like broccoli, nuts, seeds and many other vegetables and fruits.
- Eat healthy protein-Protein is believed to produce a chemical that in the brain that helps mood regulation. Healthy protein can be found in yogurt, beans, soy and poultry.

- Eat food with selenium-low selenium has been liked to low moods. Selenium can be found in beans, legumes, nuts, seeds, seafood and lean meat. Getting your selenium from food would be best as too much selenium can be toxic. If you were getting it from supplements you'd have to be very careful.
- Eat food with Omega-3 Fatty Acids-of all the nutrients; this has been the most closely linked with depression specifically. Higher rates of depression have been associated with those who consume low levels of omega-3 fatty acids. Omega-3 fatty acids can be obtained from fish, flaxseed, nuts, soybean and leafy vegetables (4).

Exercise

We talked about food, so what comes next? Correct. Exercise! The benefits of exercise, like the benefits of good food, cannot be denied. It is often prescribed by doctors and other health care professionals for numerous conditions and its advantages are backed by extensive evidence. In fact, in the UK, doctors are actively prescribing exercise as treatment for various conditions including depression. It is even paid for by the National Health Service, an organization that provides healthcare for residents of the UK (5).

Coming back to depression, many studies have shown that there is a correlation between regular exercise and lower rates of depression. There are theories as to how this may come about and we will explore this briefly.

When we exercise, our bodies release chemicals called endorphins. These chemicals affect or bodies in that they help reduce our perception of pain, they help produce a positive feeling in us, and even act as mild sedatives. Aside from these specific actions, a lifestyle of regular exercise has shown to reduce stress, improve sleep, increase self-esteem, lower blood pressure, and increase cardiovascular strength.

There are various forms of exercise you can engage in, ranging from swimming to yoga. Looking at the multitude of benefits that it produces and the clear link between exercise and lowered rates of depression, this is one treatment recommendation that is not to be ignored, and your doctor will definitely agree!

To those of us who exercise is relatively new, it can seem very daunting. Many of us may have an exaggerated perception of exercise as being strenuous, time consuming and therefore demanding a strong commitment, one that you are not ready for. Get rid of this idea. Exercise does not have to fit into this mould. In fact, it often doesn't. Even 10 minutes a day can yield positive results. As with everything else, aim to start small and work from there. Begin with activities that you may already enjoy or have enjoyed in the past such as dancing, walking your dog, or yoga. When you exercise through activities that you enjoy, the chances are higher that you will stick with it. As you get more comfortable and are able to incorporate this into your routine, slowly add more activities or do the same for longer periods. It is highly likely that the positive feelings that you experience from exercising will itself motivate you to do more.

CHAPTER 104:

Cure For Recovery From Depression

Let's face it, nobody likes to open up about their feelings to strangers, but sometimes it is necessary, especially when you're depressed.

Sometimes you just need somebody to talk to.

Therapy is a very popular way to cope with depression because you can tell your feelings to someone who can give you an unbiased opinion, and doesn't know you personally so that they won't judge you.

It may be challenging at first, but after a few sessions you will get comfortable with your therapist and be able to tell them everything!

If you are not satisfied with your therapist, you can always switch until you find a person that you are most comfortable talking to. For example, many women feel more comfortable talking to another woman of a close age, because it feels like they are having a casual conversation with a friend instead of receiving a therapy treatment. Say that you are a middle aged man, you may not feel comfortable talking to a young woman, therefore your therapy sessions will not be effective because you will be hesitant to open up to that therapist. Most people claim that they look for a therapist that is the same sex as they are, but that might not be the case for you. Everybody is different, that is why you shouldn't give up on therapy on your first session if you feel that it's not beneficial. Keep going until you find a practice and a doctor that works for you.

But isn't therapy expensive?

No, not necessarily! You may be surprised to learn that many insurance companies cover therapy sessions entirely, so you more than likely won't have to pay to see someone. Always call your insurance company to find out if they cover therapy before you go!

Statistical polls show that nearly 1 out of 5 Americans have visited a therapist at some point in their lives. Almost all of these people found their therapy sessions helpful and beneficial in improving their mental health. Why not give it a try? Out of this group of Americans, a large majority of them (over half) suffer from depression.

In addition, if you are feeling sad, overwhelmed, stressed, or annoyed, therapy is a great way to sort out your emotions before you develop depression. In fact, many people find out that they have depression after talking with a therapist. Patients that receive therapeutic treatment often leave their treatment sessions feeling empowered, because psychologists and psychiatrists offer expert advice on how to take charge of their lives and cope with sadness.

When you go to therapy, you might even learn about new strengths and talents that you didn't know you had prior to your visit!

Therapy sessions are always confidential too, so nobody will know what you talked about with your psychiatrist or psychologist. Nobody will even know that you attend therapy unless you tell them. A breach of your security and private information by the therapist is illegal.

Many people aren't sure about the difference between psychiatrist and psychologists, and while these differentiating therapists typically offer the same effectiveness of treatment, there is a large difference between them. The main difference between a psychiatrist and a psychologist is that psychiatrists have doctorate degrees in psychology and are able to prescribe medication for any mental illnesses they may think are present. A psychologist, on the other hand, typically has certificates and licenses in psychology instead of a doctorate degree. Psychologists aren't able to prescribe medications either.

Much like regular medical doctors, many therapists specialize in different aspects of mental processes. For example, there are therapists that specifically specialize in treating and managing depression! A quick Google search will show if any of these specialists are available in your area.

One specific type of therapy that you may consider trying is cognitive therapy. A fairly new alternative method to battling depression, cognitive therapy surrounds the concept that negative thoughts create negative illnesses such as depression.

When you attend cognitive therapy, your mentor will teach you mindful and positive ways to think in order to ensure positive outcomes in your life. This will include not blaming yourself when things go wrong, and not beating yourself up about your faults. Cognitive therapy will change your thought pattern from bleak and hopeless, to optimistic and productive.

The reason for this therapy is that many people with depression have feelings of worthlessness and self-doubt. In order to change your depression, you must first change the way in which you view and think about yourself. From there, your depression will subside and your self-esteem will soar!

In addition to cognitive therapy, you may consider trying marital therapy or group therapy if you feel that your depression may stem from an unsupportive spouse or relative. That way, you can both (or all) receive treatment to manage your relationship together so that everyone can be happy, productive, and more loving towards one another. You may also consider trying group therapy if one of more of your relatives suffers from depression as you do. That way, you can both work towards eliminating the cause of your sadness, which may turn out to be the same shared problem. A shared solution would be desirable if both of your depression symptoms are damaging your relationship in a negative way. That way, you can improve your relationship while overcoming your depression. It's a win-win situation for everyone involved!

Contrary to many people's beliefs, going to a therapist does not mean that you are crazy! You just need someone to talk to, everyone does at some point, and therapists are only here to help! You are only crazy if you deny the chance to improve your life and overcome your sadness. The journey to happiness may be difficult, but it will make you a stronger and more resilient person afterwards. If you have a chance to attend therapy sessions, you should take full advantage of it!

The next time you are feeling depressed, you might want to consider receiving a professional opinion, and you might leave your therapy session feeling more inspired!

Overcoming Depression Through Healthy Eating

When we are depressed, we typically reach in the cabinet for our comfort foods. You know, the cheesecake, potato chips, or fudge brownies that you tend to buy on impulse. These unhealthy foods make us feel good in the moment, but many people don't realize that these unhealthy foods may be contributing to their depression.

Along with exercise, a healthy diet can dramatically shift your symptoms of depression. While there is not a specific diet to essentially "cure" your depression, there are certain foods that you can eat that are proven to better your mood and make you feel good.

For example, a diet that is rich in Vitamin C and Vitamin E can positively affect your mood and make you feel more energetic. Vitamin C is released by the sun's rays and is what causes your brain to releaser Serotonin, so eating food that is rich in Vitamin C helps you feel as great as sitting in the sun.

In general, fruits and vegetables are high in vitamins and minerals that ease depression when eaten in large amounts. Fruits such as Blueberries, Grapefruit, and Kiwi are high in Vitamin C; while Broccoli, Potatoes, and Green Peppers are vegetables that are high in Vitamin C as well.

Furthermore, good Carbohydrates are also known to release high amounts of Serotonin in the brain. Be sure to eat whole grains and stay away from foods that are high in sugar. For example, whole grain bread is healthy and cakes and cookies are not. Sugary foods that are full of bad carbohydrates typically leave you feeling run down and depressed, hence the name "junk food."

Not only does junk food contribute to depression, but it can also cause obesity, skin problems, heart problems, and diabetes. Generally, it is best to stay away from these unhealthy choices because they may taste delicious and yummy in the moment, but down the road you may regret eating them if you acquire serious health problems, in addition to the already looming depression.

In addition, these healthier food options allow you to think quicker and make more practical decisions. Overall, your brain will feel more refreshed and alert when you switch to a healthier diet. If your brain is happy, you will be happy too.

When it comes to living a healthy lifestyle, it's fine to cheat once in a while and order those French fries or eat that cookie, but don't go overboard! As the age old saying goes, everything is okay in moderation.

Many people may think that eating healthy is impossible if you are living on a budget, but it actually isn't! For an extra hobby and to save a few bucks, you could always grow your own mini-garden in your backyard and eat what you grow! Alternatively, you could try shopping at local markets instead of food store chains to avoid large markups on produce.

Healthy food, healthy lifestyle, healthy you!

Step Outside, Smell The Roses

When we are depressed, we tend to isolate ourselves to dark and secluded locations. We may find ourselves spending most of our time lying in bed with the lights off and the blinds drawn, or lying on the couch in a dark living room. We don't often realize that doing this only enables our depression because our body needs sunlight to make us feel happy and healthy. Without light, we won't recover from depression anytime soon.

Physical activity can do wonders for alleviating symptoms of depression. Many professionals note that even sitting in the sun can help increase the levels of neurotransmitters such as melatonin and serotonin in the brain, which brightens your mood and helps you feel more energetic. If you get out more, you will no longer have to deal with some symptoms of depression such as fatigue, headaches, lack of motivation, or sleepiness.

Still need convincing?

What's more, these neurotransmitters that are released by the sun are so powerful that they mimic the strong effects of taking antidepressants such as Prozac for the same symptoms. Prozac is known to release high amounts of serotonin in the brain which alleviates negative emotions, but who knew that instead of turning to medication, you could just step outside and enjoy the nice weather?

If you suffer from Post-Partum Depression, as some new mothers do after giving birth, or seasonal depression as some do during winter time, this method is just as effective for all. If the sun's not out because of winter, you can always turn on bright lights inside your home and sit near them for a few hours in order to boost your mood as well. These lights will act as the sun does to boost your serotonin and melatonin levels and have you feeling better.

If you don't want to exercise outside, you can always join a local gym or YMCA in order to exercise and start feeling better as well. Here you might also acquire friendships with more active people that will keep you on track and keep you looking great, and feeling great too!

Exercising doesn't only alleviate depression though; it also works wonders with anxiety, stress, suicidal thoughts, and feelings of low self-esteem.

Sunlight, the miracle drug! You should try it sometime.

CHAPTER 105:

Do You Have Anxiety?

There are two types of anxiety that take control of our day, including occasional and permanent anxiety. The second one is what you don't want to have. In most cases, it cannot be helped unless one treats it intentionally. Occasional anxiety, on the other hand, is not permanent. It only occurs every now and again, usually when you experience some stress or are placed under pressure to perform. It is often accompanied by fear, predominantly the fear of failing in some sense, like in the workplace or any other professional and social environment. Whether you're prone to stress or lucky enough to be comfortable in stressful situations, occasional anxiety is normal. It is something everybody experiences at some point, if not regularly, in their lives. Even though one thinks it cannot be dealt with due to the extensity of its severity, in some cases, it can.

Anxiety makes you feel helpless and as though you can't seem to fight the fight. I know what you feel like when you are having an anxiety attack, whether it's temporary and haunt you for a few hours, or linger with you for weeks on end. It's real, and don't worry, you are not imagining it. It is as bad as you think it is, and if you know you have it or feel resistant to the very idea that you could have it, you should know that what you are going through is okay. You are the only one who knows the truth. You know deep down inside whether you do or don't have it. Not wanting to accept it and suppressing it is unhealthy, but quite normal for the average individual. Nobody wants to accept that there's something wrong with them, and that is the truth for a number of reasons. Anxiety is intense, and if you are suffering from the permanent type, and don't just feel anxious every now and again when you have to write a test or go for a job interview, accepting that you have this disorder is hard. It is the same with depression— something that can be very difficult to accept. You don't want to shout to the world or even believe it for yourself. Nobody wants to admit that they have depression, but yet so many people are living with it among us, including our friends, family members, coworkers, and perfect strangers. The irony is that anxiety and depression go hand-in-hand. These disorders cause intense emotions, and because anxiety feels as though it is impossible to control, people often develop depression as a result thereof. Whether you want to acknowledge and accept the fact that you have an anxiety disorder is entirely up to you. You don't have to scream it out loud for the whole world to know, but you do have to recognize it and try to treat and improve your symptoms so that you could have a better experience altogether. Accepting it means that you have an answer to what you've been feeling or experiencing for what's probably been a long time, as permanent anxiety doesn't just come and go. When you know what's wrong, you can address the problem and face your fears with ease in knowing that you're trying to achieve something.

Consider whether you have the following symptoms:

- Hyperventilating
- Lack of concentration
- Sweating and trembling
- Feelings nervous, tense, weak, tired, or restless
- experiencing a sense of danger and panic
- Nightmares or trouble sleeping
- Feeling the urge to avoid anxiety triggers
- Gastrointestinal problems
- Inability to control worry
- Concern about things you can't control

Anxiety disorders include:

- Agoraphobia - The fear and avoidance of places or situations that can cause panic or make you feel helpless, trapped, or embarrassed.

- Generalized anxiety disorder - Persistent or excessive anxiety that includes worrying about events or activities. The worry is uncontrollable and affects how one feels physically. It is accompanied by other anxiety disorders and sometimes, depression.

- Panic disorder - Experiencing repeated episodes of sudden intense fear, terror, or anxiety that causes panic attacks within minutes. Feelings include shortness of breath, heart fluttering or palpitations, chest pain, and thoughts about impending doom. It causes worry about these emotions happening or avoiding situations where they could occur.

- Medical anxiety - Experiencing intense panic or anxiety directly linked to physical health issues.

- Selective mutism - The consistent failure when children, and sometimes adults, can't speak in certain settings or situations, including school, social settings, and work, which could affect their functioning in these environments.

- Social anxiety disorder - The fear and avoidance of any kind of social situation as a result of feeling self-conscious, embarrassed, or concerned about judgment or being perceived in the wrong way. This disorder involves extremely high levels of anxiety.

- Phobias - A major anxiety that gets triggered by an object or situation one may be fearful of. The number one priority for someone with a phobia is to avoid it at all costs, as it may cause panic attacks.

- Separation anxiety disorder - A disorder that occurs in children and began during a child's developmental phase due to experiencing separation from their parents or guardians.

- Substance-induced disorders - Accompanied by symptoms of severe panic and anxiety caused by the misuse of medication, drugs, or alcohol. With this disorder, one generally feels hopeless without it or fearful of not being able to access it at any given time.

- Additional disorders (specified and unspecified) - Anxiety or phobias that are both distressing and disruptive to one's life.

Should you see a doctor or diagnose yourself?

If you feel as though you are worried about having a potential anxiety disorder or are feeling helpless about your feelings or behavior consistently, you should think of seeking medical treatment immediately. A good indication to see a doctor about a potential disorder or treatment thereof is to consider whether it affects your work, relationships, and different aspects of your life. If you are constantly experiencing fear, depression, are concerned about your mental health due to experiencing uncontrollable anxiety, have trouble with substance abuse, or fear that your anxiety could be linked to medical health issues, you should also seek help.

If you are not worried about your anxiety and feel as though it is temporary and could go away on its own without getting worse with time, you don't necessarily need to seek immediate professional support. Everybody experiences some degree of stress and panic in their lives. When it gets persistent or worse, however, that's when you should take it seriously and address it.

The top treatment options for anxiety disorders by medical professionals are as follows:

- Antidepressant medications - A prescribed medication, usually combined with Cognitive Behavioral Therapy (CBT) to combat anxiety and depression together. This medication contains selective serotonin reuptake inhibitors (SSRIs), which works by producing fewer side effects than previously prescribed medications, like beta-blockers.

- Cognitive Behavioral Therapy (CBT) - Therapy used in conjunction with antidepressant medications to treat anxiety, depression, and fear. It works by helping patients learn how to regain control of their feelings and life.

- Exercise - A simple, yet very effective solution for releasing good-feeling hormones, including serotonin, in the brain. Exercise always makes you feel better, whether it's mentally or physically. It can provide a sense of feeling as though you are in control of something in your life, such as becoming stronger in your body or improving your fitness level. It allows you to push yourself and rely on your ability to complete the exercise routine you planned out for yourself daily. It keeps you accountable, aids in relaxation, and promotes overall well-being.

- Relaxation methods - If anxiety and depression make you feel out of control, and then taking time to implement a relaxation method can only do you good. It includes meditation and the practice of mindfulness. You can also engage in a hobby that makes you feel more relaxed like reading, practicing yoga, deep breathing, or listening to calming music. It can relax your nervous system and improve the quality of your daily life.

To prevent the development of anxiety disorders altogether, the key is to get help early as it is a mental health disorder that requires treatment. The longer you wait to have it treated, the more difficult it will be to treat it as it can continue to get worse. You should also avoid alcohol and drug use, including smoking cigarettes, to prevent your anxiety from getting worse. If your condition is still underdeveloped, you should try your best to stay active, engage in social interaction, and build as many caring relationships as you possibly can.

The Presence Of Anxiety In Relationships

Imagine if you overreact to stressful situations and do something as simple as over-preparing for a social event or perhaps even a storm. It may seem harmless. I mean, it's better to be over-prepared than underprepared, right?

Yes.

However, it does indicate the presence of anxiety. When you have anxious thoughts about anything that is entirely simple to those around you, you could be suffering from an anxiety disorder. How does that affect the people in your life, especially your spouse? Well, even though it seems like an underlying issue to you, it can be draining for your partner or anybody else in your life. Overestimating events or considering everything that could go wrong, is not something everybody wants to hear about. Even though it is necessary to accept the people around us, including your people accepting you, this is just one sign of anxiety that could negatively affect a relationship.

Now imagine overreacting to, not only potentially stressful events, like a big event or an upcoming storm, but situations at work. If you are prone to stressing more than normal about work, including your responsibilities or deadlines, you could be bringing a lot of your stress and anxiety home. It can hurt your family. It can also cause you to find it difficult to relax, which could also make the people around you tense. I don't know about you, but a tense relationship is an uneasy and unpleasant one.

If you're startled easily, your response to anything unexpected can affect your life and those around you. That includes having difficulty concentrating, struggling to sleep, overthinking unrealities, getting headaches, muscle aches, or experiencing tension, throat issues, sweating, trembling, and becoming nauseous. These symptoms of anxiety all have an effect on your physical and mental health and disrupt your quality of life, which then also disrupts the lives of the people around you.

So, if you ever thought that anxiety was something you could shake off or hide from the people around you, you're wrong. It affects you in more ways than you can imagine, and probably more ways than you would like to admit. Anxiety causes panic, fear, uneasy, and tense feelings. It affects all areas of your life because it affects you.

CHAPTER 106:

Habits

The Sleep Habit

For most people with depression, believe it or not, the number one cause of that depression is "sleep." I don't care if you are going through divorce, lost a loved one, going through bankruptcy, or just been feeling down for a very long time with no apparent reason.

Sleep is an absolute necessity for your body to self-regulate all of your pent-up emotions. Yes, I know that some of you feel like you sleep too much, but the problem is you are just not sleeping in a restful manner.

Here are the sleep habits you need to do to get ready for sleep:

1. Go to bed at the same time every night. Even on weekends to regulate your body's natural sleep schedule.

2. Get up at the same time, even on weekends. Schedule the amount of sleep you need based on how many hours you need to feel well rested. If you need caffeine to feel awake, then you have not rested well.

Start with purposefully planning for 9 hours of sleep. If you wake on your own and feel rested, then you've gotten the right amount of sleep.

Experiment on reducing the hours of sleep, but be aware that most people need about 7 to 9 hours of sleep. Chances are you would benefit from the full 9 hours.

By the way, dump the caffeine use altogether.

Don't confuse sleepiness with caffeine withdrawals. Caffeine has terrible withdrawal effects, making you feel headaches and a drowsy feeling that begs for that shot of caffeine--Don't!

3. Do not take naps during the day. Naps will disrupt your sleeping schedule tremendously. If you nap during the day, even on your days

off, your sleep schedule goes out of whack and your body becomes confused.

If you feel the urge, sit up straight or even better, stand or do something on your feet. Be thankful that you feel sleepy and delay that sleeping energy until your designated bed time.

4. Dim the lights around your house about an hour before your bedtime. Turn down the lights, get some low intensity bulbs and put it on select lamps if necessary.

Your body needs to feel the night is coming, otherwise it will feel like you're trying to sleep while it's still sunlight out.

5. Do your bed time routine about half an hour before bed. Shut your lights off, brush your teeth, take a warm shower or bath, or whatever else you do before bed. Stay consistent, and your body and mind will recognize the routine and automatically start feeling sleepy.

6. Keep your bedroom for sleeping and sex only. Don't watch TV or have your computer devices in the bedroom. Stop watching TV or surfing the internet about half an hour before bed time.

TV, computer, and smart phone screens fake your body into being awake and stimulate your brain too much to feel sleepy.

You may read in bed, only if you go to bed early.

Do not read if it's past your actual bed time, as reading will also stimulate your brain.

What if I Can't Fall Asleep?

If you can't sleep while in bed there could be a couple of reasons. You may have napped during the day and your body now has the extra energy, or your mind may still be racing through all the experiences you've had during the day.

Learn from that mistake and don't repeat it.

If you consistently have trouble sleeping, you should also check with your doctor, as some people may go for years with sleep disorders, such as sleep apnea, and never know they have it or get treated.

There is a good set of habits that is very helpful for insomnia!

If you can't sleep do the following:

1. No matter what, stay in bed.

2. Do some meditation. Meditation in bed is an excellent alternative to sleeping as your mind will still rest (a lot more rest than staying up reading or watching TV), even when you're awake.

Self-guided meditations are excellent for this. Here's an example of a simple but very good self-guided meditation to rest your mind:

Close your eyes, take some deep slow breaths as you purposely relax your body from head to toe.

Then imagine your bed starting to float. Imagine floating gently out of your window and into the sky. Take yourself to a beautiful place that you like.

An example is going to a beautiful place on a mountain, near a peaceful running stream where you can lay listening to the water run, and the gentle birds singing.

Notice and feel the beautiful surroundings. Do this as long as you please.

Then float back up and gently sail through the sky back toward your home. Notice all the beauty of the view, and sense the fresh air, as you float through the sky.

Float back into your bedroom window and softly land.

Express gratefulness, and feel it, for all the things you saw and felt in your beautiful journey.

Repeat as much as you want and until you fall sleep. The advantage of guided meditation is that you choose to focus on what you want rather than let your mind drift in possibly unwanted thoughts. Even if you don't sleep, your mind is still resting while doing this relaxing, guided meditation.

Now that we've established the importance of sleep, we'll look at what to do when you wake up.

The Daily Movement Habit

Here's another daily must-do habit. You prep your body and mind for the day. Remember that the mind is connected to your body. So you're going to jump start your body first thing in the morning.

Once you start doing this and feel the benefits, you will feel that your day is missing out when you miss this routine.

Plan to wake up an hour early to go for a walk or a jog for at least 20 minutes; 40 minutes is even better.

NOTE: You should consult your doctor before starting any new exercise routine.

Exercise is the key for anybody helping themselves with fitness of the mind as well as the body. Not only does exercise improve your body but for our purposes it's a toxin cleanser.

Most people build up tension and stress throughout the day accumulating adrenaline and other toxins. These toxins have a special purpose for us to give us the extra energy to deal with difficult situations and help us react even when we're in danger.

The problem is that at the end of the day, the toxins are accumulated in our body. The only known way to get rid of these toxins is exercise. Cardio exercise to be more specific.

So the simplest thing that you can do, is to do some cardio exercise keeping your heart rate at a recommended pace for at least 20 minutes. Do this at least 3 days a week if not all 7 days.

NOTE: Depending on any health issues or injuries you may have, always consult your doctor before starting any exercise activity.

Empowering Exercise

When you awake, get up right away, put your shoes on and go for a power walk. This is an excellent way to charge your body and truly energize your day.

The following is adopted by the Tony Robbins method on how to start your day and I've used this for a long time with some minor changes. Feel free to adopt this to your own needs. The important thing is that you get some good cardio exercise.

Remember, the body and mind is connected. Here's what you do:

1. Start A Brisk Walk And Take Power Breaths. As you start walking to warm up your body, take in lots of oxygen by breathing in through your nose four times in a row in quick succession, then breathing out four times in a row.

So for every brisk step you take a breath in through your nose loudly. Fill your stomach rather than your chest:

Breathe in, in, in, in.

Then exhale like blowing through your mouth four times:

Blow, blow, blow, blow.

Repeat the breathing pattern three times for 30 breaths.

2. List Three Things You Are Grateful For. For example, you can start with the fresh air hitting your face, and two more.

Really feel the grateful feelings flowing within you. It's very important you feel the gratefulness.

3. Then Feel God (Or Your Universal Power) Within You Healing You. Feel God healing you from the inside. Healing your emotions, your relationships, your body and your spirit. Really feel how God heals you from within.

If you don't believe in God, feel the forces of the universe converge within you to heal you. It's important you feel healing from the inside out.

4. Next, List Three Major Things You Are Going To Do Today. Imagine you doing these.

5. Call Cadence. If you are walking, make sure you are doing a fast enough pace to increase your heart rate. Start calling cadence to your walking or jogging in a similar pattern as your breathing.

Cadences are customary in the military for a reason. There's a power of self-talk in rhythm that energizes your body and your soul. Before I even knew about depression I used cadences to help me finish a 26-mile marathon.

Every marathoner knows that after the first 10 miles, you're battling your mind rather than your body. Cadences are a great way to keep you focused in exercise and you can combine them with any aspect in life.

Continue on walking briskly, or if you can start jogging.

These cadences should be positive in nature, describing what you want. Make sure you're in cadence with your footsteps.

Some example cadences:

"Every day in every way I get stronger and stronger."

"I give love. I get love"

"I'm feeling good. Looking good. Feeling strong."

You can repeat one or more of your cadences as you wish and what you want your goals to be.

6. Finish and Stretch. Take 5 Minutes to stretch in silence. Put on meditation music if you can.

7. You're done. Eat some breakfast, drink your water and start your day.

If you can, I also recommend you doing some weight lifting. But this is icing on the cake. You can do it right after your walk, or do it in the late afternoon after a full day's work.

Although not absolutely necessary, weight training at least three times a week for as little as 20 minutes, boosts your feel-good hormones, makes you stronger, improves your posture and can improve your general feeling of well-being and mental strength.

I recommend doing your walking/jogging routine first thing in the morning, because it really does prepare you for the day, both physically and mentally in so many ways. Your body is energized with the blood flowing through your body and mind as your heart pumps blood through your organs brain and your brain. What a way to start your day feeling invigorated.

CHAPTER 107:

Dealing A Blow To Depression And Forging Into The Future

So far, we have learned anxiety and depression are conditions that are regular and painful. The feelings can range from a normal feeling down that can last for several weeks to a severe condition that might need treatment at a hospital. CBT works with or without antidepressant medicines and has been proven to lessen relapse rates. A crucial first step in helping you deal with depression is getting to know what thoughts are running through your head and to get more familiar with how you react to problematic situations. Many of us are locked into automatic, negative paths. Something happens, and then the thoughts, feelings, and behaviors cascade over one another inevitably.

Careful self-analysis is critical to be able to pause this process and analyze what is going on.

People who have depression often have what is known as a cognitive triad. This is the set of three negative views that characterize depression: negative views about you, negative views about the world, and negative views about the future. It is useful to look for any of these negative thought patterns in your life. The first part of the cognitive triad, negative views about yourself, is somewhat easy to recognize. These are the automatic thoughts that include the personal pronouns I, me, or my. You might find yourself saying things like this:

I am a bad person.

Nobody likes me.

I am terrible at my job.

As an exercise, take some time to write down the thoughts you repeatedly have that are negative about yourself. What is the way that you beat yourself up?

How do you speak to yourself when a mistake happens? These negative statements are global and seem to come automatically.

Do not take time to evaluate whether or not the statements are true. Simply write them down.

The second element of the cognitive triad is viewing that is related to the world at large. These are sometimes more difficult to spot because many people mistakenly think their negative views are accurate descriptions of the world. Many people with thought disturbances have a vague sense that it is the rest of the world that is disturbed, and only they see things accurately.

A good clue that it is a negative outlook as opposed to an accurate description of the world is that it is absolute: If you think something never works out or is always bad, you probably are overstating the case.

Either way, take some time to write down the negative thoughts you have about the world. Do not evaluate whether or not the statements are true; at this point, look for thoughts that are negative and directed outward. Some examples include:

- All men are jerks.
- The powerful are corrupt.
- Life is unfair.

The last part of the negative cognitive triad is the negative thoughts you have about the future. You might say to yourself:

- My life will get worse.
- Nothing will work out.
- The world is going to destroy itself.

These thoughts are predictions about how things are going to turn out, and they are generally negative. Without stopping to determine whether or not they are true, write down all the thoughts you have about the future that are negative. Do you focus on the fact things will be bad? Are you continually predicting negative results for things you might try?

Look at your lists. To what extent are you generally negative? In what category are your thoughts the most negative? It is crucial that you have a good sense of how your negative thoughts manifest and where you should focus your time.

Many people come to therapy with the general knowledge that they are gloomy, worried, or cynical. On the other hand, your thoughts feel true and accurate. You aren't a pessimist. You might think you are a realist. Evaluating all your negative thoughts as a collective is one way to realize there is a general pattern of negative thoughts.

The act of thinking about your thoughts is a skill in itself, and it needs to be developed. Sometimes it will be difficult for people to establish the ability to analyze their thoughts.

What effect did the negative thought have? Did it change your behavior in any way? Could you imagine your behavior changing if you had a different thought?

Make a worksheet with five columns: situation, feelings, physical reactions, behaviors, and thoughts.

With the situation you are thinking of, fill out each column. In the situation category, write down what happened who was involved, where it happened, and when it happened. In the feelings column, write down what you felt and rank the intensity of that emotion from 1 to 10. In the physical reactions column, write down how your body reacted and rank that from 1 to 10. In the behaviors column, write down the actions you did. Finally, in the thoughts column, write down the thoughts you had in that situation.

Analyze what the relationship between these columns is and how they interacted with each other. As you go through your life, develop the habit of viewing things from the outside and analyze them in this way.

It might be helpful to fill out this form every day. Make a point to spend time analyzing your situations and behavior so that you develop an awareness of patterns.

Every person has specific types of situations that set their automatic negative paths in motion. You have triggers things that spur you on into the thoughts, feelings, and behaviors that lead you to change. To address your problems, you have to know what type of situations is difficult for you and trigger your negative patterns.

Sometimes you will already be aware of your triggers. For other people, it is difficult to identify the specific situations that provoke problematic emotions. You might think that you are always sad or always drink too much and be unable to identify specific situations that become problems.

A helpful first step can be to monitor problematic feelings or behaviours. You can check and see if there are some situations where the feelings are worse or the behaviors more problematic. Imagine someone who thinks she is always angry. At first, this person might think they are angry all the time. But if she carefully monitored her feelings and determined when they were the strongest, she will begin to see patterns. Perhaps, in this particular case, she gets most angry at her teenage daughter when she doesn't do her homework or breaks curfew. She might find that her anger toward her daughter was overflowing into the rest of her life.

You can use a simple monitoring worksheet like the one below. When you use it, note what situations are the most difficult and rate your feelings from 1 to 10. When you do that, you will often start to see patterns. Imagine Richard, who feels like he is unhappy at his new school all the time. If he filled out the worksheet, it might look something like this.

When Richard looks at the worksheet, all filled out, he discovers he was unhappy in social situations. He was not listing unhappiness when he was in class or answering questions. He was only unhappy when he felt like he was excluded socially. It helped him realize that academically, the school was going well. Maybe he was in a band and didn't feel unhappy in the band. That could be going well. The problem was when he felt socially rejected.

Situations can involve interpersonal events, solitary things, or even things that are imagined. They can be memories, partial images, or mental pictures to which you are responding. They are often locked into certain times of day, so make sure to ask yourself questions about contextual aspects of the situation.

As you are identifying situations, ask yourself the W questions. What happened? Who was involved? Where did it happen? When did it happen? It is in some ways similar to being a journalist, figuring out the facts of the matter. You need to have a sense of what events caused the negative feelings or behaviors you are targeting.

Sometimes, if you are struggling to figure out what is important about a particular situation, describe the situation in vivid detail. Events exist in multiple senses, including sounds, smells, and touch. When you use multiple sensations, you can help yourself visualize the space you occupied and identify the sights, sounds, and sensations to help yourself trigger your memory. If the situation involved another person, you could ask trusted confidants to role-play the situation with you. They can take the place of the other person, and then you can analyze the situation again.

One common thing that happens is that the situation that causes negative feelings and thoughts is not just one discrete situation or a single moment. Situations that trigger us can evolve. A dispute with a friend can start as a fairly minor insult or hurt and then quickly escalate into mutual insults before you leave hurt and wounded. Your thoughts and feelings will likely evolve throughout the entire interaction. In those cases, it is useful to break the set of events down into specific moments with various stages of the interaction.

Always be as specific and concrete as possible when describing these situations to you. When you identify trigger situations in vague terms, you won't get a full sense of what happened. Instead of saying, "My wife does not respect my work," it would be better to say, "My wife told me she thought her work was more important than mine."

When you get more specific and concrete, you move forward in the process of describing the world without interpretation. Sometimes our thoughts color what our memories are. In the previous example, your wife tells you she thought her work was more important. But what if she had said that she does not want to miss a work event for her company to attend a work event for yours? It is possible to remember this interaction as her thinking her work is more important. But on closer analysis of that thought, it is not justified by the situation. Her being unwilling to prioritize your work over hers does not mean she thinks your work is unimportant.

One way to think of this is that the facts of a situation are different than the meaning of a situation. The goal is to separate the facts from the thoughts and feelings about the situation. An example might be that you might describe your child as being rude to her teacher. Rude is an adjective and describes what you think about your child's actions, but it does not describe what your child did. What was the action?

Do not record situations with your thoughts and feelings embedded in them. Instead of thinking, "I was so angry at my mother when she was late," separate those two elements. Your mother was late, and you were angry. The event happened without the feelings, and then the feelings happened.

CHAPTER 108:

Changing Your Focus — You Change Your Life

Sometimes what you need to reduce anxiety and stress is to change your focus. When you try to prevent thoughts from entering your mind, the more such thoughts keep popping right back up.

One of the best ways to avoid that thought cycle is simply to change channels. You change your focus you change the quality of your life.

Macy's Story

Macy (not her real name) felt like her life was stuck in the rut. She felt like she lived a not-so remarkable life and it was like she drifted away from her early teens to this day moving from one relationship to the other.

Now she is almost 40 and she's still in the same nine to five job that she's worked in for the past 10 years. No career change and no promotions. If anything, the biggest thing that happened in her life was having her first born son.

She was suffering from PCOS and she felt like her son Dean, now age 3, was nothing less than a miracle. Her husband had no job and stayed home to care for the kid. She felt like the lioness every day going out to provide for her pride.

It's not that they never tried to make a better life for themselves. Three years ago she and her husband started a takeout counter where Macy went to work. Her husband was a remarkable cook.

People loved the home cooked meals that they provided. Their lunches were a big hit. Everyone wanted what they served. The management was happy to have them too since they get free lunches every now and then.

When things started to pick up, Macy learned that she was pregnant. It was a miracle but she felt that it was ill timed.

Since someone had to stay home to care for their newborn, Macy's husband eventually had to stay home and she went back to work after her maternity leave was up. What happened to their food cart business?

Well, they had to put that on hold until further notice.

Fast forward to today Macy felt the pressure coming on. They have no capital to start a new business and she was turning 40. Her friends who were her age were either already well off and had grown children.

They were still in the rat race so to speak and Macy was having a mid-life crisis. She never thought that she could feel that way since she thought that it was a guy thing. But life was catching up on her and things didn't look so good.

She began to feel that her husband wasn't really pulling his weight since it seems that he wasn't really taking any initiative to try and provide for their family. What's worse is that she was beginning to get attracted to men who seemed more successful than her husband.

At one time she admitted to a close friend that somehow she wanted to change things and that maybe she chose the wrong guy after all. But now she felt that she was stuck in a dead end job and a relationship that seemed like it was something that was holding her back.

But something changed since the last time I talked to her. She learned to live by changing her focus. Today she enjoys her life.

Her husband was still trying to find work—he's still working on it. Now that there is country-wide community quarantine in place due to the corona virus pandemic, finding work as a repair man will not be that easy.

They spend all of their time together as a family—they've been doing that for an entire month now since the lockdown due to the spread of COVID-19. So, what changed in the last 4 weeks since? It was her perspective, her focus changed.

Spending time with the people who mattered most to her helped to change her perspective. Sure he was a "house husband" (that's what they called guys like him) and she did all the bread winning. But she appreciated the work her husband was doing taking care of their young boy.

She was able to change how she saw things, she learned to value her son more than her career, and she realized how big an influence her family is to her. And that motivated her to do better.

It's a Skill That Can Be Acquired

Changing one's focus is actually a skill—it's something that people can acquire. It gives you a big bump in the right direction. When you're in an up against the wall situation, you can turn things around by simply changing your focus.

Stephen Covey calls this a paradigm shift. You change the way you see things and change the center of your perspective, and you're up for something that is truly game changing. In the case of Macy, she found a way to work from home and enlist the help of her husband to work as well when the couple is not busy caring for their kids. Sometimes they take turns caring for their baby boy.

The increased cooperation both in child care and providing for the family has dramatically changed how the couple views their present condition and how they now plan for a better future despite these gloomy days.

The Key To Life Changing Focus

According to life coach JD Meier, one of the keys to changing your focus is to ask yourself the right questions. Ask yourself the right questions and it provokes your thoughts. It also makes you explore other possibilities.

Asking and answering better thought provoking questions forces you to take a stand. It also compels you to reason and it challenges your current mindset allowing you to see things in a new light.

Why Changing Your Focus is Crucial

By asking questions and challenging your current bias and assumptions increases your chances of improving your results. You can't expect to improve your lot in life unless you change your mindset.

If you're still stuck in a one income employee mindset, then don't expect it ahead in life and reel in some riches like high stakes entrepreneurs. To live like a millionaire entrepreneur you need to change your mind set to that of an entrepreneur.

Improve the questions and ask better ones and you get better results. The better you will be. The better are your chances at changing your focus for the better.

Tips on Changing Your Focus

Before we go over the actual steps that you can take to change your focus, here are some tips and guidelines that you can use to help you get a paradigm shift.

- Don't make statements, ask questions

People usually try to tell themselves what to focus on. I would like to tell you right now that that never works. You can tell yourself to be a better person but that it might take a while before such a mantra would take into effect in your life. The better option is to ask thought provoking questions that challenge your current point of view. For instance, instead of telling yourself to stop focusing on the current company layoffs, ask yourself what is the better option for you—to stay in this company or look for better opportunities elsewhere.

- Ask less why questions and more how questions

Instead of asking yourself why you're always late for everything—work, picking up your kids, etc. ask yourself how you can be on time from this time on. How questions motivate your creative side to come up with solutions.

- Ask questions that make you focus on what you truly want

Life will always give you obstacles. There will always be rocks on the ground that can make you trip. There is that scary height when you're walking on a tightrope. When you're skiing, you shouldn't look at the trees.

Ask questions that will lead you to focus on your desires and life goals. Ask yourself how much money you want to see in your bank account. Ask what kind of car you really want to drive. Ask yourself where you want to go for a dream vacation.

- Ask future related questions

The future is where the adventure is. Instead of asking why things went bad, ask yourself how you can make the most of this situation to improve your future.

Ask what your next move is for tomorrow. These questions help you to increase your resolve and focus on what you can do in the future.

- Ask questions that trigger positive emotions

Do you notice how forgiving people are when they attend someone's funeral?

It's obvious that not every person who dies is an actual saint.

But when you listen to the eulogy or testimonies given by people who attend funerals, they usually recall the good things about the deceased. They bring up happy memories and things that trigger happy memories and good feelings.

You can use the same principle when you're asking questions that will help you change your focus.

If you had a bad day, don't ask yourself what's the worst thing that happened that day. Instead, you should ask yourself instead what your favorite part of that day was.

Changing Your Focus One Question At A Time

Here's a step by step guide on how you can ask the right questions and improve your focus.

1. Step back and take a back seat so you can see the issue a lot better

It's hard to focus on solutions if you have the problem right in your face blocking your view of everything else. Stop problem solving and sit down. Grab a pen a paper answer the following questions:

- What are the things that make you feel unhappy?
- What is going and isn't going right?
- What is your major concern today?

2. Next, ask more positive and supporting questions this time

- What makes you happy at this time?
- What is going right in your life right now?
- What are the things that are going well for you now?

3. Don't ask questions that tend to be disempowering.

Avoid questions that have no clear answers. Avoid questions that are too speculative or ones that dwell on vague topics. Don't ask about situations that are beyond your control. Here are some questions that you should avoid:

- Why can't I get ahead in life?
- Why is this happening to me?
- Could she have loved me?

4. Ask more constructive, affirming, and supportive questions

- How can my new business benefit others and in turn create opportunities for me?
- What else can I do to boost sales even further?
- How do I show my love for my family a lot better than what I am doing now?

5. Ask questions to highlight the positive

- What opportunities are there for me during this community lockdown?
- How can I maximize online interactions to improve my business?
- What opportunities have been opened up for me because of this recession?

CHAPTER 109:

Give Yourself The Gift Of Perspective

Assuming an attitude of humility is one of the most important things you could ever do for yourself. It allows you to enjoy the gift of perspective. If you think in dramatic terms then everything eventually becomes an emergency. Every negative thing happening in your life seems like a crisis. You never run out of drama. You will always operate at peak stress levels.

A lot of people like this mental state, believe it or not. These people would rather feel pain, as long as they're feeling something. In other words, they're so afraid of the numbness or the sense of loss they think they'll feel when they let go. Accordingly, they'd rather live life at a high level of stress and pressure.

Talk about running yourself ragged; talk about living in a toxic mental soup. That's exactly what's happening when you try to hang on to the idea that life is all about you. You think that once you die, the world dies with you because there's nobody left to experience the world. Well let me tell you, the world has its own separate reality. If you die and you're not there to observe, believe me, the world is still going to go on. This is part of absolute reality. Reality doesn't die with you.

You have to let go of your addiction to the self, and the illusion of control that it gives you. Once you allow yourself a healthy dose of humility, you unleash perspective. Things aren't that serious; things aren't that big. Most certainly, things aren't as dramatic as you think they are. The sun will still rise tomorrow. People will move on and get on with their lives. This is one of the most liberating things you could do for yourself. It's a great way of resetting your mental and emotional energy levels so you can become a more effective person.

When you truly believe that you are not the center of the universe, and that there are other competing interests and objectives out there bigger and worthier than you, you will be okay. You won't lose a thing. Well, not quite. You do end up losing the delusion and false sense of security which being self-centered brings to the table. Furthermore, you end up gaining perspective regarding the following.

Your Proper Relationship to Others

When you allow yourself to feel contented being another face in the crowd, you start seeing your proper relations to other people. Instead of people existing to please you or fulfill your needs, you understand that you're part of the giant jazz concerto of life. You're playing your piece while everybody else is playing their segment, and you leave it at that. You no longer feel that everything has to flow through you or everything has to make sense to you. You just groove with the music. It was there before you showed up. It will continue to be there long after you're gone.

Knowing your place in the great scheme of things enables you to accept people for who they are. This works to reduce a lot of the unnecessary drama in your life because once you understand that the world doesn't revolve around you and your sensitivities, you become more tolerant of people. You will be able to stop expecting perfection from others, while expecting them to accept you for who you are. Your relationships (and your attitude towards them) become two-way streets.

Proper Relationship between the Past, the Present, and the Future

Another great perspective that you get is your personal relationship with your past, present, and future. Unless you have access to a time machine, there's really not much you can do about the past except for one thing: you can change your perception and reading of your past. This is one of the most empowering things you can ever do.

If you can't let go of past hurts and trauma, you are letting your past define you negatively. This can ruin your life because you can't to let go of these past perceived and real injuries. Stop being a prisoner of your past by choosing to view yourself as a historical being. You know there are many things that happened in the past that you could read in a very empowering and positive way to help you feel powerful and in control of the present.

As the old saying goes: those who control the present control the past; those who control the past control the future. This definitely applies to personal narratives. If you take ownership of the fact that you are in control of your thoughts and refuse to blame others or make excuses, you can go a long way in developing a proper mindset regarding your past. Your past doesn't have to be a place of pain, frustration, disappointment, and degradation. It doesn't have to involve mental images or psychological movies of your parents abandoning you, slapping you around, saying hurtful words or otherwise not treating you the way you feel you should have been treated.

You have to understand that even in hell; there are still small pieces of heaven. Even in heaven, there is still a way to imagine you in hell. It all depends on your perspective. This is why it's crucial to wrap your mind around the concept of humility and shift your attention from the center to the side. If you're able to define yourself this way, then you can develop a new relationship with your past. Of course, this requires you to take full control of your personhood. If you're able to do that, then you can set a different course for your future. After all, your past provides emotional fuel and inspiration for what you do in the future.

If you define yourself as a winner, a victor, and somebody who makes things happens, then chances are your future will be much brighter. At the very least, it would be much better than if you choose to constantly imagine yourself as a victim or somebody who is just continuously unlucky.

Proper Relationships between Your Potential, Your Present Capabilities, and Your Past Abilities

The most damaging thing about hanging on to any self-centered narrative is that it may seem to be an act of confidence on the surface. After all, you have to be confident to think that you are the center of the universe, right? Well, not quite. You may be occupying the center because you're trying to hang on for dear life. What you really feel is that you're at the bottom end or at the bottom of the barrel. You don’t feel you’re all that desirable. You keep hanging on to the center of your consciousness and trying to filter everything based on your needs and what you feel you deserve. The more you do that, the more depressed, discouraged, frustrated, or disappointed you become.

When you adopt an attitude of humility, you understand that your past abilities are just starting points-that's all they are. They don't define you. They don't limit how far you will go. They show where you started. Your past merely indicates what you have to work with. Your past experiences are building blocks if you choose to build on them. This is a very crucial change in mindset because your estimation of your potential, present capabilities, and past abilities has a big role to play on how much energy and focus you're going to put into making things happen in your life.

If you're constantly hobbled by your negative perception of your past abilities, it's very easy for you to define yourself by your past failures and shortcomings. Make no mistake about it, all of us have shortcomings and none of us came out of the gate 100% winners. Even Michael Jordan had problems playing high school basketball. There were many times he was disappointed by his performance and was doubtful of his ability to play basketball.

It doesn't matter how many times you failed in the past. You shouldn't define yourself by those setbacks. What you should focus on is the fact that you have started. At least, you put one foot in front of the other, and you tried your hand at something. It's your choice whether you can continue to keep on building. Adopting a certain level of humility helps you get a proper perspective on the relationship between your past abilities, your current capabilities, and your future potential.

Step Out Of Yourself To Get A Hold Of Yourself

Before we go any further, let me quickly recap what cognitive behavioral therapy is about. CBT operates under the premise that maladaptive behaviors stem from badly formed, imprecise, or incorrect thoughts arising from inaccurate or less than optimal personal narratives. In plain English, the negative actions you take and the emotions you feel come from badly formed thoughts that arise from your inaccurate or harmful personal narratives. You simply chose to read your personal reality negatively.

In other words, there's no space here for blame; there's no bad guy. Instead, this negativity you're feeling is really just a mistake. If you can adopt new narratives, you can live a more positive life because you feel more positive and powered, and this enables you to take more positive actions.

One key reason why people have maladaptive behaviors which they can't seem to stop is because they are hanging on to negative narratives. By simply shifting your focus from yourself to others, or to the world in general, you can make tremendous progress in changing your behavior and emotional states. You increase the likelihood that you will feel good about yourself and act like a positive person, which then makes you feel even better, and you repeat the process over and over again.

Instead of this process dragging you into a downward spiral, you can actually reverse the direction of the spiral. If you become more positive, you feel more empowered. This leads you into becoming more competent and having a greater and more positive impact, not only on the people immediately surrounding you, but the world in general.

It's easy to say this; it's easy to say that you just need to swap out narratives and all these amazing positive things will happen. As you probably already know, there are many things in life that are easier said than done; CBT is definitely one of those. So if you're having a tough time letting go of yourself, choose to care about others. If you really want to turbo charge the power of humility in your life, do this one thing: choose to care about others.

I'm not talking about thinking about them. I'm not talking about reducing them into some sort of icon or ritual, where you feel like you're going to feel better if you ritually think of other people. I'm talking about actually caring for others. This is where compassion comes in. If you look at caring for others as another to-do list item or something you can reduce to some sort of ritual, then you're not going to fully benefit from the power of shifting your narrative from yourself to something else.

CHAPTER 110:

Actions Plans3

Action Plan I – Dealing With Anxious Thoughts

Nothing impacts your thoughts as much as anxiety. Suddenly, the world comes to a standstill, and you begin to think that everyday things and situations may be threatening to you. The feeling that everything is going wrong and you will not be able to deal with the situation makes anxiety psychologically dangerous.

You may start finding normal situations, or situations that seemed normal to you the last time they occurred, potentially dangerous. For instance, having a cough may bother you and force you to think that you are suffering from some chronic disorder of which coughing is a symptom, or your child is 10 minutes from school may compel to think about all the wrong things that could have happened to him.

As a general rule, people with anxiety live under the constant fear that something is going wrong so much so that they may be dreadful to cope with. What they fear is not just the situation, but also the fact that they won't be able to manage them and remain calm and safe in it. This is a vicious circle of sorts. When you feel anxious, you are surrounded by thoughts of fear, thread, and avoidance. Moreover, when you consider a situation negative or threatening, you are likely to get anxious about it.

What you think directly impacts what you feel, and if you can curb your thoughts, your anxiety levels will automatically drop. Let us take the example of two ladies who are waiting for their children to come home from a party, late at night. Their children are out of a party and had promised to come back by 11 that night. The first lady begins to panic as the clock strikes 11, and every minute past that time her anxiety levels rise. To add to her misery, she begins to think about all the bad things that could have happened to her child, which adds up to the anxiety substantially.

The second lady also gets anxious to find out why her child is not back at 11. However, she remembers all the times when her child had promised to be back by a certain time and had arrived 10 minutes late, so her anxiety levels drop. In both the cases, the ladies got anxious, however, while the first lady increased her anxiety levels by thinking about things that escalated her worrying, the second lady thought of things that helped her manage her anxiety.

Suffice to say; thoughts have a significant impact on your overall anxiety levels. An increased anxiety level will translate into more serious physical symptoms. The best way to manage the scenario is to break the cycle of anxiety and thoughts by thinking from a positive and constructive perspective.

To help you manage anxious thoughts, you need to give yourself some time and follow the two-step exercise explained below.

- The first step is to identify the thoughts that are making you anxious.

- The second step is to replace these thoughts with alternative thoughts which are more constructive and realistic.

As we move forward, we shall explore these two steps in greater detail and how you can implement them in your daily life.

You cannot challenge anxiety unless you know the cause behind it. This is exactly the reason why identifying anxious thoughts is extremely important. With this said, it is equally important to state that this identification process will not be simple. Such thoughts are automatic and so quick in occurring that you don't even realize that something you thought just raised your anxiety levels. Fortunately, the effects of anxiety are such that you will know that you are anxious and applying reverse psychology to the cause is the only option that lies with you. As you practice, identifying such thoughts will get easier.

Typically, anxious thoughts that take the shape of "what if" and "I don't think I can cope." If you have experienced an embarrassing situation lately or an incident that has had a grave impact on your mind, then reliving the incident in our mind may also be a cause of anxiety. The most challenging aspect of identifying anxious thoughts is that you are already anxious, and being able to control your mind at such time for an analysis of this degree may be tough.

Therefore, it is a good idea to have well-framed questions to ask yourself every single time you get anxious. For instance, as yourself:

Why am I feeling this way?

How did it start?

Are all my assumptions actually true?

Will this actually happen?

How will this impact me, my life, and the people around me?

How To Start?

Prepare a Thought Record

As you perform this activity, be sure to record your thoughts and answers in a diary. This will help you analyze the situation with a calmer mind at a later time. In this way, you will be able to use the information gathered for managing anxiety for future instances even if you weren't able to help yourself at the time it occurred. This record sheet can also play an instrumental role in determining behavioral patterns and identifying underlying issues if any are present.

Determine the Realistic Thoughts from Anxious Thoughts

Once the identification process is complete, and you know how to do it at the time of anxiety, as well as analyze the pattern over a period, the next step is to determine if the thoughts are realistic. Most anxious thoughts are actually assumptions or exaggerated reactions to situations. Therefore, this step will help you get a realistic and objective viewpoint on the situation.

Moving forward, you must look for an alternative way of thinking about the situation. This is a daunting task, and you will need to practice this method a lot before you can be an expert on it. In the previous step, you had prepared a thoughts record. In this step, you may add two columns to the thought record, one each for evidence that supports your thought and the one that goes against it. Use these columns to create a new column, which gives an alternative and realistic thought for replacing the existing anxious thought.

Assessment

No method is effective unless it can help you reduce your anxiety levels. Therefore, before you sit down to write the record: assess your anxiety levels by rating it on a scale from 1 to 10; 1 being the least anxious and ten being the highest. Rate your anxiety levels after you have evaluated the alternative thought using the method illustrated here. This will help you know if the method is working for you.

Action Plan II – Dealing With Worries

Understanding Worrying

People who complain of anxiety usually worry a lot; much more than they should. If you are a chronic worrier, the chances are that you worry about a lot of things. Therefore, you will have some worry topics in your mind. As you learn to manage one worry topic, your mind may switch to another one. It soon becomes evident that it is not the incidents that are the problem; the problem lies within you, and in the manner you perceive things. Therefore, what you need to manage is not the situation, but you need to manage yourself.

Another thing that you need to understand is that worrying is a maze on its own. If you believe that worrying is a bad thing, then you will worry about the very fact that you worry. On the other hand, if you start believing that worrying is good, you will continue to worry about that isn't worrying enough.

Touching on the negative effects of worrying, some people worry about worrying and how it can negatively affect them. Such thoughts usually take the shape of worrying about the fact that worrying has gone out of control or realizing that it is harmful and not being able to do anything about it.

What is even worse is that you don't even know if you are worrying too much about worrying. To determine this, you must question yourself about whether worrying is a problem, and what is the worst that can happen to you if you continue to worry. Lastly, a quick assessment on whether it is possible for you to stop worrying can bring about decent closure to the situation.

While talking about concerns and worries, the one thing about worrying that is most disturbing for people is the fact that they are not able to control their feelings. To challenge this belief, the first thing that you must do is to look for an incident where you were worried. Assess the situation to answer a few questions:

- Was it possible for you to control your worries?
- How hard did you stop worrying?
- Were you able to stop these worries in the end?
- If yes, what was the cause of this stoppage?

Now that you have analyzed the situation in full look at your anxiety situation in its entirety.

How often have you worried, and were you able to control your worries?

Can you think of any incident when you were able to get rid of your worries successfully? If yes, how?

By the end of this analysis, you will be in a better position to assess whether your worrying situation is controllable or not. The biggest problem with controlling worries and anxious thoughts is that it is rather difficult to control or suppress them. When you try to get rid of such thoughts and think of something else, you are likely to come back to these anxiety-provoking thoughts no matter what. It is a human psychology to think more of the things that you think of not thinking. For example, if I ask you to not think about the white pigeon, images of it will keep coming back to your mind all the time.

How Can You Manage Worrying?

Controlled worry periods is a concept that shall help you in this endeavor. When you can accomplish this a couple of times, you will get a sense of control over the situation, which shall be instrumental in helping you break the vicious cycle of uncontrollable worrying.

The concept of controlled worry period is based on the fact that you must set out a fixed length of time, place, and exact time for worrying. This should be the same time every single day.

Whenever you feel like worrying, you must tell yourself that you will take up this issue at a later time that is set aside for worrying. While you can choose any time of the day for this activity, it is best to avoid performing this right before bedtime.

- Spend no more than 15 minutes worrying about all the things you had marked during the day.

- Brainstorm by yourself, and the moment you decide that the matter is not worth worrying about anymore, you must stop worrying about it.

To ensure that you don't forget any of the agendas that you had set for the worrying period, jot down your worries for the day on a notepad set aside for this purpose. The beauty of this method is that it plays with your psychology in such a way that you are not asking your mind to stop worrying. You are actually giving it a later appointment for worrying.

CHAPTER 111:

Fear And Anxiety

Fear and anxiety may seem similar, but they are two different feelings. While they both involve stress responses, they are reactions to different types of threats. Anxiety is a reaction to the idea of a threat: It keeps you on edge and apprehensive, heightening your awareness so you can respond to any potential threats quickly. Fear, on the other hand, is a response to a known threat. Rather than being alert in case of a threat, you are reacting to something that is actively threatening you. This distinction is necessary to make, as fear can cause anxiety symptoms, but true anxiety does not directly cause fear. Understanding these nuances, as well as how to identify when what you are feeling is fear or anxiety, will help you learn to control your feelings of anxiety or worry. After all, anxiety is a feeling that something bad is coming, not that there is a threat present.

Symptoms of Anxiety

Anxiety attacks can take many forms. Sometimes, they present as aggression, while other times, you may feel frozen in fear. In the heart of an anxiety attack, you feel as though there is danger surrounding you, and no matter how hard you try to tell yourself that you are okay, you cannot shake the fear of imminent danger. This often manifests in any combination of the following symptoms:

Overwhelming fear or sense of imminent death or harm

Chest pain and heart palpitations

Shortness of breath and choking sensation

Trembling, numbness, and chills or hot flashes

Nausea and dizziness

Feeling as though you have detached from the world around you (called derealization)

All of these symptoms could be signs of anxiety, but also signs of serious medical problems. If you feel these symptoms on a regular basis, even when there is no apparent danger around you, it is worth getting a checkup from a licensed medical professional. Your primary physician can eliminate any worry that there is a more sinister cause of your symptoms, and also refer you to a mental health expert that will be able to accurately pinpoint your diagnosis and help you on the journey toward wellness.

Healthy Fear and Anxiety

As unpleasant as both fear and anxiety are, they are both healthy and normal feelings, in moderation. They each have very specific biological purposes, and when they are functioning normally, they both work together to keep you alive. Imagine if you are faced with someone with a gun: Your gut reaction will likely be fear. This fear will start your fight-or-flight reflex, and with adrenaline coursing through your veins, you will be able to focus better on what is happening, so you can react to movements quicker. Your muscles will be ready to burst into action to either fight or run. Your endurance will increase. All of these together will prepare you to protect your life, allowing you to either flee from the person with a gun or fight off the person with the gun.

Likewise, with anxiety, your body is reacting to the idea of a threat. There may not be a threat that you are aware of, but you do feel as though there might be one somewhere nearby, and that apprehension is enough to keep you prepared to launch into fight-or-flight mode at a moment's notice. If you are walking through a dark alley at night, you may feel that anxiety gnawing at you, telling you to stay on edge and alert for any threats that may potentially jump out at you.

That anxiety warns you that there may be a danger, and in healthy individuals, this anxiety is balanced. It will keep you alert when your surroundings dictate that you probably should be, such as when you walk through a dark alley, or when you are hiking through a mountain trail at dusk: Your alertness keeps you prepared to react to any sort of threat that may arise. For people without anxiety disorders, it is smart to pay attention to these gut feelings, as we often have them for good reasons, such as knowing that we are in a dangerous area or that we are doing something risky.

When Fear and Anxiety Become Problematic

For someone with an anxiety disorder, however, they frequently feel that state of constant alertness and worry even when it is unwarranted, leaving them unable to relax and constantly hyper-aware of their surroundings. These symptoms can begin to spill into other aspects of your life, and when they become a persistent problem, repeatedly impeding on your ability to function or enjoy day-to-day life, your anxiety may be problematic. Only you, yourself, can identify when your anxiety has become a problem. In order to do so, you can ask yourself a few questions. Is your anxiety unwarranted? Is it persistent, no matter how irrational you may think it is? Is there a recurring cause to it? Is it becoming so bad that you have to rearrange your life to accommodate your negative feelings? Is it becoming so bad that your friends and family have begun to mention or question your anxiety? When your anxiety is triggered, are you unable to react in a rational or healthy way? Do you feel as though you cannot cope with your anxiety at this moment? Is your anxiety becoming overwhelming?

These questions can help you gauge your own opinions or feelings about your anxiety. If you can answer yes to any of them, you may have a problem with anxiety, and speaking to a medical professional would likely be beneficial to you. The professionals are trained to walk you through the steps of healing and controlling your anxiety, so please do not feel intimidated or afraid of seeking help. If your anxiety is truly problematic, you will likely need some intervention, whether through self-help books such as this one or through working with a therapist, to begin coping in a healthy, efficient manner.

Common Anxiety Disorders

Anxiety on its own does not refer to one specific problem. It is a spectrum of disorders that can come in a wide range of shapes and forms, some of which seem to be intuitively related to anxiety, while others may not seem obviously related. Ultimately, each of the following disorders you will read about is related to anxiety, though each present in different ways. All of them relate to stress or anxiety responses for a variety of reasons. Here is a general overview of the most common anxiety disorders.

Generalized Anxiety Disorder (GAD)

GAD is characterized primarily by chronic feelings of anxiety, even when unprovoked. The sufferer often feels a persistent, extreme sense of worry or fear, as well as the symptoms listed above. In order to be persistent, the feelings of anxiety must be present for the majority of days for at least half a year. They are general, meaning there is no particular trigger, and the symptoms of anxiety are felt in a wide range of circumstances, such as work, driving, or even social interactions with friends or family. Oftentimes, they are pervasive, meaning they are significantly impacting the sufferer's life in a negative fashion. For example, if the sufferer feels anxiety when forced to present for class, she may fail multiple assignments that involve public speaking. Likewise, a person who is anxious when talking to strangers may intentionally avoid all unnecessary contact with other people, costing him his social life. Since the anxiety occurs in a wide range of situations, it may lead to the sufferer attempting to avoid everything possible in hopes of avoiding the feelings of anxiety they fear, which really only exacerbates the situation. When fearing anxiety itself, the sufferer may find him or herself becoming anxious at the thought of becoming anxious.

Panic Disorder

Panic disorder is typically described as unexpected episodes of intense, overwhelming terror along with the physical symptoms typically experienced with anxiety in the absence of any true danger.

This must happen more than once, and the sufferer will often feel an overwhelming sense of losing control and a sense of impending demise. There is no known cause for these panic attacks, though it is believed that genetics, stress, and a sensitive temperament can leave an individual at a higher risk for suffering from these debilitating symptoms. Without seeking treatment, panic disorder can impact most aspects of daily life.

The fear of another panic attack can cause you to develop phobias, feel as though you must seek medical treatment for other issues that may not even be present, because issues with work and school increase the occurrence of suicidal thoughts, and risk of substance abuse and addiction.

Obsessive-Compulsive Disorder (OCD)

OCD, as the name implies, involves a series of obsessions and compulsions. Obsessions are thoughts that repeatedly occur, even why you try to stop or resist them and are typically disturbing in some way, shape, or form. Even if the person suffering from OCD is aware that the obsessions are irrational, they cannot be stopped. They are time-consuming and inhibit the individual's ability to function normally.

Compulsions are repetitive behaviors or thoughts that are acted upon with the intention of stopping the obsessions. The person with OCD recognizes that the relief from the obsession will only occur temporarily and that it may be irrational, but cannot help him or herself. This coping mechanism is frequently time-consuming, and the repetitiveness of the compulsions detracts from day-to-day life. Compulsions are context-dependent, meaning that for some, a specific action might be compulsion when it is entirely normal or even expected for others. For example, someone who works in a kitchen may wash his or her hand's hundreds of times throughout the day due to touching different substances whereas someone with an obsession over cleanliness might compulsively wash his or her hand's hundreds of times for no real reason other than to get rid of the obsessive thoughts. Clearly, the person working in a restaurant needs to keep hands clean and will be handling multiple products that will call for hands to be clean, but a person sitting at home has no reason to wash his or her hands so often.

Phobias

Phobias are intense fears or aversion to specific stimuli. For example, one of the most common phobias is toward spiders. Someone with a spider phobia may have irrational reactions to seeing or even thinking about possibly seeing a spider, and will likely intentionally change how things are done or where he or she goes to avoid spiders.

If a spider is encountered, people with arachnophobia typically experience intense anxiety. Phobias can be specific or general, and range from fearing a specific animal or insect to fearing a concept, or even fearing specific shapes and smells. There is even a phobia specific to tiny holes on a background, such as the pitting of strawberry seeds on fruit, called trypophobia. These phobias can be somewhat innocuous, such as fearing spiders, but also completely overwhelming and detrimental to day-to-day life, such as fearing to drive, fly, or even leave the house.

Social Anxiety Disorder

Social anxiety disorder is quite similar to a phobia and used to be referred to as social phobia instead of social anxiety disorder. People with this form of anxiety feel an intense fear toward being judged or rejected in social situations. Oftentimes, they fear that their behaviors rooted in anxiety, such as blushing or stuttering, will be seen negatively by others, and because of this, they avoid social interactions as much as possible.

CHAPTER 112:

Cognitive Beliefs Of Obsessive-Compulsive Disorder

Obsessive compulsive disorder is often misinterpreted. It's actually pretty common for people to be stuck in a routine they have a tough time breaking.

Typically, the person performs the behavior to get rid of the obsessive ideas, but this only gives him or her short-lived relief. Not carrying out these obsessive rituals can trigger a lot of anxiety.

An individual's level of OCD can be anywhere from mild to extreme, but if it is serious and doesn't get treated, it can wipe out somebody's capability to function.

About 1 in 150 people struggle with this compulsive disorder. Eating conditions, sadness, anxiety conditions, and other personality conditions can, in some cases, accompany OCD. The percentage of males and females who have this disorder is about the exact same.

Normally, it already becomes apparent in somebody's youth, and research has pointed out that it is genetic. Although those signs can be noticeable in someone's childhood, studies indicate that the majority of the signs typically begin in the teen years. Treatment and testing are essential when such signs are observed.

As the words indicate, Obsessive Compulsive disorder is a disorder that is based on obsessions, which originate from fears and anxiety. As a response to these obsessions, a person will usually perform rituals and repeat the same action over and over again, which is specified as compulsive behavior. The behavior, however, is only a noticeable result of the root. The root is the fears that trigger the obsessive thoughts.

Understanding OCD

It has been said that Obsessive Compulsive Disorder is most likely caused by adverse effects to brain activity and its ability to link information to all the parts that need it. The onset of OCD most commonly begins in childhood up to early adulthood with an average diagnosis age of 19 years old.

Attempting to Understand a Compulsion:

The following is a sample from an OCD sufferer. In order to understand the complexities of compulsions, he shares his specific ritual style experienced from early childhood up to diagnosis at age 11:

"This compulsion was all about counting. Everything had to be an even number with the exception of 6 and 25. Which means that 6 were not allowed, but 25 was?

The number 25 played a part in daily life by having to do such things as: drinking water and keeping the glass on my lips for 25 seconds, chewing 25 times, scratching 25 times etc.

Even numbers played a role by everything having to be, well, even. Simple things such as if I was eating potato chips, I count and have to end at an even number. Both hands would have to be even in holding chips and with the same amount of chips per hand. Which means that I have to eat an even amount of chis per hand, and alternate hands an even amount of times so one doesn't go first or second.

If I accidentally fall and scrape my knee, I purposely make myself fall again to scrape the other knee and make it even."

Germaphobia:

The fear of Germs. People with this type of OCD feel inclined to wash whenever they can, or to clean their homes, office spaces, and whatever else may be around them. When normal people see dirt on their hands or a bit of spilt milk on the table, they wash and clean up, easy as that. But people with OCD wash their hands, dry, and then have lingering thoughts about microscopic bacteria with claws clinging onto their skin and refusing to let go. They wash again, this time with a scrub, and then dry; their hands are finally clean. But wait, they used the same towel to dry. The bacteria with claws are back. Towel gets thrown away, washes again, dry's with a new towel and thrown it and the scrub away.

Germaphobia can be expensive. Often times, people replace household items much too often and use unnecessary things that could amount to extra expenses such as gloves or face masks.

Hoarding:

Lots of people are messy, so what is the difference between messy, avid collector of strange items, and hoarding? Hoarding is when a person collects items to the point that the functionality and quality of their lives become impaired. When someone's home is so filled with clutter or 'collections' that they can no longer live comfortably and do normal household chores such as cook, use the coat rack, or sleep with enough space. The frontal lobe of the brain which is in charge of balancing options and a rational thought process has proven to be inhibited in people who hoard. As a result of this, a hoarder's priorities drift from one point to the point of hoarding.

Symptoms of Hoarding:

1. Placing sentimental value on objects even of this may prove irrational. Some of these objects might include things that would classify as trash.

2. Seeing value and usefulness in any and all items.

3. Believing that everything in their collection will be of some use someday.

Signs

Which symptoms are common for a person with Obsessive Compulsive disorder? Here is a list of possible signs:

1. They have repeated ideas or pictures in their heads about several things, like a worry of germs, trespassers, hurting others, violence, sexual acts, or being overly cool.

2. They repeat the exact same behavior again and again, like locking doors, washing hands, counting, or hoarding.

3. They continuously have unwanted ideas and habits that they find tough to control.

4. They do not find satisfaction in the routines or behaviors, but they find a short remedy for stress and anxiety if they perform them.

5. They spend an hour or more each day on these ideas and routines, which interfere with life.

6. Consistent paranoid, typically impractical fears; hard to reason with them concerning those aspects of their perception of reality.

7. Monitoring and washing are the most common reported compulsions of OCD. Others have a compulsive propensity to reorganize things or mentally repeating expressions or list making.

8. Those with OCD usually fear losing control. This doesn't mean that all "control-freaks" have OCD or that all people with OCD are "control-freaks," but the fear that they may harm someone with their compulsive conduct exists.

9. Spiritual fixations could happen, like the fear of offending God.

10. Undesirable sexual ideas, like homosexuality or incest ideas, could accompany OCD. This does not mean that these people should be considered a hazard, homosexual, pedophilic, or unrestrained, but it's normal for some with OCD to just have those thoughts troubling them.

11. Perfectionism is usually typical, to the point that they fret over exactness or a need to recall things or track things.

Some people with Obsessive Compulsive disorder partly or totally comprehend the uselessness of their obsessiveness and compulsive habits, though they find it hard to stop doing it. Others lack an understanding of their unusual thinking patterns and are blinded to the simple fact that they are overdoing things.

A lot of people with OCD have the ability to keep their obsessive-compulsive symptoms under control when they are working, studying, or feeling relaxed enough to function well. Resistance may deteriorate occasionally, however, particularly when they're restless, when abrupt changes in their lives occur, or when they panic on the inside.

It's typical for symptoms to come and go, to become simpler to manage or worse. If the symptoms be too terrible, their obligations and everyday tasks may experience the consequences. Another way some people with OCD manage their compulsive disorder is by taking alcohol or drugs, which undoubtedly isn't a good resolution at all.

Drug Therapies That Exist

What will your doctor need to know before recommending anti-obsessive-compulsive drugs? Some of the information your medical professional will really need consists of:

Your medical history-Do you have other medical conditions? For instance, you might be asked about cardiovascular disease, liver disease, epilepsy, anemia or other blood diseases, glaucoma, and diabetes. Any medical condition you have might be very important, so be sure to give the physician a total listing. Aside from that, you will be asked if you have ever had any allergies to medication. For instance, have you ever had an allergic reaction to an antidepressant?

Any medications you are taking-For example, are you taking medication for your heart or blood pressure, contraceptive pills, blood thinners, antibiotics, or antidepressants? Any medication you are taking might be essential, so make certain to inform your medical professional of all medications including non-prescription drugs.

Your typical diet-Do you drink big quantities of coffee or tea? How much alcohol do you take in? Are you on a special diet of any type? Do you prepare to begin any unique diets in the future?

Your profession and activities-Do you really need to operate dangerous equipment or drive a car? Sometimes anti-obsessive-compulsive drugs cause sedation or hinder coordination, but these are usually momentary negative effects.

Or special note to women-Are you pregnant? Is there a possibility of pregnancy while taking anti-obsessive-compulsive drugs? Are you or will you be breast-feeding your child? (Anti-obsessive-compulsive drugs may be harmful to coming or breast-fed infants, though there's no evidence to date that they are). The above lists of questions aren't complete but should give some idea of what your physician will want to know before you start anti-obsessive-compulsive drug treatment. The bottom line is to inform your doctor about any medical conditions, medications, etcetera even if you are being treated by some medical professionals. If you are unsure whether certain facts should be brought out, mention them and let your doctor choose how crucial they are in your treatment. Without such information, a physician would have trouble treating you safely and effectively. Are anti-obsessive-compulsive drugs damaging during pregnancy? There is no proof from substantial screening in lab animals that clomipramine or fluvoxamine cause problems with breeding, fertility, pregnancy, or fetal development. Some women have taken clomipramine at different stages of pregnancy without obvious ill impacts for themselves or their kids. Nevertheless, it is risky for women to take any medications while conceiving or carrying babies unless the dangers of not taking the medication are substantial. Anti-obsessive-compulsive drugs in some cases produce anorgasmia inability to achieve orgasm or climax) in males and women but this adverse effects is reversible when the dosage of the drug is minimized or the medication is terminated completely. Are laboratory tests needed before beginning anti-obsessive-compulsive drugs? Since all anti-obsessive-compulsive drugs currently readily available in the United States are investigational or speculative, they remain under close examination by the Food and Drug Administration. All clients receiving them in speculative programs are currently needed to have blood and other tests (typically a urinalysis, electrocardiogram, physical examination, and eye examination) at routinely set up intervals to examine the drug's effect on the body and to guarantee that patients taking part in experiments with the drugs can do so safely. If the FDA (FDA authorizes these drugs for prescription by doctors right across the United States, there may or might not be specific requirements for lab tests.

CHAPTER 113:

The Self-Love Model And Breaking The Negative Cycle

This part of the book has a singular focus: to provide you with a better path toward self-love. Yes, loving yourself. The journey toward self-love is a balance of doing things for yourself, and what is expected of you. But what matters isn't what the world expects but what you expect from yourself.

Only you can make yourself happy. I know this sound like arbitrary common sense or old-style advice, but have you ever placed your dependence for love on someone else and ended up disappointed? Exactly. Nobody can provide you 24/7 with that feeling of self-compassion. Nobody can feed you the thoughts that go into creating your positive mindset.

You are the master gardener of your mind, body, and spirit. So, live that way. Do nice things for yourself. Treat your life as if it is a great creation. Do this and you'll always be working on the self-love model.

Throughout this book we looked at the negative mindsets that destroy self-esteem, damage confidence and keep us trapped in our own hell. But these are not who we are. They are obstacles preventing us from becoming what we have always wanted to be. Remove the patterns of defeat and you rise up to be unbeatable.

The Self-Love Model

When you hear the words self-love, what do you think about? I used to hate these words because I didn't love myself very much, and I certainly wasn't going to tell people about it if I did. But what is it, really?

Well, we know what it means not to love ourselves. You may have had deep resentments toward those who harmed you. You probably had many things about yourself you didn't like. Maybe you hated yourself so much you were near self-loathing. Self-loathing is powerful, but so is love. In fact, it's much stronger. Yet, we are conditioned to believe in fear from a very young age.

Self-love is everywhere, but we have kept it hidden all these years. Why? Vulnerability. Shame. Fear. We hide ourselves to protect ourselves. But in protecting who we are we fail to express who we are. It's a powerful catch-22.

Before you can truly have any balance in your life, and build healthy relationships with the people around you, you need first to develop your relationship with yourself.

But what does it mean to love yourself?

Many people never do. They are taught that success is about getting high grades, achievement, getting a job to justify our purpose, and then finding that special someone who is going to complete us. I have done all these things and I can tell you with confidence, while it helped to contribute to my self-esteem and success, there wasn't much focus on self-actualization. When it came to my achievements, they didn't make me happy for long. Soon I was looking for something else. Relationships provided me with some sense of what love was supposed to be like, but they eventually failed when they couldn't live up to my needy expectations. I had success in business one day, but a bad day the next. What was missing was a sustainable system that wasn't dependent on some level of success that had to be achieved first.

When I was rejected or told, "Sorry, it just isn't working out," I was convinced that I was unlovable. It turns out that external validation isn't a formula for happiness.

Unconditional Love

I have met many people over the years who confessed that they couldn't feel any sense of love for themselves. When I asked why, they would say, "Nobody ever made me feel lovable." Unconditional love is what we learn from our parents. We either get it or we don't, and if we don't get it, the love we grow up with becomes largely conditional.

We go through life believing that we can be loved once we have achieved a certain level of status. If you were raised in an environment that favors your achievements, you may have received this validation for doing well in school or sports.

"If I get the highest grades in class, my parents will be happy."

"If I do as my wife says, she'll show her appreciation for me."

"If I can live up to the expectations of my boss, she'll respect me."

When we lack that connection to unconditional love, we spend our lives in pursuit of it. It is the quest for validation that we are lovable. But when it comes to unconditional love, aside from our parents, the greatest source comes from within you. Yes, what you are looking for is what you already have. Hollywood has it all wrong.

We are made to think that once we find that special someone, we'll discover that source of power that can provide us with the love we need. But it isn't so. People are people, and what you mistake for unconditional love in the beginning is really the start of a relationship that has yet to discover the flaws of each individual. Many relationships fail because, as they move forward, these flaws become more evident and the illusion is shattered.

We have to stay grounded in reality.

Doesn't it feel at times that we are always trying to please, validate, or fulfill someone else's expectations based on performance or achievement? There are some societies that are driven this way. Many individuals as well. Here is what happens in this case: We move through life focused on doing and not being. Instead of being a person who is worthy of love, you put all your efforts into getting recognition for achievements. But these achievements are short-lived. As soon as you are done with one achievement, you're on to the next one. It becomes a never-ending cycle without any sustainable results. You're only as good as your last achievement.

Most people lose the concept of unconditional love because they believe love is based on a results-driven achievement.

Here's what Marty said about his childhood:

I'd be expected to perform well in school. When I did well, my parents bought me a computer. If I did exceptionally well, it was a reward of some sort. But once, I got sick and couldn't go to school. My grades and performance went down, and when that happened, I didn't get any rewards. I realized years later that the love I was getting only stuck around as long as I was doing something. It was based on doing and not being.

Imagine two doors:

Door #1: Conditional love is the love we get when we please others. In fulfilling their expectations, we feel like we are doing what is expected of us. Love is justified.

Door #2: Unconditional love is the love we receive when we are just being ourselves. No validation or justification is needed.

I'll take door number two.

Conditional love is based on doing. You have to show that you are worthy before someone will give you the recognition you want. But unconditional love is being, and this is what you can give yourself. The single mistake people make is expecting to draw this from the world.

We have spent a large part of our lives dependent on other people for our needs, and this is especially true when it comes to attracting love. Of course, we can get love from our friends and family. But the real respect is self-respect, self-reliance, and self-trust.

One of the best things you can do for yourself is to live with self-compassion. You've heard of showing compassion to others in hopes that you receive some in return? Well, try turning that around and showing compassion to yourself. After everything you've been through so far, don't you think you deserve it?

You don't have to earn compassion. It is yours. Decide to treat yourself with respect. Make it a daily reprieve. You don't need permission from anyone, and if someone in your life is a compassion thief and holding you back from experiencing your full potential, then it is time to take a look at that relationship.

I have a motto that works for me: Anything or anyone who is a consistent negative force in my life is either helping me to grow as an individual or isn't. If it isn't contributing to your life, it gets the boot. We can only feel good about ourselves when we learn to treat ourselves like valuable human beings.

Open Honesty

You should be honest with yourself as much as possible. Stay true to what you know is right. Accept it when you are feeling fearful and uncertain. This is a sign that you could be slipping back into an old routine. Accept yourself with all your flaws and don't criticize yourself or others for theirs. People will make mistakes and screw things up. That's what we do.

Work your empathy and have an open mind toward people who are trying their best. But be aware of the people you meet who need to be kept at a distance. If they are dragging you down with their issues and problems, but they seem unwilling to do anything about it, cut the ties, and step away from them. To be true to yourself is also about being aware of the people you can't help right now. They are not ready.

Expressing Self-Compassion

As I mentioned earlier, compassion is a positive driving force that we need. Without compassion in your life, specifically directed toward yourself, it's as if you're trying to sail a boat with a hole in it.

Compassion is a weapon against tyranny, animosity, and selfishness. When we create compassion and tap in to the love we can find there, it shifts your mindset from fear-based to empowerment. Self-compassion is the ability to love yourself, not in a narcissistic way, but out of genuine concern for self.

An example would be a person who spent a lifetime stuck in destructive addictions. Bill was trapped in feeling sorry for himself, mired in self-pity, and felt that he had little to contribute to society or humanity.

But Bill also had a strong will to change. When he finally decided to turn it around, he quit most of his addictions and developed his compassion through helping others recover, as well. You can find your compassion by sharing your story with others. There isn't any better way to contributing than helping other person overcome obstacles.

Self-Acceptance

It is hard for many of us to like ourselves as we are. As we truly are, right now, without having an attachment to the past or the future. This is the area in which we fight to balance our lives.

CHAPTER 114:

Emotional Control

Using Mastery

Using the mastery skills in this section will help you achieve Wise Mind. If you practice Wise Mind when the seas of life are calm, it will be easier to bring to mind those skills during times of turbulence. Doing something that makes you feel a little better every day helps relieve stress and inspire confidence. Attaining confidence helps reduce stress in stressful situations as well as in everyday situations.

Taking care of yourself helps you stay grounded so that when difficulties arise, and they will, you can keep your cool and maintain a consistent level of emotions.

Build Positive Experiences

Building positive experiences is necessary for emotion regulation in that we need a well of positives to draw from when we're running on empty. Many experiences are wonderful at the time, and then we later may not be friends with the people we had the experience with. Do not let that mar the memory. Remember who they were when you had the experience together. There are two important categories in which to build positive experiences: the short term and the long term.

Short Term

Short-term memories include talking to a good friend, taking a walk, noticing a beautiful area, going to the dog park, reading a good book, watching a show or movie you love, dining out, having a picnic, and laughing on a break with a coworker. Most of us already do something to create short-term positive experiences daily without thinking about it.

This exercise asks you to create more short-term positive experiences and do it deliberately. Call up an old friend. Stay off social media after work for a few days. Make a concerted effort to tell ridiculous, silly stories with your kids. Send your nieces and nephews presents from the clearance aisle. Do something that will create positive experiences deliberately.

When you deliberately practice making and noticing positive experiences, you'll begin to make and notice more as part of your daily life. When positivity is a part of your daily life, you feel better emotionally and physically.

Do at least one of these things, or choose something else that makes you happy, every day for a week. Go out of your way to do it for a week. After that, try to make it IN your way. Do something you've never tried before. There are probably a few things you've never thought of trying:

- Reading a good book
- Writing a good story
- Going out for drinks midweek
- Going to a movie midweek
- Sex
- Eating a good meal
- Going out just for dessert
- Going to a poetry jam
- Going to a karaoke bar
- Joining pub trivia with friends
- Learning to make sushi or another exotic dish

- Trying a new exotic dish
- Jogging
- Kickboxing
- Swimming
- Watching a children's movie in the theater and focusing on the laughter
- Stopping on the dog's walking route to smell the flowers
- Doing something nice for a stranger
- Doing something nice for a friend
- Playing a carnival game
- Getting the expensive, full inside and out car wash
- Completing your to-do list
- Writing a ridiculously easy to-do list so you can complete it
- Taking pictures with a real camera
- Going down a waterslide
- Playing board games with friends
- Playing interactive games, like "How to Host a Murder"
- Going to a movie or concert in the park
- Going to a new hobby class like painting or writing or learning to skate
- Organizing your bookshelf or closet

- Buying a new article of clothing, jewelry or book for yourself
- Visiting a nursing home to sing or play bingo with the residents
- Letting your kids teach you how to play their favorite video game
- Getting a massage
- Going to the chiropractor
- Going to a play or the opera
- Going to a high school play
- Going to a college football game
- Driving to a different city for dinner with a friend
- Going sightseeing
- Joining Toastmasters
- Volunteering at a homeless shelter during the months they really need it: January-October
- carrying "homeless packs" in your cars: gallon Ziploc bags with personal hygiene materials, feminine hygiene products, smokes, granola bars, bottles of water, socks, candy bars, stuffed animals, cash, gift cards to McDonald's, etc. Put them with blankets, coats, and clothes you would've given away. Drive around the areas where there are homeless people and give these out.
- Gardening
- Planning a party
- Getting your hair done

- Talking in a different accent for an evening
- Dedicating a song on the radio to someone
- Writing in your journal
- Spending some time alone without the television, radio, or internet; just you and a cup of the beverage of your choice
- Going out to lunch with a friend
- Playing volleyball
- Playing hide and seek with your coworkers (and trying not to go home when their eyes are closed)
- Singing in the car
- Driving to the mountains
- Roasting marshmallows
- Going to the sauna
- Sitting in a hot tub
- Sitting in a cold tub
- Making a fort in the elevator at work with a sign that says, 'No bosses allowed!'
- Silently challenging the driver in the car next to you at a stoplight to a dance-off in your cars
- Keeping a box of fruit snacks in your desk for anyone having a bad day
- Having a song fight with your spouse

- Convincing a stranger you think you're a vampire
- Calling a radio station and telling them a funny story
- Doing a jigsaw puzzle
- Riding a unicycle
- Going to a museum or aquarium
- Going to a psychic, just for giggles
- Getting a Reiki session done
- Taking a stuffed animal for a walk, pretending to cry when anyone points out it's not real
- Calling a radio station and pretending to be psychic. Google the DJ while you're talking and tell them all about themselves so they'll believe you.
- Going to a belly dancing class

Long Term

Long-term positive experiences are more goal-oriented, creating a life worth living. What are some goals that you would like to achieve? Write down a few specific goals. Break them down into subcategories.

Money

Many people have goals that are money-oriented. Write down how much you'd like to save each month or put towards your debt. If you put it in a place you'll forget or an IRA (Individual Retirement Account) you can't touch, you're less likely to spend it.

Learn how to budget. Keep track of how much you spend versus how much you make. Keep track of all your expenses. See where you can cut back. Itemize your spending as you go – keep it on your phone until you put it into a spreadsheet. When tax time comes, you will already know how much you have spent on medical supplies or work-related expenses. Use your debit card instead of your credit card. Then you're only spending what you have, and if you don't keep your receipts, everything is on your bank statement anyway. Get out of debt as much as possible. You may always have debt for education, health, and home, but you can pay off your credit cards and chip away at the others. Save as much as possible. Save by packing your own lunch instead of eating out. Put that in a jar. Use those coins when your kid needs shoelaces or something. After a while of paying with change, you forget you ever had any dignity; it's cool.

If your job offers a 401(k), take it. Immediately. The 401(k) follows the person, not the job. If your job offers overtime, do it. Pick up shifts. Show up in your uniform and ask who wants to go home. When a couple complains that they don't know where their waitress is, promise to take care of them yourself because she clearly doesn't value her customers. Then pocket that $20 tip. Find little tricks to make your job, and your screw-ups work FOR you.

Relationships

Repair a relationship.

If you have a relationship in your life that you feel must be repaired in order for you to move on with your life, you may have to take the initiative. You may have to make the first move, offer the first apology. Not a fake "I'm sorry you feel that way" apology, but a sincere "I'm sorry I treated you that way" apology. Not even a half-sincere apology – "I'm sorry I treated you that way, but you deserved it and here's why…" Let that second half come about if they accept your apology and you can open a discussion.

End a relationship.

Not all relationships can be saved, and not all should be. If you have offered a sincere apology and have been rebuffed, it may be time to cut your losses and move on. It may be sad for both of you, but some relationships over time become toxic for one or both parties. If this is the case, you might try one last-ditch effort, and then you should actually ditch it. If they come back, you can see how you feel at that time, and whether it's something you want to renew. Some relationships are better off dead. Reviving those is the true zombie apocalypse.

Create new relationships.

The older we get, the harder it is to create new relationships. We have to actually go out of our comfort zone to meet new people. Talk to people at your bowling league. Start a bowling league. Talk to new people at functions you attend regularly, like church or kayaking or suing people. Or even family reunions.

Go to weekly things. Join Toastmasters. You'll migrate towards the same people each week, but how much do you really talk to them? Get to know someone, more than at just surface level. Ask probing questions like, "If you invented a superpower, what would it be?" None of this already-invented superpower business. That's boring. "You can travel to the past, before a huge disaster, with the ability to warn people, but you might get stoned or burned as a witch, or you can travel twenty seconds into the future every day. Which do you choose?"

Work on current relationships.

Work on maintaining the relationships you have. Develop deeper bonds with people. Do you really know their hopes and fears, wishes, and dreams?

Go out of your way to stay in touch. Most friendships are built on convenience – when it's convenient for both or all parties to talk or hang out. Texting is a great way to let them know you're thinking about them, and they'll respond when they can. It's also a great way to miscommunicate, but that can be done in any medium.

Conclusion

I hope that you have been able to follow the advice and exercises throughout the book and that these have already begun to take effect. Now, we need to deal with something that all depressed people go through. The reason depression is so powerful is because your mind is telling you negative things on an ongoing basis. You reinforce your own depression. If you tell yourself that you are depressed all the time, then depression is the only thing that you can experience. You block out the possibility of feeling any different.

All through our lives, we take note of negative comments. Perhaps your parents were negative about things that you did as a child. Perhaps your partner is critical of the way you do things. Perhaps you critique yourself all of the time as a result of the experiences that you have been through in your life. One particular client I worked with had the potential to do so much in her life, but held herself back because she continually told herself that:

a) She wasn't good enough

b) She would never achieve anything

c) She didn't have what it took to succeed

None of these facts were actually true, but she convinced herself that they were true by continually repeating them to herself. If she did something wrong, she would call herself names and repeatedly tell herself that she was stupid. The problem with this kind of feedback is that it feeds depression.

Thus, when you hear yourself negative talking, switch it off and let it go. Say something positive to yourself and repeat the positive mantras often so that you are letting your mind know that you don't actually have to be perfect. For each negative comment you hear yourself make, follow it with a positive statement, so that you stave off becoming depressed and start to normalize your state of mind. It will make you much happier.

During the course of this book, you have learned different ways to deal with your depression and anxiety. You need to go back into the chapters of the book and follow the exercises again, making them part of your everyday life. Once you are able to do this, you will feel better about who you are. You will have better interactions with others and will also be able to find your own level of happiness.

Remember that you have the choice. Become a statistic or respect yourself and work toward not becoming a statistic. Depression all happens when you turn your mind into a source of negativity, or you let your surroundings dictate who you are. By employing the tips I have shown you in this book, you can walk away from depression and feel much healthier and happier on a permanent basis because you will have the mental tools to deal with bad events that happen in life and will be able to take them in your stride.

Flip this page to leave a review for this book on Amazon!

Thank you and good luck!

CPSIA information can be obtained
at www.ICGtesting.com
Printed in the USA
LVHW020602191020
669133LV00009B/278